THE 8051 MICROCONTROLLER
AND EMBEDDED SYSTEMS

Muhammad Ali Mazidi
Janice Gillispie Mazidi

Prentice Hall
Upper Saddle River, New Jersey   *Columbus, Ohio*

ISBN: 0-13-861022-3

Publisher: Charles E. Stewart, Jr.
Production Editor: Alexandrina Benedicto Wolf
Cover art and design: Alan Bumpus
Cover Design Coordinator: Karrie Converse-Jones
Production Manager: Matt Ottenweller
Marketing Manager: Ben Leonard

This book was set in Times Roman by Janice Mazidi and was printed and bound by Courier/Westford. The cover was printed by Phoenix Color Corp.

© 2000 by Prentice-Hall, Inc.
Pearson Education
Upper Saddle River, New Jersey 07457

Printed in the United States of America

10 9 8 7 6 5 4 3 2

ISBN: 0-13-861022-3

Prentice-Hall International (UK) Limited, *London*
Prentice-Hall of Australia Pty. Limited, *Sydney*
Prentice-Hall of Canada, Inc., *Toronto*
Prentice-Hall Hispanoamericana, S.A., *Mexico*
Prentice-Hall of India Private Limited, *New Delhi*
Prentice-Hall of Japan, Inc., *Tokyo*
Prentice-Hall (Singapore) Pte. Ltd., *Singapore*
Editora Prentice-Hall do Brasil , Ltda., *Rio de Janeiro*

... man's glory lieth in his knowledge,
his upright conduct, his praiseworthy character,
his wisdom, and not in his nationality or rank.

Baha'u'llah

CONTENTS AT A GLANCE

CHAPTERS

0:	Introduction to Computing	1
1:	The 8051 Microcontrollers	23
2:	8051 Assembly Language Programming	35
3:	Jump, Loop, and Call Instructions	65
4:	I/O Port Programming	83
5:	8051 Addressing Modes	95
6:	Arithmetic Instructions and Programs	109
7:	Logic Instructions and Programs	127
8:	Single-bit Instructions and Programming	143
9:	Timer/Counter Programming in the 8051	157
10:	8051 Serial Communication	183
11:	Interrupts Programming	209
12:	Real-world Interfacing I: LCD, ADC, and Sensors	235
13:	Real-world Interfacing II: Stepper Motor, Keyboard, DAC	255
14:	8051/31 Interfacing to External Memory	273
15:	8031/51 Interfacing to the 8255	303

APPENDICES

A:	8051 Instructions, Timing, and Registers	325
B:	8051-Based Systems: Wire-Wrapping and Testing	365
C:	IC Technology and System Design Issues	375
D:	Flowcharts and Pseudocode	395
E:	8051 Primer for X86 Programmers	400
F:	ASCII Codes	401
G:	Assemblers, Development Resources, and Suppliers	402
H:	Data Sheets	404

CONTENTS

CHAPTER 0: INTRODUCTION TO COMPUTING   1
    Section 0.1: Numbering and Coding Systems   2
    Section 0.2: Digital Primer   9
    Section 0.3: Inside the Computer   13

CHAPTER 1: THE 8051 MICROCONTROLLERS   23
    Section 1.1: Microcontrollers and Embedded Processors   24
    Section 1.2: Overview of the 8051 Family   28

CHAPTER 2: 8051 ASSEMBLY LANGUAGE PROGRAMMING   35
    Section 2.1: Inside the 8051   36
    Section 2.2: Introduction to 8051 Assembly Programming 39
    Section 2.3: Assembling and Running an 8051 Program   42
    Section 2.4: The Program Counter and ROM Space in the 8051   44
    Section 2.5: Data Types and Directives   47
    Section 2.6: 8051 Flag Bits and the PSW Register   50
    Section 2.7: 8051 Register Banks and Stack   53

CHAPTER 3: JUMP, LOOP, AND CALL INSTRUCTIONS   65
    Section 3.1: Loop and Jump Instructions   66
    Section 3.2: Call Instuctions 71
    Section 3.3: Time Delay Generation and Calculation   76

CHAPTER 4: I/O PORT PROGRAMMING   83
    Section 4.1: Pin Description of the 8051   84
    Section 4.2: I/O Programming; Bit Manipulation   91

CHAPTER 5: 8051 ADDRESSING MODES   95
    Section 5.1: Immediate and Register Addressing Modes   96
    Section 5.2: Accessing Memory Using Various Addressing Modes 98

CHAPTER 6: ARITHMETIC INSTRUCTIONS AND PROGRAMS 109
    Section 6.1: Unsigned Addition and Subtraction   110
    Section 6.2: Unsigned Multiplication and Division 117
    Section 6.3: Signed Number Concepts and Arithmetic Operations 119

CHAPTER 7: LOGIC INSTRUCTIONS AND PROGRAMS   127

Section 7.1: Logic and Compare Instructions   128
Section 7.2: Rotate and Swap Instructions   134
Section 7.3: BCD and ASCII Application Programs   137

CHAPTER 8: SINGLE-BIT INSTRUCTIONS AND PROGRAMMING 143

Section 8.1: Single-Bit Instruction Programming   144
Section 8.2: Single-Bit Operations with CY   150
Section 8.3: Reading Input Pins vs. Port Latch   152

CHAPTER 9: TIMER/COUNTER PROGRAMMING IN THE 8051 157

Section 9.1: Programming 8051 Timers   158
Section 9.2: Counter Programming   173

CHAPTER 10: 8051 SERIAL COMMUNICATION   183

Section 10.1: Basics of Serial Communication   184
Section 10.2: 8051 Connection to RS232   191
Section 10.3: 8051 Serial Communication Programming   193

CHAPTER 11: INTERRUPTS PROGRAMMING   209

Section 11.1: 8051 Interrupts   210
Section 11.2: Programming Timer Interrupts   212
Section 11.3: Programming External Hardware Interrupts   216
Section 11.4: Programming the Serial Communication Interrupt   223
Section 11.5: Interrupt Priority in the 8051   227

**CHAPTER 12: REAL-WORLD INTERFACING I: LCD, ADC,
AND SENSORS   235**

Section 12.1: Interfacing an LCD to the 8051   236
Section 12.2: 8051 Interfacing to ADC, Sensors   243

**CHAPTER 13: REAL-WORLD INTERFACING II: STEPPER MOTOR,
KEYBOARD, DAC 255**

Section 13.1: Interfacing a Stepper Motor   256
Section 13.2: 8051 Interfacing to the Keyboard   261
Section 13.3: Interfacing a DAC to the 8051   266

CHAPTER 14: 8051/31 INTERFACING TO EXTERNAL MEMORY   **273**

    Section 14.1: Semiconductor Memory   274
    Section 14.2: Memory Address Decoding   284
    Section 14.3: 8031/53 Interfacing with External ROM   287
    Section 14.4: Data Memory Space   292

CHAPTER 15: 8031/51 INTERFACING TO THE 8255   **303**

    Section 15.1: Programming the 8255   304
    Section 15.2: 8255 Interfacing   312
    Section 15.3: Other Modes of the 8255   316

APPENDIX A: 8051 INSTRUCTIONS, TIMING, AND REGISTERS   325

APPENDIX B: 8051-BASED SYSTEMS: WIRE-WRAPPING AND TESTING 365

APPENDIX C: IC TECHNOLOGY AND SYSTEM DESIGN ISSUES   375

APPENDIX D: FLOWCHARTS AND PSEUDOCODE   395

APPENDIX E: 8051 PRIMER FOR X86 PROGRAMMERS   400

APPENDIX F: ASCII CODES   401

APPENDIX G: ASSEMBLRS, DEVELOPMENT RESOURCES, AND SUPPLIERS   402

APPENDIX H: DATA SHEETS   404

INTRODUCTION

Products using microprocessors generally fall into two categories. The first category uses high-performance microprocessors such as the Pentium in applications where system performance is critical. We have an entire book dedicated to this topic, *The 80x86 IBM PC and Compatible Computers, Volumes I and II*, from Prentice Hall. In the second category of applications, performance is secondary; issues of space, power, and rapid development are more critical than raw processing power. The microprocessor for this category is often called a microcontroller.

This book is for the second category of applications. The 8051 is a widely used microcontroller. There are many reasons for this, including the existence of multiple producers and its simple architecture. This book is intended for use in college-level courses teaching microcontrollers and embedded systems. It not only establishes a foundation of assembly language programming, but also provides a comprehensive treatment of 8051 interfacing for engineering students. From this background, the design and interfacing of microcontroller-based embedded systems can be explored. This book can be also used by practicing technicians, hardware engineers, computer scientists, and hobbyists. It is an ideal source for those building stand-alone projects, or projects in which data is collected and fed into a PC for distribution on a network.

Prerequisites

Readers should have had an introductory digital course. Knowledge of a programming language would be helpful but is not necessary. Although the book is written for those with no background in assembly language programming, students with prior assembly language experience will be able to gain a mastery of 8051 architecture very rapidly and start on their projects right away.

Overview

A systematic, step-by-step approach is used to cover various aspects of 8051 Assembly language programming and interfacing. Many examples and sample programs are given to clarify the concepts and provide students with an opportunity to learn by doing. Review questions are provided at the end of each section to reinforce the main points of the section.

Chapter 0 covers number systems (binary, decimal, and hex), and provides an introduction to basic logic gates and computer terminology. This is designed especially for students, such as mechanical engineering students, who have not taken a digital logic course or those who need to refresh their memory on these topics.

Chapter 1 discusses 8051 history and features of other 8051 family members such as the 8751, 89C51, DS5000, and 8031. It also provides a list of various producers of 8051 chips.

Chapter 2 discusses the internal architecture of the 8051 and explains the use of an 8051 assembler to create ready-to-run programs. It also explores the stack and the flag register.

In Chapter 3 the topics of loop, jump, and call instructions are discussed, with many programming examples.

Chapter 4 is dedicated to the discussion of I/O ports. This allows students who are working on a project to start experimenting with 8051 I/O interfacing and start the project as soon as possible.

Chapter 5 covers the 8051 addressing modes and explains how to use the code space of the 8051 to store data, as well as how to access data.

Chapter 6 is dedicated to arithmetic instructions and programs.

Logic instructions and programs are covered in Chapter 7.

In Chapter 8 we discuss one of the most important features of the 8051, bit manipulation, as well as single-bit instructions of the 8051.

Chapter 9 describes the 8051 timers and how to use them as event-counters.

Chapter 10 is dedicated to serial data communication of the 8051 and its interfacing to the RS232. It also shows 8051 communication with COM ports of the IBM PC and compatible computers.

Chapter 11 provides a detailed discussion of 8051 interrupts with many examples on how to write interrupt handler programs.

Chapter 12 shows 8051 interfacing with real-world devices such as LCDs, ADCs, and sensors.

Chapter 13 shows 8051 interfacing with real world devices such as the keyboard, stepper motors, and DAC devices.

In Chapter 14 we cover 8031/51 interfacing with external memories, both ROM and RAM.

Finally, in Chapter 15 the issue of adding more ports to the 8031/51 is discussed, and the interfacing of an 8255 chip with the microcontroller is covered in detail.

The appendices have been designed to provide all reference material required for the topics covered in the book. Appendix A describes each 8051 instruction in detail, with examples. Appendix A also provides the clock count for instructions, 8051 register diagrams, and RAM memory maps. Appendix B describes wire wrapping, and how to design your own 8051 trainer board based on 89C51 or DS5000 chips. Appendix C covers IC technology and logic families, as well as 8051 I/O port interfacing and fan-out. Make sure you study this before connecting the 8051 to an external device. In Appendix D, the use of flowcharts and psuedocode is explored. Appendix E is for students familiar with x86 architecture who need to make a rapid transition to 8051 architecture. Appendix F provides the table for ASCII characters. Appendix G lists resources for assembler shareware, and electronics parts. Appendix H contains data sheets for the 8051 and other IC chips.

Diskette contents

The diskette attached to the book contains the lab manual, which has many experiments for software programming and hardware interfacing of the 8051. These are in Microsoft Word 97 format. In addition, the diskette contains the source code for all the programs in the book (in ASCII files). Also on the diskette are two guides for using 8051 assemblers and simulators from Franklin Software and Keil Corporation.

Acknowledgments

This book is the result of the dedication and encouragement of many individuals. Our sincere and heartfelt appreciation goes to all of them.

First, we would like to thank Professor Danny Morse, the most knowledgeable and experienced person on the 8051 that we know. He felt a strong need for a book such as this, and due to his lack of time he encouraged us to write it. He is the one who introduced us to this microcontroller and was always there, ready to discuss issues related to 8051 architecture.

Also we would like to express our sincere thanks to Professor Clyde Knight of Devry Institute of Technology for his helpful suggestions on the organization of the book.

In addition, the following professors and students found errors while using the book in its pre-publication form in their microcontroller course, and we thank them sincerely: Professor Phil Golden and John Berry of DeVry Institute of Technology, Robert Wrightson, Priscilla Martinez, Benjamin Fombon, David Bergman, John Higgins, Scot Robinson, Jerry Chrane, James Piott, Daniel Rusert, Michael Beard, Landon Hull, Jose Lopez, Larry Hill, David Johnson, Jerry Kelso, Michael Marshall, Marc Hoang, Trevor Isra.

Mr. Rolin McKinlay, an excellent student of the 8051, made many valuable suggestions, found many errors, and helped to produce the solution manual for the end-of-chapter problems. We sincerely appreciate his enthusiasm for this book.

Finally, we would like to thank the people at Prentice Hall, in particular our publisher, Mr. Charles Stewart, who continues to support and encourage our writing, and our production editor Alex Wolf who made the book a reality.

We enjoyed writing this book, and hope you enjoy reading it and using it for your courses and projects. Please let us know if you have any suggestions or find any errors.

Assemblers

The following gives two sites where you can download assemblers:

www.fsinc.com	for Franklin Software, Inc.
www.keil.com	for Keil Corporation

Another interesting web site is www.8052.com for more discussion on the microcontroller. Finally, the following site provides useful Intel manuals:

http://developer.intel.com/design/auto/mcs51/manuals

ABOUT THE AUTHORS

Muhammad Ali Mazidi holds Master's degrees from both Southern Methodist University and the University of Texas at Dallas, and currently is completing his Ph.D. in the Electrical Engineering Department of Southern Methodist University. He is a co-founder and chief researcher of Microprocessor Education Group, a company dedicated to bringing knowledge of microprocessors to the widest possible audience. He also teaches microprocessor-based system design at DeVry Institute of Technology in Dallas, Texas.

Janice Gillispie Mazidi has a Master of Science degree in Computer Science from the University of North Texas. After several years experience as a software engineer in Dallas, she co-founded Microprocessor Education Group, where she is the chief technical writer and production manager, and is responsible for software development and testing.

The Mazidis have been married since 1985 and have two sons, Robert Nabil and Michael Jamal.

The authors can be contacted at the following address if you have any comments or suggestions, or if you find any errors.

Microprocessor Education Group
P.O. Box 381970
Duncanville, TX 75138
U.S.A.

mmazidi@dal.devry.edu

This volume is dedicated to the memory of Dr. A. Davoodi, Professor of Tehran University, who in the tumultuous years of my youth taught me the importance of an independent search for truth. -- Muhammad Ali Mazidi

CHAPTER 0

INTRODUCTION TO COMPUTING

OBJECTIVES

Upon completion of this chapter, you will be able to:

≫ Convert any number from base 2, base 10, or base 16 to any of the other two bases
≫ Add and subtract hex numbers
≫ Add binary numbers
≫ Represent any binary number in 2's complement
≫ Represent an alphanumeric string in ASCII code
≫ Describe logical operations AND, OR, NOT, XOR, NAND, NOR
≫ Use logic gates to diagram simple circuits
≫ Explain the difference between a bit, a nibble, a byte, and a word
≫ Give precise mathematical definitions of the terms *kilobyte*, *megabyte*, *terabyte*, and *gigabyte*
≫ Explain the difference between RAM and ROM and describe their use
≫ Describe the purpose of the major components of a computer system
≫ List the three types of buses found in computers and describe the purpose of each type of bus
≫ Describe the role of the CPU in computer systems
≫ List the major components of the CPU and describe the purpose of each

To understand the software and hardware of a microcontroller-based system, one must first master some very basic concepts underlying computer design. In this chapter (which in the tradition of digital computers can be called Chapter 0), the fundamentals of numbering and coding systems are presented. After an introduction to logic gates, an overview of the workings inside the computer is given. Finally, in the last section we give a brief history of CPU architecture. Although some readers may have an adequate background in many of the topics of this chapter, it is recommended that the material be scanned, however briefly.

SECTION 0.1: NUMBERING AND CODING SYSTEMS

Whereas human beings use base 10 *(decimal)* arithmetic, computers use the base 2 *(binary)* system. In this section we explain how to convert from the decimal system to the binary system, and vice versa. The convenient representation of binary numbers, called *hexadecimal,* also is covered. Finally, the binary format of the alphanumeric code, called *ASCII*, is explored.

Decimal and binary number systems

Although there has been speculation that the origin of the base 10 system is the fact that human beings have 10 fingers, there is absolutely no speculation about the reason behind the use of the binary system in computers. The binary system is used in computers because 1 and 0 represent the two voltage levels of on and off. Whereas in base 10 there are 10 distinct symbols, 0, 1, 2, ..., 9, in base 2 there are only two, 0 and 1, with which to generate numbers. Base 10 contains digits 0 through 9; binary contains digits 0 and 1 only. These two binary digits, 0 and 1, are commonly referred to as *bits*.

Converting from decimal to binary

One method of converting from decimal to binary is to divide the decimal number by 2 repeatedly, keeping track of the remainders. This process continues until the quotient becomes zero. The remainders are then written in reverse order to obtain the binary number. This is demonstrated in Example 0-1.

Example 0-1

Convert 25_{10} to binary.

Solution:

	Quotient	Remainder	
25/2 =	12	1	LSB (least significant bit)
12/2 =	6	0	
6/2 =	3	0	
3/2 =	1	1	
1/2 =	0	1	MSB (most significant bit)

Therefore, $25_{10} = 11001_2$.

Converting from binary to decimal

To convert from binary to decimal, it is important to understand the concept of weight associated with each digit position. First, as an analogy, recall the weight of numbers in the base 10 system, as shown in the diagram. By the same token, each digit position in a number in base 2 has a weight associated with it:

$$
\begin{array}{rcl}
740683_{10} & = & \\
3 \times 10^0 & = & 3 \\
8 \times 10^1 & = & 80 \\
6 \times 10^2 & = & 600 \\
0 \times 10^3 & = & 0000 \\
4 \times 10^4 & = & 40000 \\
7 \times 10^5 & = & \underline{700000} \\
& & 740683
\end{array}
$$

$$
\begin{array}{lll}
110101_2 = & & \textit{Decimal} \quad \textit{Binary} \\
1 \times 2^0 = & 1 \times 1 = & 1 \qquad\qquad\quad 1 \\
0 \times 2^1 = & 0 \times 2 = & 0 \qquad\qquad\quad 00 \\
1 \times 2^2 = & 1 \times 4 = & 4 \qquad\qquad\quad 100 \\
0 \times 2^3 = & 0 \times 8 = & 0 \qquad\qquad\quad 0000 \\
1 \times 2^4 = & 1 \times 16 = & 16 \qquad\qquad 10000 \\
1 \times 2^5 = & 1 \times 32 = & \underline{32} \qquad\qquad \underline{100000} \\
& & 53 \qquad\qquad 110101
\end{array}
$$

Knowing the weight of each bit in a binary number makes it simple to add them together to get its decimal equivalent, as shown in Example 0-2.

Example 0-2

Convert 11001_2 to decimal.

Solution:

Weight:	16	8	4	2	1
Digits:	1	1	0	0	1
Sum:	16 +	8 +	0 +	0 +	1 = 25_{10}

Knowing the weight associated with each binary bit position allows one to convert a decimal number to binary directly instead of going through the process of repeated division. This is shown in Example 0-3.

Example 0-3

Use the concept of weight to convert 39_{10} to binary.

Solution:

Weight:	32	16	8	4	2	1
	1	0	0	1	1	1
	32 +	0 +	0 +	4 +	2 +	1 = 39

Therefore, $39_{10} = 100111_2$.

Hexadecimal system

Base 16, the *hexadecimal* system as it is called in computer literature, is used as a convenient representation of binary numbers. For example, it is much easier for a human being to represent a string of 0s and 1s such as 100010010110 as its hexadecimal equivalent of 896H. The binary system has 2 digits, 0 and 1. The base 10 system has 10 digits, 0 through 9. The hexadecimal (base 16) system has 16 digits. In base 16, the first 10 digits, 0 to 9, are the same as in decimal, and for the remaining six digits, the letters A, B, C, D, E, and F are used. Table 0-1 shows the equivalent binary, decimal, and hexadecimal representations for 0 to 15.

Converting between binary and hex

To represent a binary number as its equivalent hexadecimal number, start from the right and group 4 bits at a time, replacing each 4-bit binary number with its hex equivalent shown in Table 0-1. To convert from hex to binary, each hex digit is replaced with its 4-bit binary equivalent. See Examples 0-4 and 0-5.

Table 0-1: Base 16 Number Systems

Decimal	Binary	Hex
0	0000	0
1	0001	1
2	0010	2
3	0011	3
4	0100	4
5	0101	5
6	0110	6
7	0111	7
8	1000	8
9	1001	9
10	1010	A
11	1011	B
12	1100	C
13	1101	D
14	1110	E
15	1111	F

Example 0-4

Represent binary 100111110101 in hex.

Solution:
First the number is grouped into sets of 4 bits: 1001 1111 0101.
Then each group of 4 bits is replaced with its hex equivalent:

          1001   1111   0101
            9   F   5

Therefore, 100111110101_2 = 9F5 hexadecimal.

Example 0-5

Convert hex 29B to binary.

Solution:

              2   9   B
        =   0010   1001   1011

Dropping the leading zeros gives 1010011011.

Converting from decimal to hex

Converting from decimal to hex could be approached in two ways:
1. Convert to binary first and then convert to hex. Example 0-6 shows this method of converting decimal to hex.
2. Convert directly from decimal to hex by repeated division, keeping track of the remainders. Experimenting with this method is left to the reader.

Example 0-6

(a) Convert 45_{10} to hex.

32	16	8	4	2	1	First, convert to binary.
1	0	1	1	0	1	$32 + 8 + 4 + 1 = 45$

$45_{10} = 0010\ 1101_2 = 2D$ hex

(b) Convert 629_{10} to hex.

512	256	128	64	32	16	8	4	2	1
1	0	0	1	1	1	0	1	0	1

$629_{10} = (512 + 64 + 32 + 16 + 4 + 1) = 0010\ 0111\ 0101_2 = 275$ hex

(c) Convert 1714_{10} to hex.

1024	512	256	128	64	32	16	8	4	2	1
1	1	0	1	0	1	1	0	0	1	0

$1714_{10} = (1024 + 512 + 128 + 32 + 16 + 2) = 0110\ 1011\ 0010_2 = 6B2$ hex

Converting from hex to decimal

Conversion from hex to decimal can also be approached in two ways:
1. Convert from hex to binary and then to decimal. Example 0-7 demonstrates this method of converting from hex to decimal.
2. Convert directly from hex to decimal by summing the weight of all digits.

Example 0-7

Convert the following hexadecimal numbers to decimal.

(a) $6B2_{16} = 0110\ 1011\ 0010_2$

1024	512	256	128	64	32	16	8	4	2	1
1	1	0	1	0	1	1	0	0	1	0

$1024 + 512 + 128 + 32 + 16 + 2 = 1714_{10}$

(b) $9F2D_{16} = 1001\ 1111\ 0010\ 1101_2$

32768	16384	8192	4096	2048	1024	512	256	128	64	32	16	8	4	2	1
1	0	0	1	1	1	1	1	0	0	1	0	1	1	0	1

$32768 + 4096 + 2048 + 1024 + 512 + 256 + 32 + 8 + 4 + 1 = 40,749_{10}$

Table 0-2: Counting in Bases

Decimal	Binary	Hex
0	00000	0
1	00001	1
2	00010	2
3	00011	3
4	00100	4
5	00101	5
6	00110	6
7	00111	7
8	01000	8
9	01001	9
10	01010	A
11	01011	B
12	01100	C
13	01101	D
14	01110	E
15	01111	F
16	10000	10
17	10001	11
18	10010	12
19	10011	13
20	10100	14
21	10101	15
22	10110	16
23	10111	17
24	11000	18
25	11001	19
26	11010	1A
27	11011	1B
28	11100	1C
29	11101	1D
30	11110	1E
31	11111	1F

Counting in bases 10, 2, and 16

To show the relationship between all three bases, in Table 0-2 we show the sequence of numbers from 0 to 31 in decimal, along with the equivalent binary and hex numbers. Notice in each base that when one more is added to the highest digit, that digit becomes zero and a 1 is carried to the next-highest digit position. For example, in decimal, $9 + 1 = 0$ with a carry to the next-highest position. In binary, $1 + 1 = 0$ with a carry; similarly, in hex, $F + 1 = 0$ with a carry.

Table 0 - 3: Binary Addition

A + B	Carry	Sum
0 + 0	0	0
0 + 1	0	1
1 + 0	0	1
1 + 1	1	0

Addition of binary and hex numbers

The addition of binary numbers is a very straightforward process. Table 0-3 shows the addition of two bits. The discussion of subtraction of binary numbers is bypassed since all computers use the addition process to implement subtraction. Although computers have adder circuitry, there is no separate circuitry for subtractors. Instead, adders are used in conjunction with *2's complement* circuitry to perform subtraction. In other words, to implement "$x - y$", the computer takes the 2's complement of y and adds it to x. The concept of 2's complement is reviewed next. Example 0-8 shows the addition of binary numbers.

Example 0-8

Add the following binary numbers. Check against their decimal equivalents.

Solution:

	Binary	*Decimal*
	1101	13
+	1001	9
	10110	22

2's complement

To get the 2's complement of a binary number, invert all the bits and then add 1 to the result. Inverting the bits is simply a matter of changing all 0s to 1s and 1s to 0s. This is called the *1's complement*. See Example 0-9.

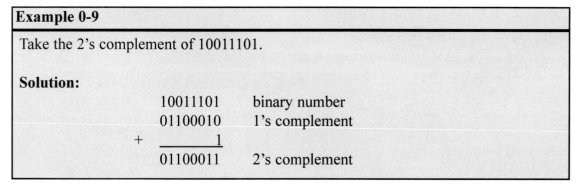

Example 0-9

Take the 2's complement of 10011101.

Solution:

	10011101	binary number
	01100010	1's complement
+	1	
	01100011	2's complement

Addition and subtraction of hex numbers

In studying issues related to software and hardware of computers, it is often necessary to add or subtract hex numbers. Mastery of these techniques is essential. Hex addition and subtraction are discussed separately below.

Addition of hex numbers

This section describes the process of adding hex numbers. Starting with the least significant digits, the digits are added together. If the result is less than 16, write that digit as the sum for that position. If it is greater than 16, subtract 16 from it to get the digit and carry 1 to the next digit. The best way to explain this is by example, as shown in Example 0-10.

Example 0-10

Perform hex addition: 23D9 + 94BE.

Solution:

	23D9	LSD: $9 + 14 = 23$	$23 - 16 = 7$ with a carry
+	94BE	$1 + 13 + 11 = 25$	$25 - 16 = 9$ with a carry
	B897	$1 + 3 + 4 = 8$	
		MSD: $2 + 9 = B$	

Subtraction of hex numbers

In subtracting two hex numbers, if the second digit is greater than the first, borrow 16 from the preceding digit. See Example 0-11.

ASCII code

The discussion so far has revolved around the representation of number systems. Since all information in the computer must be represented by 0s and 1s, binary patterns must be assigned to letters and other characters. In the 1960s a standard representation called *ASCII* (American Standard Code for Information Interchange) was established. The ASCII (pronounced "ask-E") code assigns binary patterns for numbers 0 to 9, all the letters of the English alphabet, both uppercase (capital) and lowercase, and many control codes and punctuation marks. The great advantage of this system is that it is used by most computers, so that information can be shared among computers. The ASCII system uses a total of 7 bits to represent each code. For example, 100 0001 is assigned to the uppercase letter "A" and 110 0001 is for

Hex	Symbol	Hex	Symbol
41	A	61	a
42	B	62	b
43	C	63	c
44	D	64	d
...	...	...	...
59	Y	79	y
5A	Z	7A	z

Figure 0-1. Selected ASCII Codes

the lowercase "a". Often, a zero is placed in the most significant bit position to make it an 8-bit code. Figure 0-1 shows selected ASCII codes. A complete list of ASCII codes is given in Appendix F. The use of ASCII is not only standard for keyboards used in the United States and many other countries but also provides a standard for printing and displaying characters by output devices such as printers and monitors.

Notice that the pattern of ASCII codes was designed to allow for easy manipulation of ASCII data. For example, digits 0 through 9 are represented by ASCII codes 30 through 39. This enables a program to easily convert ASCII to decimal by masking off the "3" in the upper nibble. Also notice that there is a relationship between the uppercase and lowercase letters. The uppercase letters are represented by ASCII codes 41 through 5A while lowercase letters are represented by codes 61 through 7A. Looking at the binary code, the only bit that is different between the uppercase "A" and lowercase "a" is bit 5. Therefore, conversion between uppercase of lowercase is as simple as changing bit 5 of the ASCII code.

Example 0-11

Perform hex subtraction: 59F - 2B8.

Solution:

```
      59F       LSD:  8 from 15 = 7
    - 2B8             11 from 25 (9 + 16) = 14 (E)
      2E7             2 from 4 (5 - 1) = 2
```

Review Questions

1. Why do computers use the binary number system instead of the decimal system?
2. Convert 34_{10} to binary and hex.
3. Convert 110101_2 to hex and decimal.
4. Perform binary addition: 101100 + 101.
5. Convert 101100_2 to its 2's complement representation.
6. Add 36BH + F6H.
7. Subtract 36BH - F6H.
8. Write "80x86 CPUs" in its ASCII code (in hex form).

SECTION 0.2: DIGITAL PRIMER

This section gives an overview of digital logic and design. First, we cover binary logic operations, then we show gates that perform these functions. Next, logic gates are put together to form simple digital circuits. Finally, we cover some logic devices commonly found in microcontroller interfacing.

Binary logic

As mentioned earlier, computers use the binary number system because the two voltage levels can be represented as the two digits 0 and 1. Signals in digital electronics have two distinct voltage levels. For example, a system may define 0 V as logic 0 and +5 V as logic 1. Figure 0-2 shows this system with the built-in tolerances for variations in the voltage. A valid digital signal in this example should be within either of the two shaded areas.

Figure 0-2. Binary Signals

Logic gates

Binary logic gates are simple circuits that take one or more input signals and send out one output signal. Several of these gates are defined below.

AND gate

The AND gate takes two or more inputs and performs a logic AND on them. See the truth table and diagram of the AND gate. Notice that if both inputs to the AND gate are 1, the output will be 1. Any other combination of inputs will give a 0 output. The example shows two inputs, x and y. Multiple outputs are also possible for logic gates. In the case of AND, if all inputs are 1, the output is 1. If any input is 0, the output is zero.

OR gate

The OR logic function will output a 1 if one or more inputs is 1. If all inputs are 0, then and only then will the output be 0.

Tri-state buffer

A buffer gate does not change the logic level of the input. It is used to isolate or amplify the signal.

Logical AND Function

Inputs	Output
X Y	X AND Y
0 0	0
0 1	0
1 0	0
1 1	1

X —⊐D— X AND Y
Y

Logical OR Function

Inputs	Output
X Y	X OR Y
0 0	0
0 1	1
1 0	1
1 1	1

X —⊐D— X OR Y
Y

Buffer

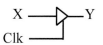

X —▷— Y

Clk

Inverter

The inverter, also called NOT, outputs the value opposite to that input to the gate. That is, a 1 input will give a 0 output, while a 0 input will give a 1 output.

XOR gate

The XOR gate performs an exclusive-OR operation on the inputs. Exclusive-OR produces a 1 output if one (but only one) input is 1. If both operands are 0, the output is zero. Likewise, if both operands are 1, the output is also zero. Notice from the XOR truth table, that whenever the two inputs are the same, the output is zero. This function can be used to compare two bits to see if they are the same.

NAND and NOR gates

The NAND gate functions like an AND gate with an inverter on the output. It produces a zero output when all inputs are 1; otherwise, it produces a 1 output. The NOR gate functions like an OR gate with an inverter on the output. It produces a 1 if all inputs are 0; otherwise, it produces a 0. NAND and NOR gates are used extensively in digital design because they are easy and inexpensive to fabricate. Any circuit that can be designed with AND, OR, XOR, and INVERTER gates can be implemented using only NAND and NOR gates. A simple example of this is given below. Notice in NAND, that if any input is zero, the output is one. Notice in NOR, that if any input is one, the output is zero.

Logic design using gates

Next we will show a simple logic design to add two binary digits. If we add two binary digits there are four possible outcomes:

	Carry	Sum
0 + 0 =	0	0
0 + 1 =	0	1
1 + 0 =	0	1
1 + 1 =	1	0

Logical Inverter

Input	Output
X	NOT X
0	1
1	0

X ——▷o— NOT X

Logical XOR Function

Inputs	Output
X Y	X XOR Y
0 0	0
0 1	1
1 0	1
1 1	0

X, Y ——)D— X XOR Y

Logical NAND Function

Inputs	Output
X Y	X NAND Y
0 0	1
0 1	1
1 0	1
1 1	0

X, Y ——Do— X NAND Y

Logical NOR Function

Inputs	Output
X Y	X NOR Y
0 0	1
0 1	0
1 0	0
1 1	0

X, Y ——Do— X NOR Y

Notice that when we add 1 + 1 we get 0 with a carry to the next higher place. We will need to determine the sum and the carry for this design. Notice that the sum column above matches the output for the XOR function, and that the carry column matches the output for the AND function. Figure 0-3 (a) shows a simple adder implemented with XOR and AND gates. Figure 0-3 (b) shows the same logic circuit implemented with AND and OR gates.

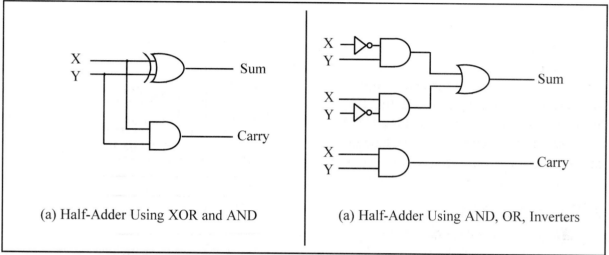

(a) Half-Adder Using XOR and AND   (a) Half-Adder Using AND, OR, Inverters

Figure 0-3. Two Implementations of a Half-Adder

Figure 0-4 shows a block diagram of a half-adder. Two half-adders can be combined to form an adder that can add three input digits. This is called a full-adder. Figure 0-5 shows the logic diagram of a full adder, along with a block diagram which masks the details of the circuit. Figure 0-6 shows a 3-bit adder using 3 full-adders.

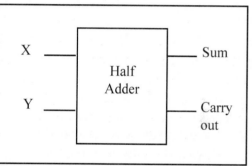

Figure 0-4. Block Diagram of a Half-Adder

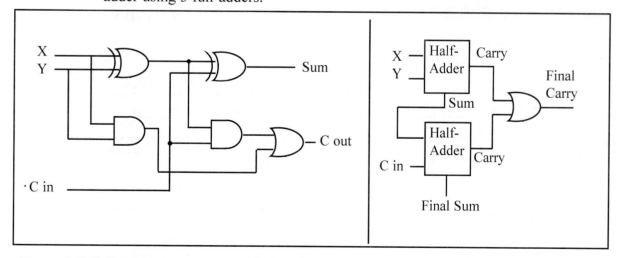

Figure 0-5. Full-Adder Built From a Half-Adder

Decoders

Another example of the application of logic gates is the decoder. Decoders are widely used for address decoding in computer design. Figure 0-7 shows decoders for 9 (1001 binary), and 5 (0101) using inverters and AND gates.

Flip-flops

A widely used component in digital systems is the flip-flop. Frequently, flip-flops are used to store data. Figure 0-8 shows the logic diagram, block diagram, and truth table for a flip-flop.

The D flip-flop is widely used to latch data. Notice from the truth table that a D-FF grabs the data at the input as the clock is activated. A D-FF holds the data as long as the power is on.

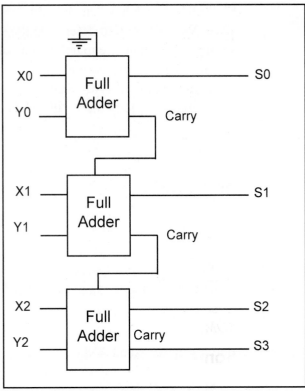

Figure 0-6. 3-Bit Adder Using 3 Full-Adders

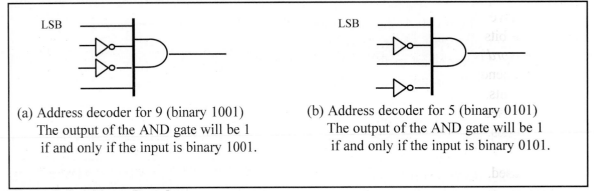

(a) Address decoder for 9 (binary 1001)
The output of the AND gate will be 1
if and only if the input is binary 1001.

(b) Address decoder for 5 (binary 0101)
The output of the AND gate will be 1
if and only if the input is binary 0101.

Figure 0-7. Address Decoders

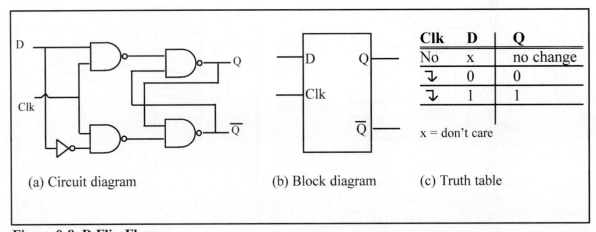

Clk	D	Q
No	x	no change
⤵	0	0
⤵	1	1

x = don't care

(a) Circuit diagram   (b) Block diagram   (c) Truth table

Figure 0-8. D Flip-Flops

Review Questions

1. The logical operation _____ gives a 1 output when all inputs are 1.
2. The logical operation _____ gives a 1 output when 1 or more of its inputs is 1.
3. The logical operation _____ is often used to compare if two inputs have the same value.
4. A _____ gate does not change the logic level of the input.
5. Name a common use for flip-flops.
6. An address _____ is used to identify a pre-determined binary address.

SECTION 0.3: INSIDE THE COMPUTER

In this section we provide an introduction to the organization and internal working of computers. The model used is generic, but the concepts discussed are applicable to all computers, including the IBM PC, PS/2, and compatibles. Before embarking on this subject, it will be helpful to review definitions of some of the most widely used terminology in computer literature, such as *K*, *mega*, *giga*, *byte*, *ROM*, *RAM*, and so on.

Some important terminology

One of the most important features of a computer is how much memory it has. Next we review terms used to describe amounts of memory in IBM PCs and compatibles. Recall from the discussion above that a *bit* is a binary digit that can have the value 0 or 1. A *byte* is defined as 8 bits. A *nibble* is half a byte, or 4 bits. A *word* is two bytes, or 16 bits. The display is intended to show the relative size of these units. Of course, they could all be composed of any combination of zeros and ones.

```
Bit                            0
Nibble                      0000
Byte              0000  0000
Word  0000  0000  0000  0000
```

A *kilobyte* is 2^{10} bytes, which is 1024 bytes. The abbreviation K is often used. For example, some floppy disks hold 356K bytes of data. A *megabyte*, or meg as some call it, is 2^{20} bytes. That is a little over 1 million bytes; it is exactly 1,048,576 bytes. Moving rapidly up the scale in size, a *gigabyte* is 2^{30} bytes (over 1 billion), and a *terabyte* is 2^{40} bytes (over 1 trillion). As an example of how some of these terms are used, suppose that a given computer has 16 megabytes of memory. That would be 16×2^{20}, or $2^4 \times 2^{20}$, which is 2^{24}. Therefore 16 megabytes is 2^{24} bytes.

Two types of memory commonly used in microcomputers are *RAM*, which stands for "random access memory" (sometimes called *read/write memory*), and *ROM*, which stands for "read-only memory." RAM is used by the computer for temporary storage of programs that it is running. That data is lost when the computer is turned off. For this reason, RAM is sometimes called *volatile memory*. ROM contains programs and information essential to operation of the computer. The information in ROM is permanent, cannot be changed by the user, and is not lost when the power is turned off. Therefore, it is called *nonvolatile memory*.

Internal organization of computers

The internal working of every computer can be broken down into three parts: CPU (central processing unit), memory , and I/O (input/output) devices (see Figure 0-9). The function of the CPU is to execute (process) information stored in memory. The function of I/O devices such as the keyboard and video monitor is to provide a means of communicating with the CPU. The CPU is connected to memory and I/O through strips of wire called a *bus*. The bus inside a computer carries information from place to place just as a street bus carries people from place to place. In every computer there are three types of buses: address bus, data bus, and control bus.

For a device (memory or I/O) to be recognized by the CPU, it must be assigned an address. The address assigned to a given device must be unique; no two devices are allowed to have the same address. The CPU puts the address (of course, in binary) on the address bus, and the decoding circuitry finds the device. Then the CPU uses the data bus either to get data from that device or to send data to it. The control buses are used to provide read or write signals to the device to indicate if the CPU is asking for information or sending it information. Of the three buses, the address bus and data bus determine the capability of a given CPU.

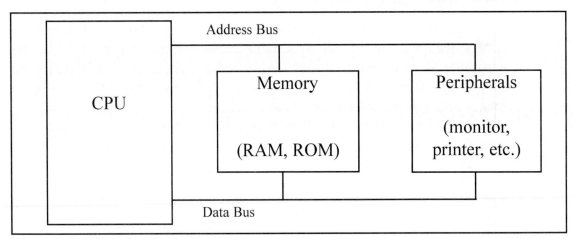

Figure 0-9: Inside the Computer

More about the data bus

Since data buses are used to carry information in and out of a CPU, the more data buses available, the better the CPU. If one thinks of data buses as highway lanes, it is clear that more lanes provide a better pathway between the CPU and its external devices (such as printers, RAM, ROM, etc.; see Figure 0-10). By the same token, that increase in the number of lanes increases the cost of construction. More data buses mean a more expensive CPU and computer. The average size of data buses in CPUs varies between 8 and 64. Early computers such as Apple 2 used an 8-bit data bus, while supercomputers such as Cray use a 64-bit data bus. Data buses are bidirectional, since the CPU must use them either to receive or to send data. The processing power of a computer is related to the size of its buses, since an 8-bit bus can send out 1 byte a time, but a 16-bit bus can send out 2 bytes at a time, which is twice as fast.

More about the address bus

Since the address bus is used to identify the devices and memory connected to the CPU, the more address buses available, the larger the number of devices that can be addressed. In other words, the number of address buses for a CPU determines the number of locations with which it can communicate. The number of locations is always equal to 2^x, where x is the number of address lines, regardless of the size of the data bus. For example, a CPU with 16 address lines can provide a total of 65,536 (2^{16}) or 64K bytes of addressable memory. Each location can have a maximum of 1 byte of data. This is due to the fact that all general-purpose microprocessor CPUs are what is called *byte addressable*. As another example, the IBM PC AT uses a CPU with 24 address lines and 16 data lines. In this case the total accessible memory is 16 megabytes (2^{24} = 16 megabytes). In this example there would be 2^{24} locations, and since each location is one byte, there would be 16 megabytes of memory. The address bus is a *unidirectional* bus, which means that the CPU uses the address bus only to send out addresses. To summarize: The total number of memory locations addressable by a given CPU is always equal to 2^x where x is the number of address bits, regardless of the size of the data bus.

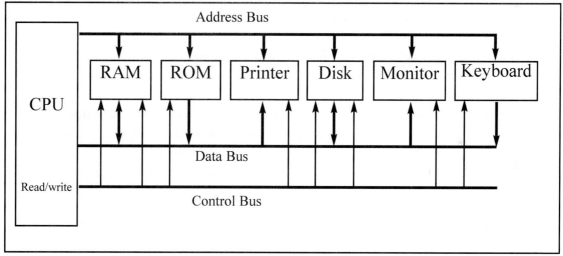

Figure 0-10: Internal Organization of Computers

CPU and its relation to RAM and ROM

For the CPU to process information, the data must be stored in RAM or ROM. The function of ROM in computers is to provide information that is fixed and permanent. This is information such as tables for character patterns to be displayed on the video monitor, or programs that are essential to the working of the computer, such as programs for testing and finding the total amount of RAM installed on the system, or programs to display information on the video monitor. In contrast, RAM is used to store information that is not permanent and can change with time, such as various versions of the operating system and application packages such as word processing or tax calculation packages. These programs are loaded into RAM to be processed by the CPU. The CPU cannot get the information from the disk directly since the disk is too slow. In other words, the CPU gets the information to be processed, first from RAM (or ROM). Only if it is not there does the CPU seek it from a mass storage device such as a disk, and then it transfers the information to RAM. For this reason, RAM and ROM are sometimes referred to as *primary memory* and disks are called *secondary memory*. Figure 0-11 shows a block diagram of the internal organization of the PC.

Inside CPUs

A program stored in memory provides instructions to the CPU to perform an action. The action can simply be adding data such as payroll data or controlling a machine such as a robot. It is the function of the CPU to fetch these instructions from memory and execute them. To perform the actions of fetch and execute, all CPUs are equipped with resources such as the following:

1. Foremost among the resources at the disposal of the CPU are a number of *registers*. The CPU uses registers to store information temporarily. The information could be two values to be processed, or the address of the value needed to be fetched from memory. Registers inside the CPU can be 8-bit, 16-bit, 32-bit, or even 64-bit registers, depending on the CPU. In general, the more and bigger the registers, the better the CPU. The disadvantage of more and bigger registers is the increased cost of such a CPU.
2. The CPU also has what is called the *ALU* (arithmetic/logic unit). The ALU section of the CPU is responsible for performing arithmetic functions such as add, subtract, multiply, and divide, and logic functions such as AND, OR, and NOT.
3. Every CPU has what is called a *program counter*. The function of the program counter is to point to the address of the next instruction to be executed. As each instruction is executed, the program counter is incremented to point to the address of the next instruction to be executed. It is the contents of the program counter that are placed on the address bus to find and fetch the desired instruction. In the IBM PC, the program counter is a register called IP, or the instruction pointer.
4. The function of the *instruction decoder* is to interpret the instruction fetched into the CPU. One can think of the instruction decoder as a kind of dictionary, storing the meaning of each instruction and what steps the CPU should take upon receiving a given instruction. Just as a dictionary requires more pages the more words it defines, a CPU capable of understanding more instructions requires more transistors to design.

Internal working of computers

To demonstrate some of the concepts discussed above, a step-by-step analysis of the process a CPU would go through to add three numbers is given next. Assume that an imaginary CPU has registers called A, B, C, and D. It has an 8-bit data bus and a 16-bit address bus. Therefore, the CPU can access memory from addresses 0000 to FFFFH (for a total of 10000H locations). The action to be performed by the CPU is to put hexadecimal value 21 into register A, and then add to register A values 42H and 12H. Assume that the code for the CPU to move a value to register A is 1011 0000 (B0H) and the code for adding a value to register A is 0000 0100 (04H). The necessary steps and code to perform them are as follows.

Action	Code	Data
Move value 21H into register A	B0H	21H
Add value 42H to register A	04H	42H
Add value 12H to register A	04H	12H

If the program to perform the actions listed above is stored in memory locations starting at 1400H, the following would represent the contents for each memory address location:

Memory address	Contents of memory address
1400	(B0) code for moving a value to register A
1401	(21) value to be moved
1402	(04) code for adding a value to register A
1403	(42) value to be added
1404	(04) code for adding a value to register A
1405	(12) value to be added
1406	(F4) code for halt

The actions performed by the CPU to run the program above would be as follows:

1. The CPU's program counter can have a value between 0000 and FFFFH. The program counter must be set to the value 1400H, indicating the address of the first instruction code to be executed. After the program counter has been loaded with the address of the first instruction, the CPU is ready to execute.

2. The CPU puts 1400H on the address bus and sends it out. The memory circuitry finds the location while the CPU activates the READ signal, indicating to memory that it wants the byte at location 1400H. This causes the contents of memory location 1400H, which is B0, to be put on the data bus and brought into the CPU.

3. The CPU decodes the instruction B0 with the help of its instruction decoder dictionary. When it finds the definition for that instruction it knows it must bring into register A of the CPU the byte in the next memory location. Therefore, it commands its controller circuitry to do exactly that. When it brings in value 21H from memory location 1401, it makes sure that the doors of all registers are closed except register A. Therefore, when value 21H comes into the CPU it will go directly into register A. After completing one instruction, the program counter points to the address of the next instruction to be executed, which in this case is 1402H. Address 1402 is sent out on the address bus to fetch the next instruction.

4. From memory location 1402H it fetches code 04H. After decoding, the CPU knows that it must add to the contents of register A the byte sitting at the next address (1403). After it brings the value (in this case 42H) into the CPU, it provides the contents of register A along with this value to the ALU to perform the addition. It then takes the result of the addition from the ALU's output and

puts it in register A. Meanwhile the program counter becomes 1404, the address of the next instruction.

5. Address 1404H is put on the address bus and the code is fetched into the CPU, decoded, and executed. This code is again adding a value to register A. The program counter is updated to 1406H.

6. Finally, the contents of address 1406 are fetched in and executed. This HALT instruction tells the CPU to stop incrementing the program counter and asking for the next instruction. In the absence of the HALT, the CPU would continue updating the program counter and fetching instructions.

Now suppose that address 1403H contained value 04 instead of 42H. How would the CPU distinguish between data 04 to be added and code 04? Remember that code 04 for this CPU means move the next value into register A. Therefore, the CPU will not try to decode the next value. It simply moves the contents of the following memory location into register A, regardless of its value.

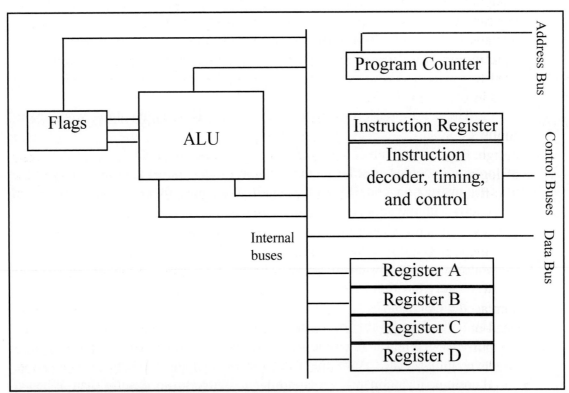

Figure 0-11: Internal Block Diagram of a CPU

Review Questions

1. How many bytes is 24 kilobytes?
2. What does "RAM" stand for? How is it used in computer systems?
3. What does "ROM" stand for? How is it used in computer systems?
4. Why is RAM called volatile memory?
5. List the three major components of a computer system.
6. What does "CPU" stand for? Explain its function in a computer.

7. List the three types of buses found in computer systems and state briefly the purpose of each type of bus.
8. State which of the following is unidirectional and which is bidirectional.
   (a) data bus   (b) address bus
9. If an address bus for a given computer has 16 lines, what is the maximum amount of memory it can access?
10. What does "ALU" stand for? What is its purpose?
11. How are registers used in computer systems?
12. What is the purpose of the program counter?
13. What is the purpose of the instruction decoder?

SUMMARY

The binary number system represents all numbers with a combination of the two binary digits, 0 and 1. The use of binary systems is necessary in digital computers because only two states can be represented: on or off. Any binary number can be coded directly into its hexadecimal equivalent for the convenience of humans. Converting from binary/hex to decimal, and vice versa, is a straightforward process that becomes easy with practice. The ASCII code is a binary code used to represent alphanumeric data internally in the computer. It is frequently used in peripheral devices for input and/or output.

The logic gates AND, OR, and Inverter are the basic building blocks of simple circuits. NAND, NOR, and XOR gates are also used to implement circuit design. Diagrams of half-adders and full-adders were given as examples of the use of logic gates for circuit design. Decoders are used to detect certain addresses. Flip-flops are used to latch in data until other circuits are ready for it.

The major components of any computer system are the CPU, memory, and I/O devices. "Memory" refers to temporary or permanent storage of data. In most systems, memory can be accessed as bytes or words. The terms *kilobyte*, *megabyte*, *gigabyte*, and *terabyte* are used to refer to large numbers of bytes. There are two main types of memory in computer systems: RAM and ROM. RAM (random access memory) is used for temporary storage of programs and data. ROM (read-only memory) is used for permanent storage of programs and data that the computer system must have in order to function. All components of the computer system are under the control of the CPU. Peripheral devices such as I/O (input/output) devices allow the CPU to communicate with humans or other computer systems. There are three types of buses in computers: address, control, and data. Control buses are used by the CPU to direct other devices. The address bus is used by the CPU to locate a device or a memory location. Data buses are used to send information back and forth between the CPU and other devices.

Finally, this chapter gave an overview of digital logic.

PROBLEMS

SECTION 0.1: NUMBERING AND CODING SYSTEMS

1. Convert the following decimal numbers to binary.
   (a) 12   (b) 123   (c) 63   (d) 128   (e) 1000
2. Convert the following binary numbers to decimal.
   (a) 100100   (b) 1000001   (c) 11101   (d) 1010   (e) 00100010
3. Convert the values in Problem 2 to hexadecimal.
4. Convert the following hex numbers to binary and decimal.
   (a) 2B9H   (b) F44H   (c) 912H   (d) 2BH   (e) FFFFH
5. Convert the values in Problem 1 to hex.
6. Find the 2's complement of the following binary numbers.
   (a) 1001010   (b) 111001   (c) 10000010   (d) 111110001
7. Add the following hex values.
   (a) 2CH+3FH   (b) F34H+5D6H   (c) 20000H+12FFH   (d) FFFFH+2222H
8. Perform hex subtraction for the following.
   (a) 24FH-129H   (b) FE9H-5CCH   (c) 2FFFFH-FFFFFH   (d) 9FF25H-4DD99H
9. Show the ASCII codes for numbers 0, 1, 2, 3, ..., 9 in both hex and binary.
10. Show the ASCII code (in hex) for the following string:
    "U.S.A. is a country" CR,LF
    "in North America" CR,LF
    CR is carriage return
    LF is line feed

SECTION 0.2: DIGITAL PRIMER

11. Draw a 3-input OR gate using a 2-input OR gate.
12. Show the truth table for a 3-input OR gate.
13. Draw a 3-input AND gate using a 2-input AND gate.
14. Show the truth table for a 3-input AND gate.
15. Design a 3-input XOR gate with a 2-input XOR gate. Show the truth table for a 3-input XOR.
16. List the truth table for a 3-input NAND.
17. List the truth table for a 3-input NOR.
18. Show the decoder for binary 1100.
19. Show the decoder for binary 11011.
20. List the truth table for a D-FF.

SECTION 0.3: INSIDE THE COMPUTER

21. Answer the following:
    (a) How many nibbles are 16 bits?
    (b) How many bytes are 32 bits?
    (c) If a word is defined as 16 bits, how many words is a 64-bit data item?
    (d) What is the exact value (in decimal) of 1 meg?

(e) How many K is 1 meg?

(f) What is the exact value (in decimal) of 1 giga?

(g) How many K is 1 giga?

(h) How many meg is 1 giga?

(i) If a given computer has a total of 8 megabytes of memory, how many bytes (in decimal) is this? How many kilobytes is this?

22. A given mass storage device such as a hard disk can store 2 gigabytes of information. Assuming that each page of text has 25 rows and each row has 80 columns of ASCII characters (each character = 1 byte), approximately how many pages of information can this disk store?

23. In a given byte-addressable computer, memory locations 10000H to 9FFFFH are available for user programs. The first location is 10000H and the last location is 9FFFFH. Calculate the following:

(a) The total number of bytes available (in decimal)

(b) The total number of kilobytes (in decimal)

24. A given computer has a 32-bit data bus. What is the largest number that can be carried into the CPU at a time?

25. Below are listed several computers with their data bus widths. For each computer, list the maximum value that can be brought into the CPU at a time (in both hex and decimal).

(a) Apple 2 with an 8-bit data bus

(b) IBM PS/2 with a 16-bit data bus

(c) IBM PS/2 model 80 with a 32-bit data bus

(d) CRAY supercomputer with a 64-bit data bus

26. Find the total amount of memory, in the units requested, for each of the following CPUs, given the size of the address buses.

(a) 16-bit address bus (in K)

(b) 24-bit address bus (in meg)

(c) 32-bit address bus (in megabytes and gigabytes)

(d) 48-bit address bus (in megabytes, gigabytes, and terabytes)

27. Regarding the data bus and address bus, which is unidirectional and which is bidirectional?

28. Which register of the CPU holds the address of the instruction to be fetched?

29. Which section of the CPU is responsible for performing addition?

30. List the three bus types present in every CPU.

ANSWERS TO REVIEW QUESTIONS

SECTION 0.1: NUMBERING AND CODING SYSTEMS

1. Computers use the binary system because each bit can have one of two voltage levels: on and off.

2. $34_{10} = 100010_2 = 22_{16}$

3. $110101_2 = 35_{16} = 53_{10}$

4. 1110001

5. 010100

6. 461

7. 275

8. 38 30 78 38 36 20 43 50 55 73

SECTION 0.2: DIGITAL PRIMER

1. AND
2. OR
3. XOR
4. Buffer
5. Storing data
6. Decoder

SECTION 0.3: INSIDE THE COMPUTER

1. 24,576
2. Random access memory; it is used for temporary storage of programs that the CPU is running, such as the operating system, word processing programs, etc.
3. Read-only memory; it is used for permanent programs such as those that control the keyboard, etc.
4. The contents of RAM are lost when the computer is powered off.
5. The CPU, memory, and I/O devices
6. Central processing unit; it can be considered the "brain" of the computer; it executes the programs and controls all other devices in the computer.
7. The address bus carries the location (address) needed by the CPU; the data bus carries information in and out of the CPU; the control bus is used by the CPU to send signals controlling I/O devices.
8. (a) bidirectional   (b) unidirectional
9. 64K, or 65,536 bytes
10. Arithmetic/logic unit; it performs all arithmetic and logic operations.
11. It is for temporary storage of information.
12. It holds the address of the next instruction to be executed.
13. It tells the CPU what steps to perform for each instruction.

CHAPTER 1

THE 8051 MICROCONTROLLERS

OBJECTIVES

Upon completion of this chapter, you will be able to:

≫≫ Compare and contrast microprocessors and microcontrollers
≫≫ Describe the advantages of microcontrollers for some applications
≫≫ Explain the concept of embedded systems
≫≫ Discuss criteria to consider in choosing a microcontroller
≫≫ Explain the variations of speed, packaging, memory, and
cost per unit and how these affect choosing a microcontroller
≫≫ Compare and contrast the various members of the 8051 family
≫≫ Compare 8051 microcontrollers offered by various manufacturers

This chapter begins with a discussion of the role and importance of microcontrollers in everyday life. In Section 1.1 we also discuss criteria to consider in choosing a microcontroller, as well as the use of microcontrollers in the embedded market. Section 1.2 covers various members of the 8051 family such as the 8052 and 8031, and their features. In addition, we discuss various versions of the 8051 such as the 8751, AT89C51, and DS5000.

SECTION 1.1: MICROCONTROLLERS AND EMBEDDED PROCESSORS

In this section we discuss the need for microcontrollers and contrast them with general-purpose microprocessors such as the Pentium and other x86 microprocessors. We also look at the role of microcontrollers in the embedded market. In addition, we provide some criteria on how to choose a microcontroller.

Microcontroller versus general-purpose microprocessor

What is the difference between a microprocessor and microcontroller? By microprocessor is meant the general-purpose microprocessors such as Intel's x86 family (8086, 80286, 80386, 80486, and the Pentium) or Motorola's 680x0 family (68000, 68010, 68020, 68030, 68040, etc.). These microprocessors contain no RAM, no ROM, and no I/O ports on the chip itself. For this reason, they are commonly referred to as *general-purpose microprocessors*.

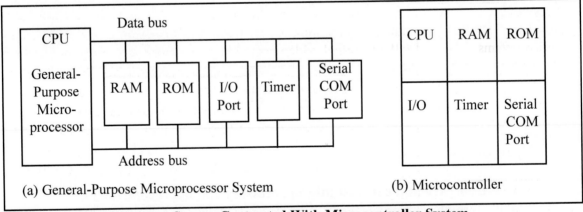

Figure 1-1. Microprocessor System Contrasted With Microcontroller System

A system designer using a general-purpose microprocessor such as the Pentium or the 68040 must add RAM, ROM, I/O ports, and timers externally to make them functional. Although the addition of external RAM, ROM, and I/O ports makes these systems bulkier and much more expensive, they have the advantage of versatility such that the designer can decide on the amount of RAM, ROM, and I/O ports needed to fit the task at hand. This is not the case with microcontrollers. A microcontroller has a CPU (a microprocessor) in addition to a fixed amount of RAM, ROM, I/O ports, and a timer all on a single chip. In other words, the processor, the RAM, ROM, I/O ports, and timer are all embedded together on one chip; therefore, the designer cannot add any external memory, I/O, or timer to it. The fixed amount of on-chip ROM, RAM, and number of I/O ports in microcontrollers makes them ideal for many applications in which cost and space are

Home
Appliances
Intercom
Telephones
Security systems
Garage door openers
Answering machines
Fax machines
Home computers
TVs
Cable TV tuner
VCR
Camcorder
Remote controls
Video games
Cellular phones
Musical instruments
Sewing machines
Lighting control
Paging
Camera
Pinball machines
Toys
Exercise equipment
Office
Telephones
Computers
Security systems
Fax machine
Microwave
Copier
Laser printer
Color printer
Paging
Auto
Trip computer
Engine control
Air bag
ABS
Instrumentation
Security system
Transmission control
Entertainment
Climate control
Cellular phone
Keyless entry

Table 1-1: Some Embedded Products Using Microcontrollers

critical. In many applications, for example a TV remote control, there is no need for the computing power of a 486 or even an 8086 microprocessor. In many applications, the space it takes, the power it consumes, and the price per unit are much more critical considerations than the computing power. These applications most often require some I/O operations to read signals and turn on and off certain bits. For this reason some call these processors IBP, "itty-bitty processors" (see "Good Things in Small Packages Are Generating Big Product Opportunities" by Rick Grehan, BYTE magazine, September 1994; www.byte.com, for an excellent discussion of microcontrollers).

It is interesting to note that some microcontroller manufacturers have gone as far as integrating an ADC (analog-to-digital converter) and other peripherals into the microcontroller.

Microcontrollers for embedded systems

In the literature discussing microprocessors, we often see the term *embedded system*. Microprocessors and microcontrollers are widely used in embedded system products. An embedded product uses a microprocessor (or microcontroller) to do one task and one task only. A printer is an example of embedded system since the processor inside it performs one task only; namely, getting the data and printing it. Contrast this with a Pentium-based PC (or any x86 IBM-compatible PC). A PC can be used for any number of applications such as word processor, print-server, bank teller terminal, video game player, network server, or internet terminal. Software for a variety of applications can be loaded and run. Of course the reason a PC can perform myriad tasks is that it has RAM memory and an operating system that loads the application software into RAM and lets the CPU run it. In an embedded system, there is only one application software that is typically burned into ROM. An x86 PC contains or is connected to various embedded products such as the keyboard, printer, modem, disk controller, sound card, CD-ROM driver, mouse, and so on. Each one of these peripherals has a microcontroller inside it that performs only one task. For example, inside every mouse there is a microcontroller to perform the task of finding the mouse position and sending it to the PC. Table 1-1 lists some embedded products.

X86 PC embedded applications

Although microcontrollers are the preferred choice for many embedded systems, there are times that a microcontroller is inadequate for the task. For this reason, in recent years many manufacturers of general-purpose microprocessors such as Intel, Motorola, AMD (Advanced Micro Devices, Inc.), and Cyrix (now a division of National Semiconductor, Inc.) have targeted their microprocessor for the high end of the embedded market. While Intel, AMD, and Cyrix push their x86 processors for both the embedded and desk-top PC markets, Motorola is determined to keep the 68000 family alive by targeting it mainly for the high end of embedded systems now that Apple no longer uses the 680x0 in their Macintosh. In the early 1990s Apple computer began using Power PC microprocessors (604, 603, 620, etc.) in place of the 680x0 for the Macintosh. The Power PC microprocessor is a joint venture between IBM and Motorola, and is targeted for the high end of the embedded market as well as the PC market. It must be noted that when a company targets a general-purpose microprocessor for the embedded market it optimizes the processor used for embedded systems. For this reason these processors are often called *high-end embedded processors*. Very often the terms *embedded processor* and *microcontroller* are used interchangeably.

One of the most critical needs of an embedded system is to decrease power consumption and space. This can be achieved by integrating more functions into the CPU chip. All the embedded processors based on the x86 and 680x0 have low power consumption in addition to some forms of I/O, COM port, and ROM all on a single chip. In high-performance embedded processors, the trend is to integrate more and more functions on the CPU chip and let the designer decide which features he/she wants to use. This trend is invading PC system design as well. Normally, in designing the PC motherboard we need a CPU plus a chip-set containing I/O, a cache controller, a flash ROM containing BIOS, and finally a secondary cache memory. New designs are emerging in industry. For example, Cyrix has announced that it is working on a chip that contains the entire PC, except for DRAM. In other words, we are about to see an entire computer on a chip.

Currently, because of MS-DOS and Windows standardization many embedded systems are using x86 PCs. In many cases using x86 PCs for the high-end embedded applications not only saves money but also shortens development time since there is a vast library of software already written for the DOS and Windows platforms. The fact that Windows is a widely used and well understood platform means that developing a Windows-based embedded product reduces the cost and shortens the development time considerably.

Choosing a microcontroller

There are four major 8-bit microcontrollers. They are: Motorola's 6811, Intel's 8051, Zilog's Z8, and PIC 16X from Microchip Technology. Each of the above microcontrollers has a unique instruction set and register set; therefore, they are not compatible with each other. Programs written for one will not run on the others. There are also 16-bit and 32-bit microcontrollers made by various chip makers. With all these different microcontrollers, what criteria do designers consider in choosing one? Three criteria in choosing microcontrollers are as follows: (1) meeting the computing needs of the task at hand efficiently and cost effectively, (2) availability of software development tools such as compilers, assemblers, and debuggers, and (3) wide availability and reliable sources of the microcontroller. Next we elaborate further on each of the above criteria.

Criteria for choosing a microcontroller

1. The first and foremost criterion in choosing a microcontroller is that it must meet the task at hand efficiently and cost effectively. In analyzing the needs of a microcontroller-based project, we must first see whether an 8-bit, 16-bit, or 32-bit microcontroller can best handle the computing needs of the task most effectively. Among other considerations in this category are:

    (a) Speed. What is the highest speed that the microcontroller supports?

    (b) Packaging. Does it come in 40-pin DIP (dual inline package) or a QFP (quad flat package), or some other packaging format? This is important in terms of space, assembling, and prototyping the end product.

    (c) Power consumption. This is especially critical for battery-powered products.

    (d) The amount of RAM and ROM on chip.

    (e) The number of I/O pins and the timer on the chip.

    (f) How easy it is to upgrade to higher-performance or lower power-consumption versions.

    (g) Cost per unit. This is important in terms of the final cost of the product in which a microcontroller is used. For example, there are microcontrollers that cost 50 cents per unit when purchased 100,000 units at a time.

2. The second criterion in choosing a microcontroller is how easy it is to develop products around it. Key considerations include the availability of an assembler, debugger, a code-efficient C language compiler, emulator, technical support, and both in-house and outside expertise. In many cases, third-party vendor (that is, a supplier other than the chip manufacturer) support for the chip is as good as, if not better than, support from the chip manufacturer.

3. The third criterion in choosing a microcontroller is its ready availability in needed quantities both now and in the future. For some designers this is even more important than the first two criteria. Currently, of the leading 8-bit microcontrollers, the 8051 family has the largest number of diversified (multiple source) suppliers. By supplier is meant a producer besides the originator of the microcontroller. In the case of the 8051, which was originated by Intel, several companies also currently produce (or have produced in the past) the 8051. These companies include: Intel, Atmel, Philips/Signetics, AMD, Siemens, Matra, and Dallas Semiconductor.

Table 1-2: Some Companies Producing a Member of the 8051 Family

Company	Web Site
Intel	www.intel.com/design/mcs51
Atmel	www.atmel.com
Philips/Signetics	www.semiconductors.philips.com
Siemens	www.sci.siemens.com
Dallas Semiconductor	www.dalsemi.com

It should be noted that Motorola, Zilog, and Microchip Technology have all dedicated massive resources to ensure wide and timely availability of their product since their product is stable, mature, and single sourced. In recent years they also have begun to sell the ASIC library cell of the microcontroller.

Review Questions

1. True or false. Microcontrollers are normally less expensive than microprocessors.
2. When comparing a system board based on a microcontroller and a general-purpose microprocessor, which one is cheaper?
3. A microcontroller normally has which of the following devices on-chip?
   (a) RAM   (b) ROM   (c) I/O   (d) all of the above
4. A general-purpose microprocessor normally needs which of the following devices to be attached to it?
   (a) RAM   (b) ROM   (c) I/O   (d) all of the above
5. An embedded system is also called a dedicated system. Why?
6. What does the term *embedded system* mean?
7. Why does having multiple sources of a given product matter?

SECTION 1.2: OVERVIEW OF THE 8051 FAMILY

In this section we first look at the various members of the 8051 family of microcontrollers and their internal features. Plus we see who are the different manufacturers of the 8051 and what kind of products they offer.

A brief history of the 8051

In 1981, Intel Corporation introduced an 8-bit microcontroller called the 8051. This microcontroller had 128 bytes of RAM, 4K bytes of on-chip ROM, two timers, one serial port, and four ports (each 8-bits wide) all on a single chip. At the time it was also referred to as a "system on a chip." The 8051 is an 8-bit processor, meaning that the CPU can work on only 8 bits of data at a time. Data larger than 8 bits has to be broken into 8-bit pieces to be processed by the CPU. The 8051 has a total of four I/O ports, each 8 bits wide. See Figure 1-2. Although the 8051 can have a maximum of 64K bytes of on-chip ROM, many manufacturers have put only 4K bytes on the chip. This will be discussed in more detail later.

The 8051 became widely popular after Intel allowed other manufacturers to make and market any flavor of the 8051 they please with the condition that they remain code-compatible with the 8051. This has led to many versions of the 8051 with different speeds and amounts of on-chip ROM marketed by more than half a dozen manufacturers. Next we review some of them. It is important to note that although there are different flavors of the 8051 in terms of speed and amount of on-chip ROM, they are all compatible with the original 8051 as far as the instructions are concerned. This means that if you write your program for one, it will run on any one of them regardless of the manufacturer.

8051 microcontroller

The 8051 is the original member of the 8051 family. Intel refers to it as MCS-51. Table 1-3 shows the main features of the 8051.

Table 1-3: Features of the 8051

Feature	Quantity
ROM	4K bytes
RAM	128 bytes
Timer	2
I/O pins	32
Serial port	1
Interrupt sources	6

Note: ROM amount indicates on-chip program space.

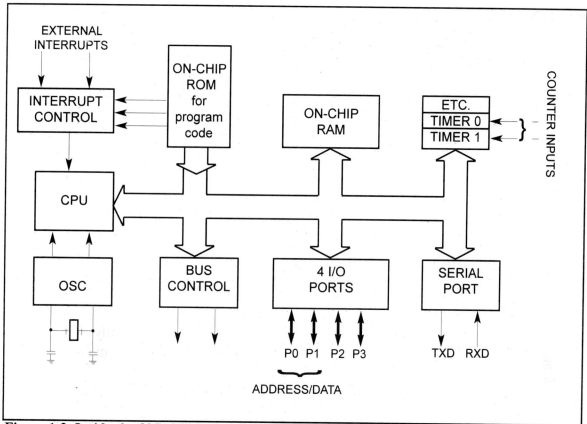

Figure 1-2. Inside the 8051 Microcontroller Block Diagram

Other members of the 8051 family

There are two other members in the 8051 family of microcontrollers. They are the 8052 and the 8031.

8052 microcontroller

The 8052 is another member of the 8051 family. The 8052 has all the standard features of the 8051 in addition to an extra 128 bytes of RAM and an extra timer. In other words, the 8052 has 256 bytes of RAM and 3 timers. It also has 8K bytes of on-chip program ROM instead of 4K bytes. See Table 1-4.

Table 1-4: Comparison of 8051 Family Members

Feature	8051	8052	8031
ROM (on-chip program space in bytes)	4K	8K	0K
RAM (bytes)	128	256	128
Timers	2	3	2
I/O pins	32	32	32
Serial port	1	1	1
Interrupt sources	6	8	6

As can be seen from Table 1-4, the 8051 is a subset of the 8052; therefore, all programs written for the 8051 will run on the 8052, but the reverse is not true.

8031 microcontroller

Another member of the 8051 family is the 8031 chip. This chip is often referred to as a ROM-less 8051 since it has 0K bytes of on-chip ROM. To use this chip you must add external ROM to it. This external ROM must contain the program that the 8031 will fetch and execute. Contrast that to the 8051 in which the on-chip ROM contains the program to be fetched and executed but is limited to only 4K bytes of code. The ROM containing the program attached to the 8031 can be as large as 64K bytes. In the process of adding external ROM to the 8031, you lose two ports. That leaves only 2 ports (of the 4 ports) for I/O operations. To solve this problem, you can add external I/O to the 8031. Interfacing the 8031 with memory and I/O ports such as the 8255 chip is discussed in Chapter 14. There are also various speed versions of the 8031 available from different companies.

Various 8051 microcontrollers

Although the 8051 is the most popular member of the 8051 family, you will not see "8051" in the part number. This is because the 8051 is available in different memory types, such as UV-EPROM, flash, and NV-RAM, all of which have different part numbers. A discussion of the various types of ROM will be given in Chapter 14. The UV-EPROM version of the 8051 is the 8751. The flash ROM version is marketed by many companies including Atmel Corp. The Atmel Flash 8051 is called AT89C51. The NV-RAM version of the 8051 made by Dallas Semiconductor is called DS5000. There is also the OTP (one-time programmable) version of the 8051 made by various manufacturers. Next we discuss briefly each of the above chips and describe applications where they are used.

8751 microcontroller

This 8751 chip has only 4K bytes of on-chip UV-EPROM. To use this chip for development requires access to a PROM burner, as well as a UV-EPROM eraser to erase the contents of UV-EPROM inside the 8751 chip before you can program it again. Due to the fact that the on-chip ROM for the 8751 is UV-EPROM, it takes around 20 minutes to erase the 8751 before it can be programmed again. This has led many manufacturers to introduce flash and NV-RAM versions of the 8051 as we will discuss next. There are also various speed versions of the 8751 available from different companies.

AT89C51 from Atmel Corporation

This popular 8051 chip has on-chip ROM in the form of flash memory. This is ideal for fast development since flash memory can be erased in seconds compared to the twenty minutes or more needed for the 8751. For this reason the AT89C51 is used in place of the 8751 to eliminate the waiting time needed to erase the chip and thereby speed up the development time. To use the AT89C51 to develop a microcontroller-based system requires a ROM burner that supports flash memory; however, a ROM eraser is not needed. Notice that in flash memory you must erase the entire contents of ROM in order to program it again. This erasing of flash is done by the PROM burner itself and this is why a separate eraser is not needed. To eliminate the need for a PROM burner Atmel is working on a version of the AT89C51 that can be programmed via the serial COM port of an IBM PC.

Table 1-5: Versions of 8051 From Atmel (All ROM Flash)

Part Number	ROM	RAM	I/O pins	Timer	Interrupt	V_{CC}	Packaging
AT89C51	4K	128	32	2	6	5V	40
AT89LV51	4K	128	32	2	6	3V	40
AT89C1051	1K	64	15	1	3	3V	20
AT89C2051	2K	128	15	2	6	3V	20
AT89C52	8K	128	32	3	8	5V	40
AT89LV52	8K	128	32	3	8	3V	40

Note: "C" in the part number indicates CMOS.

There are various speed and packaging versions of the above products. See Table 1-6. For example, notice AT89C51-12PC where "C" before the 51 is for CMOS, which has a low power consumption, "12" indicates 12 MHz, "P" is for plastic DIP package, and "C" is for commercial (vs. "M" for military). Often, the AT89C51-12PC is ideal for many student projects.

Table 1-6: Various Speeds of 8051 From Atmel

Part Number	Speed	Pins	Packaging	Use
AT89C51-12PC	12 MHz	40	DIP plastic	commercial
AT89C51-16PC	16 MHz	40	DIP plastic	commercial
AT89C51-20PC	20 MHz	40	DIP plastic	commercial

DS5000 from Dallas Semiconductor

Another popular version of the 8051 is the DS5000 chip from Dallas Semiconductor. The on-chip ROM for the DS5000 is in the form of NV-RAM. The read/write capability of NV-RAM allows the program to be loaded into the on-chip ROM while it is in the system. This can be done even via the serial port of an IBM PC. This in-system program loading of DS5000 via a PC serial port makes it an ideal home development system. Another advantage of NV-RAM is the ability to change the ROM contents one byte at a time. Contrast this with UV-EPROM and flash memory in which the entire ROM must be erased before it is programmed again.

Table 1-7: Versions of 8051 From Dallas Semiconductor's Soft Microcontroller

Part Number	ROM	RAM	I/O pins	Timers	Interrupts	V_{CC}	Packaging
DS5000-8	8K	128	32	2	6	5V	40
DS5000-32	32K	128	32	2	6	5V	40
DS5000T-8	8K	128	32	2	6	5V	40
DS5000T-8	32K	128	32	2	6	5V	40

Notes: All ROM are NV-RAM.
"T" means it has a real-time clock.

Notice that the real-time clock (RTC) is different from the timer. The real-time clock generates and keeps the time of day (hr-min-sec) and date (yr-mon-day) even when the power is off.

There are various speed and packaging versions of the DS5000 as shown in Table 1-8. For example, DS5000-8-8 has 8K NV-RAM and a speed of 8MHz. Often the DS5000-8-12 (or DS5000T-8-12) is ideal for many student projects.

Table 1-8: Versions of 8051 From Dallas Semiconductor

Part Number	NV-RAM	Speed	
DS5000-8-8	8K	8 MHz	
DS5000-8-12	8K	12 MHz	
DS5000-32-8	32K	8 MHz	
DS5000T-32-8	32K	8 MHz	(with RTC)
DS5000-32-12	32K	12 MHz	
DS5000T-8-12	8K	12 MHz	(with RTC)

OTP version of the 8051

There are also OTP (one-time-programmable) versions of the 8051 available from different sources. Flash and NV-RAM versions are typically used for product development. When a product is designed and absolutely finalized, the OTP version of the 8051 is used for mass production since it is much cheaper in terms of price per unit.

8051 family from Philips

Another major producer of the 8051 family is Philips Corporation. Indeed, they have one of the largest selections of 8051 microcontrollers. Many of their products include features such as A-to-D converters, D-to-A converters, extended I/O, and both OTP and flash.

Review Questions

1. Name three features of the 8051.
2. What is the major difference between the 8051 and 8052 microcontrollers?
3. Give the size of RAM in each of the following.
   (a) 8051   (b) 8052   (c) 8031
4. Give the size of the on-chip ROM in each of the following.
   (a) 8051   (b) 8052   (c) 8031
5. The 8051 is a(n) _____-bit microprocessor.
6. State a major difference between the 8751, the AT89C51 and the DS5000.
7. List additional features introduced in the DS5000T that are not present in the DS5000.
8. True or false. The AT89C51-12PC chip has a DIP package.
9. The AT89C51-12PC chip can handle a maximum frequency of _____ MHz.
10. The DS5000-32 has _____ K bytes of on-chip NV-RAM for programs.

SUMMARY

This chapter discussed the role and importance of microcontrollers in everyday life. Microprocessors and microcontrollers were contrasted and compared. We discussed the use of microcontrollers in the embedded market. We also discussed criteria to consider in choosing a microcontroller such as speed, memory, I/O, packaging, and cost per unit. The second section of this chapter described various family members of the 8051, such as the 8052 and 8031, and their features. In addition, we discussed various versions of the 8051 such as the AT89C51 and DS5000, which are marketed by suppliers other than Intel.

PROBLEMS

SECTION 1.1: MICROCONTROLLERS AND EMBEDDED PROCESSORS

1. True or False. A general-purpose microprocessor has on-chip ROM.
2. True or False. A microcontroller has on-chip ROM.
3. True or False. A microcontroller has on-chip I/O ports.
4. True or False. A microcontroller has a fixed amount of RAM on the chip.
5. What components are normally put together with the microcontroller into a single chip?
6. Intel's Pentium chips used in Windows PCs need external _____ and _____ chips to store data and code.
7. List three embedded products attached to a PC.
8. Why would someone want to use an x86 as an embedded processor?
9. Give the name and the manufacturer of some of the most widely used 8-bit microcontrollers.
10. In Question 9, which one has the most manufacture sources?
11. In a battery-based embedded product, what is the most important factor in choosing a microcontroller?
12. In an embedded controller with on-chip ROM, why does the size of the ROM matter?
13. In choosing a microcontroller, how important is it to have a multiple source for that chip?
14. What does the term "third-party support" mean?
15. If a microcontroller architecture has both 8-bit and 16-bit versions, which of the following statements is true.
    (a) The 8-bit software will run on the 16-bit system.
    (b) The 16-bit software will run on the 8-bit system.

SECTION 1.2: OVERVIEW OF THE 8051 FAMILY

16. The 8751 has _____ bytes of on-chip ROM.
17. The AT89C51 has _____ bytes of on-chip RAM.
18. The 8051 has ____ on-chip timer(s).

19. The 8052 has ____ bytes of on-chip RAM.
20. The ROMless version of the 8051 uses _____ as the part number.
21. The 8051 family has ____ pins for I/O.
22. The 8051 family has circuitry to support _____ serial ports.
23. The 8751 on-chip ROM is of type _____.
24. The AT8951 on-chip ROM is of type _____.
25. The DS5000 on-chip ROM is of type _____.
26. Give the speed and package type for the following chips.
    (a) AT89C51-16PC   (b) DS5000-8-12
27. In Question 26, give the amount and type of on-chip ROM.
28. Of the 8051 family, which version is the most cost effective if you are using a million of them in an embedded product?
29. What is the difference between the 8031 and 8051?
30. Of the 8051 microcontrollers, which one is the best for a home development environment? (You do not have access to a ROM burner).

ANSWERS TO REVIEW QUESTIONS

SECTION 1.1: MICROCONTROLLERS AND EMBEDDED PROCESSORS

1. True   2. A microcontroller based system   3. (d)   4. (d)
5. It is dedicated since it is dedicated to doing one type of job.
6. Embedded system means the processor is embedded into that application.
7. Having multiple sources for a given part means you are not hostage to one supplier. More importantly competition among suppliers brings about lower cost for that product.

SECTION 1.2: OVERVIEW OF THE 8051 FAMILY

1. 128 bytes of RAM, 4K bytes of on-chip ROM, four 8-bit I/O ports.
2. The 8052 has everything that the 8051 has, plus an extra timer, and the on-chip ROM is 8K bytes instead of 4K bytes. The RAM in the 8052 is 256 bytes instead of 128 bytes.
3. Both the 8051 and the 8031 have 128 bytes of RAM and the 8052 has 256 bytes.
4. (a) 4K bytes   (b) 8K bytes   (c) 0K bytes
5. 8
6. The difference is the type of on-chip ROM. In the 8751 it is UV-EPROM; in the AT89C51 it is flash; and in the DS5000 it is NV-RAM.
7. DS5000T has a real-time clock (RTC).
8. True
9. 12
10. 32

CHAPTER 2

8051 ASSEMBLY LANGUAGE PROGRAMMING

OBJECTIVES

Upon completion of this chapter, you will be able to:

≫ List the registers of the 8051 microcontroller
≫ Manipulate data using the registers and MOV instructions
≫ Code simple 8051 Assembly language instructions
≫ Assemble and run an 8051 program
≫ Describe the sequence of events that occur upon 8051 power-up
≫ Examine programs in ROM code of the 8051
≫ Explain the ROM memory map of the 8051
≫ Detail the execution of 8051 Assembly language instructions
≫ Describe 8051 data types
≫ Explain the purpose of the PSW (program status word) register
≫ Discuss RAM memory space allocation in the 8051
≫ Diagram the use of the stack in the 8051
≫ Manipulate the register banks of the 8051

In Section 2.1 we look at the inside of the 8051. We demonstrate some of the widely used registers of the 8051 with simple instructions such as MOV and ADD. In Section 2.2 we examine Assembly language and machine language programming and define terms such as mnemonics, opcode, operand, etc. The process of assembling and creating a ready-to-run program for the 8051 is discussed in Section 2.3. Step-by-step execution of an 8051 program and the role of the program counter are examined in Section 2.4. In Section 2.5 we look at some widely used Assembly language directives, pseudocode, and data types related to the 8051. In Section 2.6 we discuss the flag bits and how they are affected by arithmetic instructions. Allocation of RAM memory inside the 8051 plus the stack and register banks of the 8051 are discussed in Section 2.7.

SECTION 2.1: INSIDE THE 8051

In this section we examine the major registers of the 8051 and show their use with the simple instructions MOV and ADD.

Registers

D7	D6	D5	D4	D3	D2	D1	D0

In the CPU, registers are used to store information temporarily. That information could be a byte of data to be processed, or an address pointing to the data to be fetched. The vast majority of 8051 registers are 8-bit registers. In the 8051 there is only one data type: 8 bits. The 8 bits of a register are shown in the diagram from the MSB (most significant bit) D7 to the LSB (least significant bit) D0. With an 8-bit data type, any data larger than 8 bits must be broken into 8-bit chunks before it is processed. Since there are a large number of registers in the 8051, we will concentrate on some of the widely used general-purpose registers and cover special registers in future chapters. See Appendix A.3 for a complete list of 8051 registers.

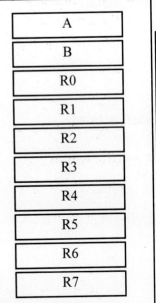

Figure 2-1 (a): Some 8-bit Registers of the 8051

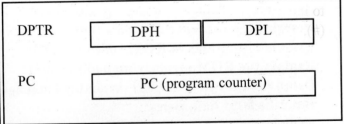

Figure 2-1 (b): Some 8051 16-bit Registers

The most widely used registers of the 8051 are A (accumulator), B, R0, R1, R2, R3, R4, R5, R6, R7, DPTR (data pointer), and PC (program counter). All of the above registers are 8-bits, except DPTR and the program counter. The accumulator, register A, is used for all arithmetic and logic instructions. To understand the use of these registers, we will show them in the context of two simple instructions, MOV and ADD.

MOV instruction

Simply stated, the MOV instruction copies data from one location to another. It has the following format:

```
MOV destination,source ;copy source to dest.
```

This instruction tells the CPU to move (in reality, copy) the source operand to the destination operand. For example, the instruction "MOV A,R0" copies the contents of register R0 to register A. After this instruction is executed, register A will have the same value as register R0. The MOV instruction does not affect the source operand. The following program first loads register A with value 55H (that is 55 in hex), then moves this value around to various registers inside the CPU. Notice the "#" in the instruction. This signifies that it is a value. The importance of this will be discussed soon.

```
MOV A,#55H       ;load value 55H into reg. A
MOV R0,A         ;copy contents of A into R0
                 ; (now A=R0=55H)
MOV R1,A         ;copy contents of A into R1
                 ; (now A=R0=R1=55H)
MOV R2,A         ;copy contents of A into R2
                 ;now A=R0=R1=R2=55H)
MOV R3,#95H      ;load value 95H into R3
                 ; (now R3=95H)
MOV A,R3         ;copy contents of R3 into A
                 ;now A=R3=95H)
```

When programming the 8051 microcontroller, the following points should be noted:

1. Values can be loaded directly into any of registers A, B, or R0 - R7. However, to indicate that it is an immediate value it must be preceded with a pound sign (#). This is shown next.

```
MOV A,#23H       ;load 23H into A (A=23H)
MOV R0,#12H      ;load 12H into R0 (R0=12H)
MOV R1,#1FH      ;load 1FH into R1 (R1=1FH)
MOV R2,#2BH      ;load 2BH into R2 (R2=2BH)
MOV B,#3CH       ;load 3CH into B (B=3CH)
MOV R7,#9DH      ;load 9DH into R7 (R7=9DH)
MOV R5,#0F9H     ;load F9H into R5 (R5=F9H)
MOV R6,#12       ;load 12 decimal (= 0CH)
                 ;into reg. R6 (R6=0CH)
```

Notice in instruction "MOV R5,#0F9H" that there is a need for 0 between the # and F to indicate that F is a hex number and not a letter. In other words "MOV R5,#F9H" will cause an error.

---

2. If values 0 to F are moved into an 8-bit register, the rest of the bits are assumed to be all zeros. For example, in "MOV A, #5" the result will be A = 05; that is, A = 00000101 in binary.

3. Moving a value that is too large into a register will cause an error.

```
MOV A,#7F2H;ILLEGAL: 7F2H > 8 bits (FFH)
MOV R2,456 ;ILLEGAL: 456 > 255 decimal (FFH)
```

4. To load a value into a register it must be preceded with a pound sign (#). Otherwise it means to load from a memory location. For example "MOV A, 17H" means to move into A the value held in memory location 17H, which could have any value. In order to load the value 17H into the accumulator we must write "MOV A, #17H" with the # preceding the number. Notice that the absence of the pound sign will not cause an error by the assembler since it is a valid instruction. However, the result would not be what the programmer intended. This is a common error for beginning programmers in the 8051.

ADD instruction

The ADD instruction has the following format:

```
ADD A,source    ;ADD the source operand
                ;to the accumulator
```

The ADD instruction tells the CPU to add the source byte to register A and put the result in register A. To add two numbers such as 25H and 34H, each can be moved to a register and then added together:

```
MOV A,#25H      ;load 25H into A
MOV R2,#34H     ;load 34H into R2
ADD A,R2        ;add R2 to accumulator
                ;(A = A + R2)
```

Executing the program above results in A = 59H (25H + 34H = 59H) and R2 = 34H. Notice that the content of R2 does not change. The program above can be written in many ways, depending on the registers used. Another way might be:

```
MOV R5,#25H     ;load 25H into R5 (R5=25H)
MOV R7,#34H     ;load 34H into R7  (R7=34H)
MOV A,#0        ;load 0 into A (A=0,clear A)
ADD A,R5        ;add to A content of R5
                ;where A = A + R5
ADD A,R7        ;add to A content of R7
                ;where A = A + R7
```

The program above results in A = 59H. There are always many ways to write the same program. One question that might come to mind after looking at the program above, is whether it is necessary to move both data items into registers before adding them together. The answer is no, it is not necessary. Look at the following variation of the same program:

```
MOV A,#25H ;load one operand into A (A=25H)
ADD A,#34H ;add the second operand 34H to A
```

In the above case, while one register contained one value, the second value followed the instruction as an operand. This is called an *immediate* operand. The examples shown so far for the ADD instruction indicate that the source operand can be either a register or immediate data, but the destination must always be register A, the accumulator. In other words, an instruction such as "ADD R2,#12H" is invalid since register A (accumulator) must be involved in any arithmetic operation. Notice that "ADD R4,A" is also invalid for the reason that A must be the destination of any arithmetic operation. To put it simply: In the 8051, register A must be involved and be the destination for all arithmetic operations. The foregoing discussion explains the reason that register A is referred to as the accumulator. The format for Assembly language instructions, descriptions of their use, and a listing of legal operand types are provided in Appendix A.1.

There are two 16-bit registers in the 8051: PC (program counter) and DPTR (data pointer). The importance and use of the program counter are covered in Section 2.3. The DPTR register is used in accessing data and is discussed in Chapter 5 when addressing modes are covered.

Review Questions

1. Write the instructions to move value 34H into register A and value 3FH into register B, then add them together.
2. Write the instructions to add the values 16H and CDH. Place the result in register R2.
3. True or false. No value can be moved directly into registers R0 - R7.
4. What is the largest hex value that can be moved into an 8-bit register? What is the decimal equivalent of the hex value?
5. The vast majority of registers in 8051 are _____ bits.

SECTION 2.2: INTRODUCTION TO 8051 ASSEMBLY PROGRAMMING

In this section we discuss Assembly language format and define some widely used terminology associated with Assembly language programming.

While the CPU can work only in binary, it can do so at a very high speed. However, it is quite tedious and slow for humans to deal with 0s and 1s in order to program the computer. A program that consists of 0s and 1s is called *machine language.* In the early days of the computer, programmers coded programs in machine language. Although the hexadecimal system was used as a more efficient way to represent binary numbers, the process of working in machine code was still cumbersome for humans. Eventually, Assembly languages were developed which provided mnemonics for the machine code instructions, plus other features which made programming faster and less prone to error. The term *mnemonic* is frequently used in computer science and engineering literature to refer to codes and abbreviations that are relatively easy to remember. Assembly language programs must be translated into machine code by a program called an *assembler*. Assembly language is referred to as a *low-level language* because it deals directly with the internal structure of the CPU. To program in Assembly language, the programmer must know all the registers of the CPU and the size of each, as well as other details.

Today, one can use many different programming languages, such as BASIC, Pascal, C, C++, Java, and numerous others. These languages are called *high-level languages* because the programmer does not have to be concerned with the internal details of the CPU. Whereas an *assembler* is used to translate an Assembly language program into machine code (sometimes also called *object code* or opcode for operation code), high-level languages are translated into machine code by a program called a *compiler*. For instance, to write a program in C, one must use a C compiler to translate the program into machine language. Now we look at 8051 Assembly language format and use an 8051 assembler to create a ready-to-run program.

Structure of Assembly language

An Assembly language program consists of, among other things, a series of lines of Assembly language instructions. An Assembly language instruction consists of a mnemonic, optionally followed by one or two operands. The operands are the data items being manipulated, and the mnemonics are the commands to the CPU, telling it what to do with those items.

```
        ORG   0H         ;start (origin) at location 0
        MOV   R5,#25H    ;load 25H into R5
        MOV   R7,#34H    ;load 34H into R7
        MOV   A,#0       ;load 0 into A
        ADD   A,R5       ;add contents of R5 to A
                         ;now A = A + R5
        ADD   A,R7       ;add contents of R7 to A
                         ;now A = A + R7
        ADD   A,#12H     ;add to A value 12H
                         ;now A = A + 12H
HERE:SJMP   HERE         ;stay in this loop
        END              ;end of asm source file
```

Program 2-1: Sample of an Assembly Language Program

A given Assembly language program (see Program 2-1) is a series of statements, or lines, which are either Assembly language instructions such as ADD and MOV, or statements called directives. While instructions tell the CPU what to do, directives (also called pseudo-instructions) give directions to the assembler. For example, in the above program while the MOV and ADD instructions are commands to the CPU, ORG and END are directives to the assembler. ORG tells the assembler to place the opcode at memory location 0 while END indicates to the assembler the end of the source code. In other words, one is for the start of the program and the other one for the end of the program.

An Assembly language instruction consists of four fields:

```
[label:]    mnemonic    [operands]    [;comment]
```

Brackets indicate that a field is optional and not all lines have them. Brackets should not be typed in. Regarding the above format, the following points should be noted.

1. The label field allows the program to refer to a line of code by name. The label field cannot exceed a certain number of characters. Check your assembler for the rule.

2. The Assembly language mnemonic (instruction) and operand(s) fields together perform the real work of the program and accomplish the tasks for which the program was written. In Assembly language statements such as

    ```
    ADD  A,B
    MOV  A,#67
    ```

    ADD and MOV are the mnemonics which produce opcodes; "A,B" and "A,#67" are the operands. Instead of a mnemonic and operand, these two fields could contain assembler pseudo-instructions, or directives. Remember that directives do not generate any machine code (opcode) and are used only by the assembler, as opposed to instructions that are translated into machine code (opcode) for the CPU to execute. In Program 2-1 the commands ORG (origin) and END are examples of directives (some 8051 assemblers use .ORG and .END). Check your assembler for the rules. More of these pseudo-instructions are discussed in detail in Section 2.5.

3. The comment field begins with a semicolon comment indicator ";". Comments may be at the end of a line or on a line by themselves. The assembler ignores comments, but they are indispensable to programmers. Although comments are optional, it is recommended that they be used to describe the program in order to make it easier for someone else to read and understand.

4. Notice the label "HERE" in the label field in Program 2-1. Any label referring to an instruction must be followed by a colon symbol, ":". In the SJMP (short jump instruction), the 8051 is told to stay in this loop indefinitely. If your system has a monitor program you do not need this line and it should be deleted from your program. In the next section we will see how to create a ready-to-run program.

Review Questions

1. What is the purpose of pseudo-instructions?
2. _____ are translated by the assembler into machine code, whereas _____ are not.
3. True or false. Assembly language is a high-level language.
4. Which of the following produces opcode?
   (a) ADD A,R2 (b) MOV A,#12 (c) ORG 2000H (d) SJMP HERE
5. Pseudo-instructions are also called _____.
6. True or false. Assembler directives are not used by the CPU itself. They are simply a guide to the assembler.
7. In question 4, which one is an assembler directive?

SECTION 2.3: ASSEMBLING AND RUNNING AN 8051 PROGRAM

Now that the basic form of an Assembly language program has been given, the next question is: How it is created, assembled and made ready to run? The steps to create an executable Assembly language program are outlined as follows.

1. First we use an editor to type in a program similar to Program 2-1. Many excellent editors or word processors are available that can be used to create and/or edit the program. A widely used editor is the MS-DOS EDIT program (or Notepad in Windows), which comes with all Microsoft operating systems. Notice that the editor must be able to produce an ASCII file. For many assemblers, the file names follow the usual DOS conventions, but the source file has the extension "asm" or "src", depending on which assembler you are using. Check your assembler for the convention. The "asm" extension for the source file is used by an assembler in the next step.

2. The "asm" source file containing the program code created in step 1 is fed to an 8051 assembler. The assembler converts the instructions into

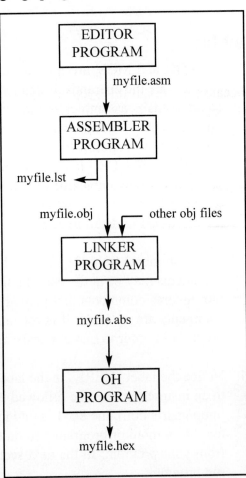

Figure 2-2. Steps to Create a Program

machine code. The assembler will produce an object file and a list file. The extension for the object file is "obj" while the extension for the list file is "lst".

3. Assemblers require a third step called *linking*. The link program takes one or more object files and produces an absolute object file with the extension "abs". This abs file is used by 8051 trainers that have a monitor program.

4. Next the "abs" file is fed into a program called "OH" (object to hex converter) which creates a file with extension "hex" that is ready to burn into ROM. This program comes with all 8051 assemblers. Recent Windows-based assemblers combine steps 2 through 4 into one step.

More about "asm" and "obj" files

The "asm" file is also called the *source* file and for this reason some assemblers require that this file have the "src" extension. Check your 8051 assembler to see which one it requires. As mentioned earlier, this file is created with an editor such as DOS EDIT or Window's Notepad. The 8051 assembler converts the asm file's Assembly language instructions into machine language and provides the obj (object) file. In addition to creating the object file, the assembler also produces the lst file (list file).

lst file

The lst (list) file, which is optional, is very useful to the programmer because it lists all the opcodes and addresses as well as errors that the assembler detected. Many assemblers assume that the list file is not wanted unless you indicate that you want to produce it. This file can be accessed by an editor such as DOS EDIT and displayed on the monitor or sent to the printer to get a hard copy. The programmer uses the list file to find syntax errors. It is only after fixing all the errors indicated in the lst file that the obj file is ready to be input to the linker program.

```
1 0000                  ORG  0H          ;start (origin) at 0
2 0000 7D25             MOV  R5,#25H     ;load 25H into R5
3 0002 7F34             MOV  R7,#34H     ;load 34H into R7
4 0004 7400             MOV  A,#0        ;load 0 into A
5 0006 2D               ADD  A,R5        ;add contents of R5 to A
                                         ;now A = A + R5
6 0007 2F               ADD  A,R7        ;add contents of R7 to A
                                         ;now A = A + R7
7 0008 2412             ADD  A,#12H      ;add to A value 12H
                                         ;now A = A + 12H
8 000A 80FE HERE:       SJMP HERE        ;stay in this loop
9 000C                  END              ;end of asm source file
```

Program 2-1: List File

Review Questions

1. True or false. The DOS program EDIT produces an ASCII file.
2. True or false. Generally, the extension of the source file is ".asm" or ".src".
3. Which of the following files can be produced by the DOS EDIT program?
   (a) myprog.asm (b) myprog.obj (c) myprog.exe (d) myprog.lst
4. Which of the following files is produced by an 8051 assembler?
   (a) myprog.asm (b) myprog.obj (c) myprog.hex (d) myprog.lst
5. Which of the following files lists syntax errors?
   (a) myprog.asm (b) myprog.obj (c) myprog.hex (d) myprog.lst

SECTION 2.4: THE PROGRAM COUNTER AND ROM SPACE IN THE 8051

In this section we examine the role of the program counter (PC) register in executing an 8051 program. We also discuss ROM memory space for various 8051 family members.

Program counter in the 8051

Another important register in the 8051 is the PC (program counter). The program counter points to the address of the next instruction to be executed. As the CPU fetches the opcode from the program ROM, the program counter is incremented to point to the next instruction. The program counter in the 8051 is 16 bits wide. This means that the 8051 can access program addresses 0000 to FFFFH, a total of 64K bytes of code. However, not all members of the 8051 have the entire 64K bytes of on-chip ROM installed, as we will see soon. Where does the 8051 wake up when it is powered? We will discuss this important topic next.

Where the 8051 wakes up when it is powered up

One question that we must ask about any microcontroller (or microprocessor) is: At what address does the CPU wake up upon applying power to it? Each microprocessor is different. In the case of the 8051 family, that is, all members regardless of the maker and variation, the microcontroller wakes up at memory address 0000 when it is powered up. By powering up we mean applying V_{CC} to the RESET pin as discussed in Chapter 4. In other words, when the 8051 is powered up, the PC (program counter) has the value of 0000 in it. This means that it expects the first opcode to be stored at ROM address 0000H. For this reason in the 8051 system, the first opcode must be burned into memory location 0000H of program ROM since this is where it looks for the first instruction when it is booted. We achieve this by the ORG statement in the source program as shown earlier. Next we discuss the step-by-step action of the program counter in fetching and executing a sample program.

Placing code in program ROM

To get a better understanding of the role of the program counter in fetching and executing a program, we examine the action of the program counter as each instruction is fetched and executed. First, we examine once more the list file of the sample program and how the code is placed in the ROM of an 8051 chip. As we can see, the opcode and operand for each instruction are listed on the left side of the list file.

```
1 0000                ORG  0H         ;start at location 0
2 0000 7D25           MOV R5,#25H     ;load 25H into R5
3 0002 7F34           MOV R7,#34H     ;load 34H into R7
4 0004 7400           MOV A,#0        ;load 0 into A
5 0006 2D             ADD A,R5        ;add contents of R5 to A
                                      ;now A = A + R5
6 0007 2F             ADD A,R7        ;add contents of R7 to A
                                      ;now A = A + R7
7 0008 2412           ADD A,#12H      ;add to A value 12H
                                      ;now A = A + 12H
8 000A 80FE HERE:     SJMP HERE       ;stay in this loop
9 000C                END             ;end of asm source file
```

Program 2-1: List File

ROM Address	Machine Language	Assembly Language
0000	7D25	MOV R5,#25H
0002	7F34	MOV R7,#34H
0004	7400	MOV A,#0
0006	2D	ADD A,R5
0007	2F	ADD A,R7
0008	2412	ADD A,#12H
000A	80FE	HERE: SJMP HERE

After the program is burned into ROM of an 8051 family member such as 8751 or AT8951 or DS5000, the opcode and operand are placed in ROM memory locations starting at 0000 as shown in the list below.

The list shows that address 0000 contains 7D which is the opcode for moving a value into register R5, and address 0001 contains the operand (in this case 25H) to be moved to R5. Therefore, the instruction "MOV R5,#25H" has a machine code of "7D25", where 7D is the opcode and 25 is the operand. Similarly, the machine code "7F34" is located in memory locations 0002 and 0003 and represents the opcode and the operand for the instruction "MOV R7,#34H". In the same way, machine code "7400" is located in memory locations 0004 and 0005 and represents the opcode and the operand for the instruction "MOV A,#0". The memory location 0006 has the opcode of 2D which is the opcode for the

Program 2-1: ROM Contents

Address	Code
0000	7D
0001	25
0002	7F
0003	34
0004	74
0005	00
0006	2D
0007	2F
0008	24
0009	12
000A	80
000B	FE

instruction "ADD A,R5" and memory location 0007 has the content 2F, which is opcode for the "ADD A,R7" instruction. The opcode for instruction "ADD A,#12H" is located at address 0008 and the operand 12H at address 0009. The memory location 000A has the opcode for the SJMP instruction and its target address is located in location 000B. The reason the target address is FE is explained in the next chapter.

Executing a program byte by byte

Assuming that the above program is burned into the ROM of an 8051 chip (or 8751, AT8951, or DS5000), the following is a step-by-step description of the action of the 8051 upon applying power to it.

1. When the 8051 is powered up, the PC (program counter) has 0000 and starts to fetch the first opcode from location 0000 of the program ROM. In the case of the above program the first opcode is 7D, which is the code for moving an operand to R5. Upon executing the opcode, the CPU fetches the value 25 and places it in R5. Now one instruction is finished. Then the program counter is incremented to point to 0002 (PC = 0002), which contains opcode 7F, the opcode for the instruction "MOV R7, . .".

2. Upon executing the opcode 7F, the value 34H is moved into R7. Then the program counter is incremented to 0004.

3. ROM location 0004 has the opcode for instruction "MOV A,#0". This instruction is executed and now PC=0006. Notice that all the above instructions are 2-byte instructions; that is, each one takes two memory locations.

4. Now PC = 0006 points to the next instruction which is "ADD A,R5". This is a 1-byte instruction. After the execution of this instruction, PC = 0007.

5. The location 0007 has the opcode 2F which belongs to the instruction "ADD A,R7". This is also a 1-byte instruction. Upon execution of this instruction, PC is incremented to 0008. This process goes on until all the instructions are fetched and executed. The fact the program counter points at the next instruction to be executed explains why some microprocessors (notably the x86) call the program counter the *instruction pointer*.

ROM memory map in the 8051 family

As we saw in the last chapter, some family members have only 4K bytes of on-chip ROM (e.g., 8751, AT8951) and some, such as the AT89C52, have 8K bytes of ROM. Dallas Semiconductor's DS5000-32 has 32K bytes of on-chip ROM. Dallas Semiconductor also has an 8051 with 64K bytes of on-chip ROM. The point to remember is that no member of the 8051 family can access more than 64K bytes of opcode since the program counter in the 8051 is a 16-bit register (0000 to FFFF address range). It must be noted that while the first location of program ROM inside the 8051 has the address of 0000, the last location can be different depending on the size of the ROM on the chip. Among the 8051 family members, the 8751 and AT8951 have 4K bytes of on-chip ROM. This 4K bytes ROM memory has memory addresses of 0000 to 0FFFH. Therefore, the first location of on-chip ROM of this 8051 has an address of 0000 and the last location has the address of 0FFFH. Look at Example 2-1 to see how this is computed.

Example 2-1

Find the ROM memory address of each of the following 8051 chips.
(a) AT89C51 (or 8751) with 4KB   (b)   DS5000-32 with 32KB

Solution:

(a) With 4K bytes of on-chip ROM memory space, we have 4096 bytes, which is 1000H
in hex (4 × 1024 = 4096 or 1000 in hex). This much memory maps to address locations of 0000 to 0FFFH. Notice that 0 is always the first location.

(b) With 32K bytes we have 32,768 (32 × 1024 = 32,768) bytes. Converting 32,768 to
hex, we get 8000H; therefore, the memory space is 0000 to 7FFFH.

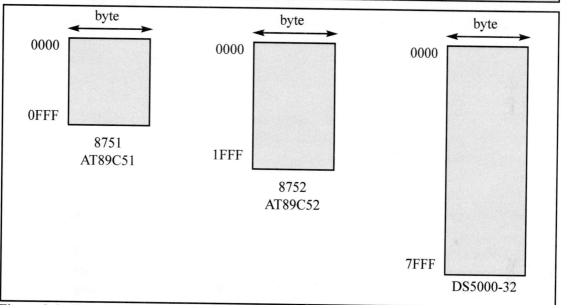

Figure 2-3. 8051 On-Chip ROM Address Range

Review Questions

1. In the 8051, the program counter is _____ bits wide.
2. True or false. Every member of the 8051 family, regardless of the maker,
   wakes up at memory 0000H when it is powered up.
3. At what ROM location do we store the first opcode of an 8051 program?
4. The instruction "MOV A, #44H" is a ____-byte instruction.
5. What is the ROM address space for the 8052 chip?

SECTION 2.5: 8051 DATA TYPES AND DIRECTIVES

In this section we look at some widely used data types and directives supported by the 8051 assembler.

8051 data type and directives

The 8051 microcontroller has only one data type. It is 8 bits, and the size
of each register is also 8 bits. It is the job of the programmer to break down data

larger than 8 bits (00 to FFH, or 0 to 255 in decimal) to be processed by the CPU. For examples of how to process data larger than 8 bits, see Chapter 6. The data types used by the 8051 can be positive or negative. A discussion of signed numbers is given in Chapter 6.

DB (define byte)

The DB directive is the most widely used data directive in the assembler. It is used to define the 8-bit data. When DB is used to define data, the numbers can be in decimal, binary, hex, or ASCII formats. For decimal, the "D" after the decimal number is optional, but using "B" (binary) and "H" (hexadecimal) for the others is required. Regardless of which is used, the assembler will convert the numbers into hex. To indicate ASCII, simply place it in quotation marks ('like this'). The assembler will assign the ASCII code for the numbers or characters automatically. The DB directive is the only directive that can be used to define ASCII strings larger than two characters; therefore, it should be used for all ASCII data definitions. Following are some DB examples:

```
        ORG   500H
DATA1:  DB    28              ;DECIMAL(1C in hex)
DATA2:  DB    00110101B       ;BINARY (35 in hex)
DATA3:  DB    39H             ;HEX
        ORG   510H
DATA4:  DB    "2591"          ;ASCII NUMBERS
        ORG   518H
DATA6:  DB    "My name is Joe";ASCII CHARACTERS
```

Either single or double quotes can be used around ASCII strings. This can be useful for strings, which contain a single quote such as "O'Leary". DB is also used to allocate memory in byte-sized chunks.

Assembler directives

The following are some more widely used directives of the 8051.

ORG (origin)

The ORG directive is used to indicate the beginning of the address. The number that comes after ORG can be either in hex or in decimal. If the number is not followed by H, it is decimal and the assembler will convert it to hex. Some assemblers use ".ORG" (notice the dot) instead of "ORG" for the origin directive. Check your assembler.

EQU (equate)

This is used to define a constant without occupying a memory location. The EQU directive does not set aside storage for a data item but associates a constant value with a data label so that when the label appears in the program, its constant value will be substituted for the label. The following uses EQU for the counter constant and then the constant is used to load the R3 register.

```
COUNT      EQU   25
...        ....
MOV        R3,#COUNT
```

When executing the instruction "MOV R3,#COUNT", the register R3 will be loaded with the value 25 (notice the # sign). What is the advantage of using EQU? Assume that there is a constant (a fixed value) used in many different places in the program, and the programmer wants to change its value throughout. By the use of EQU, one can change it once and the assembler will change all of its occurrences, rather than search the entire program trying to find every occurrence.

END directive

Another important pseudocode is the END directive. This indicates to the assembler the end of the source (asm) file. The END directive is the last line of an 8051 program, meaning that in the source code anything after the END directive is ignored by the assembler. Some assemblers use ".END" (notice the dot) instead of "END".

Rules for labels in Assembly language

By choosing label names that are meaningful, a programmer can make a program much easier to read and maintain. There are several rules that names must follow. First, each label name must be unique. The names used for labels in Assembly language programming consist of alphabetic letters in both upper and lower case, the digits 0 through 9, and the special characters question mark (?), period (.), at (@), underline (_), and dollar sign ($). The first character of the label must be an alphabetic character. In other words it cannot be a number. Every assembler has some reserved words which must not be used as labels in the program. Foremost among the reserved words are the mnemonics for the instructions. For example, "MOV" and "ADD" are reserved since they are instruction mnemonics. Aside from the mnemonics there are some other reserved words. Check your assembler for the list of reserved words.

Review Questions

1. The _____ directive is always used for ASCII strings.
2. How many bytes are used by the following?
   `DATA_1 DB "AMERICA"`
3. What is the advantage in using the EQU directive to define a constant value?
4. How many bytes are set aside by each of the following directives?
   (a) `ASC_DATA  DB "1234"` (b) `MY_DATA DB "ABC1234"`
5. State the contents of memory locations 200H - 205H for the following
   ```
            ORG  200H
   MYDATA:  DB   "ABC123"
   ```

SECTION 2.6: 8051 FLAG BITS AND THE PSW REGISTER

Like any other microprocessor, the 8051 has a flag register to indicate arithmetic conditions such as the carry bit. The flag register in the 8051 is called the program status word (PSW) register. In this section we discuss various bits of this register and provide some examples of how it is altered.

PSW (program status word) register

The program status word (PSW) register is an 8-bit register. It is also referred to as the *flag register*. Although the PSW register is 8 bits wide, only 6 bits of it are used by the 8051. The two unused bits are user-definable flags. Four of the flags are called *conditional flags*, meaning that they indicate some conditions that resulted after an instruction was executed. These four are CY (carry), AC (auxiliary carry), P (parity), and OV (overflow).

As seen from Figure 2-4, the bits PSW3 and PSW4 are designated as RS0 and RS1, and are used to change the bank registers. They are explained in the next section. The PSW.5 and PSW.1 bits are general-purpose status flag bits and can be used by the programmer for any purpose. In other words, they are user definable. See Figure 2-4 for the bits of the PSW register.

CY	AC	F0	RS1	RS0	OV	--	P

CY	PSW.7	Carry flag.
AC	PSW.6	Auxiliary carry flag.
--	PSW.5	Available to the user for general purpose.
RS1	PSW.4	Register Bank selector bit 1.
RS0	PSW.3	Register Bank selector bit 0.
OV	PSW.2	Overflow flag.
--	PSW.1	User definable bit.
P	PSW.0	Parity flag. Set/cleared by hardware each instuction cycle to indicate an odd/even number of 1 bits in the accumulator.

RS1	RS0	Register Bank	Address
0	0	0	00H - 07H
0	1	1	08H - 0FH
1	0	2	10H - 17H
1	1	3	18H - 1FH

Figure 2-4. Bits of the PSW Register

The following is a brief explanation of four of the flag bits of the PSW register. The impact of instructions on these registers is then discussed.

CY, the carry flag

This flag is set whenever there is a carry out from the d7 bit. This flag bit is affected after an 8-bit addition or subtraction. It can also be set to 1 or 0 directly by an instruction such as "SETB C" and "CLR C" where "SETB C" stands for "set bit carry" and "CLR C" for "clear carry". More about these and other bit-addressable instructions will be given in Chapter 8.

AC, the auxiliary carry flag

If there is a carry from D3 to D4 during an ADD or SUB operation, this bit is set; otherwise, it is cleared. This flag is used by instructions that perform BCD (binary coded decimal) arithmetic. See Chapter 6 for more information.

P, the parity flag

The parity flag reflects the number of 1s in the A (accumulator) register only. If the A register contains an odd number of 1s, then P = 1. Therefore, P = 0 if A has an even number of 1s.

OV, the overflow flag

This flag is set whenever the result of a signed number operation is too large, causing the high-order bit to overflow into the sign bit. In general, the carry flag is used to detect errors in unsigned arithmetic operations. The overflow flag is only used to detect errors in signed arithmetic operations and is discussed in detail in Chapter 6.

ADD instruction and PSW

Next we examine the impact of the ADD instruction on the flag bits CY, AC, and P of the PSW register. Some examples should clarify their status. Although the flag bits affected by the ADD instruction are CY (carry flag), P (parity flag), AC (auxiliary carry flag), and OV (overflow flag) we will focus on flags CY, AC, and P for now. A discussion of the overflow flag is given in Chapter 6, since it relates only to signed number arithmetic. How the various flag bits are used in programming is discussed in future chapters in the context of many applications.

See Examples 2-2 through 2-4 for the impact on selected flag bits as a result of the ADD instruction.

Table 2-1: Instructions That Affect Flag Bits

Instruction	CY	OV	AC
ADD	X	X	X
ADDC	X	X	X
SUBB	X	X	X
MUL	0	X	
DIV	0	X	
DA	X		
RRC	X		
RLC	X		
SETB C	1		
CLR C	0		
CPL C	X		
ANL C,bit	X		
ANL C,/bit	X		
ORL C,bit	X		
ORL C,/bit	X		
MOV C,bit	X		
CJNE	X		

Note: X can be 0 or 1.

Example 2-2

Show the status of the CY, AC, and P flags after the addition of 38H and 2FH in the following instructions.

```
MOV A,#38H
ADD A,#2FH ;after the addition A=67H, CY=0
```

Solution:

```
      38          00111000
    + 2F          00101111
      67          01100111
```

CY = 0 since there is no carry beyond the D7 bit
AC = 1 since there is a carry from the D3 to the D4 bit
P = 1 since the accumulator has an odd number of 1s (it has five 1s).

Example 2-3

Show the status of the CY, AC, and P flags after the addition of 9CH and 64H in the following instructions.

```
MOV A,#9CH
ADD A,#64H       ;after addition A=00 and CY=1
```

Solution:

```
      9C          10011100
    + 64          01100100
     100          00000000
```

CY=1 since there is a carry beyond the D7 bit
AC=1 since there is a carry from the D3 to the D4 bit
P=0 since the accumulator has an even number of 1s (it has zero 1s).

Example 2-4

Show the status of the CY, AC, and P flags after the addition of 88H and 93H in the following instructions.

```
MOV A,#88H
ADD A,#93H        ;after the addition A=1BH,CY=1
```

Solution:

```
      88          10001000
    + 93          10010011
     11B          00011011
```

CY=1 since there is a carry beyond the D7 bit.
AC=0 since there is no carry from the D3 to the D4 bit.
P=0 since the accumulator has an even number of 1s (it has four 1s).

Review Questions

1. The flag register in the 8051 is called _____.
2. What is the size of the flag register in the 8051?
3. Which bits of the PSW register are user-definable?
4. Find the CY and AC flag bits for the following code.

```
        MOV   A,#0FFH
        ADD   A,#01
```

5. Find the CY and AC flag bits for the following code.

```
        MOV   A,#0C2H
        ADD   A,#3DH
```

SECTION 2.7: 8051 REGISTER BANKS AND STACK

The 8051 microcontroller has a total of 128 bytes of RAM. In this section we discuss the allocation of these 128 bytes of RAM and examine their usage as registers and stack.

RAM memory space allocation in the 8051

There are 128 bytes of RAM in the 8051 (Some members, notably the 8052, have 256 bytes of RAM). The 128 bytes of RAM inside the 8051 are assigned addresses 00 to 7FH. As we will see in Chapter 5, they can be accessed directly as memory locations. These 128 bytes are divided into three different groups as follows.

1. A total of 32 bytes from locations 00 to 1F hex are set aside for register banks and the stack.
2. A total of 16 bytes from locations 20H to 2FH are set aside for bit-addressable read/write memory. A detailed discussion of bit-addressable memory and instructions is given in Chapter 8.
3. A total of 80 bytes from locations 30H to 7FH are used for read and write storage, or what is normally called a *scratch pad*. These 80 locations of RAM are widely used for the purpose of storing data and parameters by 8051 programmers. We will use them in future chapters to store data brought into the CPU via I/O ports.

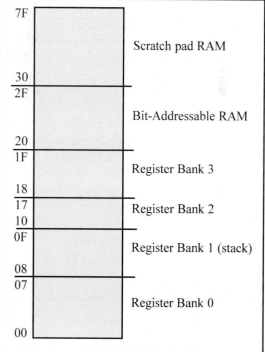

Figure 2-5. RAM Allocation in the 8051

Register banks in the 8051

As mentioned earlier, a total of 32 bytes of RAM are set aside for the register banks and stack. These 32 bytes are divided into 4 banks of registers in which each bank has 8 registers, R0 - R7. RAM locations from 0 to 7 are set aside for bank 0 of R0 - R7 where R0 is RAM location 0, R1 is RAM location 1, R2 is location 2, and so on, until memory location 7 which belongs to R7 of bank 0. The second bank of registers R0 - R7 starts at RAM location 08 and goes to location 0FH. The third bank of R0 - R7 starts at memory location 10H and goes to location 17H; finally, RAM locations 18H to 1FH are set aside for the fourth bank of R0 - R7. The following shows how the 32 bytes are allocated into 4 banks:

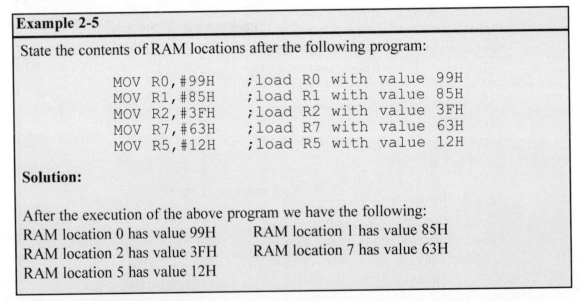

Figure 2-6. 8051 Register Banks and their RAM Addresses

As we can see from Figure 2-5, bank 1 uses the same RAM space as the stack. This is a major problem in programming the 8051. We must either not use register bank 1, or we must allocate another area of RAM for the stack. This will be discussed below.

Example 2-5

State the contents of RAM locations after the following program:

```
        MOV R0,#99H    ;load R0 with value 99H
        MOV R1,#85H    ;load R1 with value 85H
        MOV R2,#3FH    ;load R2 with value 3FH
        MOV R7,#63H    ;load R7 with value 63H
        MOV R5,#12H    ;load R5 with value 12H
```

Solution:

After the execution of the above program we have the following:
RAM location 0 has value 99H   RAM location 1 has value 85H
RAM location 2 has value 3FH   RAM location 7 has value 63H
RAM location 5 has value 12H

Default register bank

If RAM locations 00 - 1F are set aside for the four register banks, which register bank of R0 - R7 do we have access to when the 8051 is powered up? The answer is register bank 0; that is, RAM locations 0, 1, 2, 3, 4, 5, 6, and 7 are accessed with the names R0, R1, R2, R3, R4, R5, R6, and R7 when programming the 8051. It is much easier to refer to these RAM locations with names such as R0, R1, and so on, than by their memory locations. Example 2-6 clarifies this concept.

Example 2-6

Repeat Example 2-5 using RAM addresses instead of register names.

Solution:

This is called direct addressing mode and uses the RAM address location for the destination address. See Chapter 5 for a more detailed discussion of addressing modes.

```
MOV  00,#99H    ;load R0 with value 99H
MOV  01,#85H    ;load R1 with value 85H
MOV  02,#3FH    ;load R2 with value 3FH
MOV  07,#63H    ;load R7 with value 63H
MOV  05,#12H    ;load R5 with value 12H
```

How to switch register banks

As stated above, register bank 0 is the default when the 8051 is powered up. We can switch to other banks by use of the PSW (program status word) register. Bits D4 and D3 of the PSW are used to select the desired register bank as shown in Table 2-2.

The D3 and D4 bits of register PSW are often referred to as PSW.4 and PSW.3 since they can be accessed by the bit-addressable instructions SETB and CLR. For example, "SETB PSW.3" will make PSW.3=1 and select bank register 1. See Example 2-7.

Table 2.2: PSW Bits Bank Selection

	RS1 (PSW.4)	RS0 (PSW.3)
Bank 0	0	0
Bank 1	0	1
Bank 2	1	0
Bank 3	1	1

Example 2-7

State the contents of the RAM locations after the following program:

```
SETB PSW.4      ;select bank 2
MOV R0,#99H     ;load R0 with value 99H
MOV R1,#85H     ;load R1 with value 85H
MOV R2,#3FH     ;load R2 with value 3FH
MOV R7,#63H     ;load R7 with value 63H
MOV R5,#12H     ;load R5 with value 12H
```

Solution:

By default, PSW.3=0 and PSW.4=0; therefore, the instruction "SETB PSW.4" sets RS1=1 and RS0=0, thereby selecting register bank 2. Register bank 2 uses RAM locations 10H - 17H. After the execution of the above program we have the following:

RAM location 10H has value 99H   RAM location 11H has value 85H
RAM location 12H has value 3FH   RAM location 17H has value 63H
RAM location 15H has value 12H

Stack in the 8051

The stack is a section of RAM used by the CPU to store information temporarily. This information could be data or an address. The CPU needs this storage area since there are only a limited number of registers.

How stacks are accessed in the 8051

If the stack is a section of RAM, there must be registers inside the CPU to point to it. The register used to access the stack is called the SP (stack pointer) register. The stack pointer in the 8051 is only 8 bits wide, which means that it can take values of 00 to FFH. When the 8051 is powered up, the SP register contains value 07. This means that RAM location 08 is the first location being used for the stack by the 8051. The storing of a CPU register in the stack is called a PUSH, and loading the contents of the stack back into a CPU register is called a POP. In other words, a register is pushed onto the stack to save it and popped off the stack to retrieve it. The job of the SP is very critical when push and pop actions are performed. To see how the stack works, let's look at the PUSH and POP instructions.

Pushing onto the stack

In the 8051 the stack pointer (SP) is pointing to the last used location of the stack. As we push data onto the stack, the stack pointer (SP) is incremented by one. Notice that this is different from many microprocessors, notably x86 processors in which the SP is decremented when data is pushed onto the stack. Examining Example 2-8, we see that as each PUSH is executed, the contents of the register are saved on the stack and SP is incremented by 1. Notice that for every byte of data saved on the stack, SP is incremented only once. Notice also that to push the registers onto the stack we must use their RAM addresses. For example, the instruction "PUSH 1" pushes register R1 onto the stack.

Example 2-8

Show the stack and stack pointer for the following. Assume the default stack area.
```
        MOV   R6,#25H
        MOV   R1,#12H
        MOV   R4,#0F3H
        PUSH  6
        PUSH  1
        PUSH  4
```

Solution:

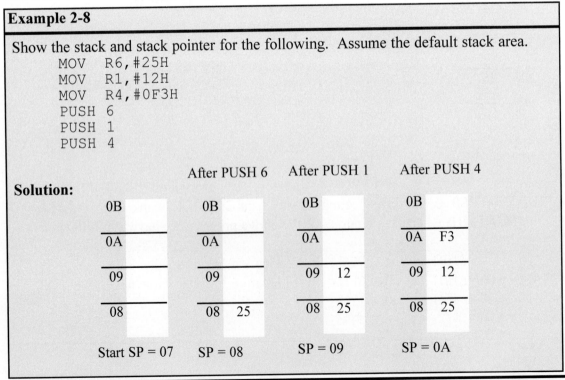

	Start	After PUSH 6	After PUSH 1	After PUSH 4
0B				
0A				F3
09			12	12
08		25	25	25
	Start SP = 07	SP = 08	SP = 09	SP = 0A

Popping from the stack

Popping the contents of the stack back into a given register is the opposite process of pushing. With every pop, the top byte of the stack is copied to the register specified by the instruction and the stack pointer is decremented once. Example 2-9 demonstrates the POP instruction.

The upper limit of the stack

As mentioned earlier, in the 8051 RAM locations 08 to 1F can be used for the stack. This is due to the fact that locations 20 - 2FH of RAM are reserved for bit-addressable memory and must not be used by the stack. If in a given program we need more than 24 bytes (08 to 1FH = 24 bytes) of stack, we can change the SP to point to RAM locations 30 - 7FH. This is done with the instruction "MOV SP, #xx".

Example 2-9

Examining the stack, show the contents of the registers and SP after execution of the following instructions. All values are in hex.

```
POP   3      ;POP stack into R3
POP   5      ;POP stack into R5
POP   2      ;POP stack into R2
```

Solution:

			After POP 3		After POP 5		After POP 2
0B	54		0B		0B		0B
0A	F9		0A	F9	0A		0A
09	76		09	76	09	76	09
08	6C		08	6C	08	6C	08	6C

Start SP = 0B   SP = 0A   SP = 09   SP = 08

CALL instruction and the stack

In addition to using the stack to save registers, the CPU also uses the stack to save the address of the instruction just below the CALL instruction. This is how the CPU knows where to resume when it returns from the called subroutine. More information on this will be given in Chapter 3 when we discuss the CALL instruction.

Example 2-10

Show the stack and stack pointer for the following instructions.

```
        MOV  SP,#5FH    ;make RAM location 60H
                        ;first stack location
        MOV  R2,#25H
        MOV  R1,#12H
        MOV  R4,#0F3H
        PUSH 2
        PUSH 1
        PUSH 4
```

Solution:

		After PUSH 2		After PUSH 1		After PUSH 4	
63		63		63		63	
62		62		62		62	F3
61		61		61	12	61	12
60		60	25	60	25	60	25
Start SP = 5F		SP = 60		SP = 61		SP = 62	

Stack and bank 1 conflict

Recall from our earlier discussion that the stack pointer register points to the current RAM location available for the stack. As data is pushed onto the stack, SP is incremented. Conversely, it is decremented as data is popped off the stack into the registers. The reason that the SP is incremented after the push is to make sure that the stack is growing toward RAM location 7FH, from lower addresses to upper addresses. If the stack pointer were decremented after push instructions, we would be using RAM locations 7, 6, 5, etc. which belong to R7 to R0 of bank 0, the default register bank. This incrementing of the stack pointer for push instructions also ensures that the stack will not reach location 0 at the bottom of RAM, and consequently run out of space for the stack. However, there is a problem with the default setting of the stack. Since SP = 07 when the 8051 is powered up, the first location of the stack is RAM location 08 which also belongs to register R0 of register bank 1. In other words, register bank 1 and the stack are using the same memory space. If in a given program we need to use register banks 1 and 2, we can reallocate another section of RAM to the stack. For example, we can allocate RAM locations 60H and higher to the stack as shown in Example 2-10.

Review Questions

1. What is the size of the SP register?
2. With each PUSH instruction, the stack pointer register, SP, is _____ (incremented, decremented) by 1.
3. With each POP instruction, the SP is _____ (incremented, decremented) by 1.
4. On power-up, the 8051 uses RAM location _____ as the first location of the stack.
5. On power up, the 8051 uses bank ___ for registers R0 - R7.
6. On power up, the 8051 uses RAM locations _____ to _____ for registers R0 - R7 (register bank 0).
7. Which register bank is used if we alter RS0 and RS1 of the PSW by the following two instructions?
   ```
   SETB PSW.3
   SETB PSW.4
   ```
8. In Question 7, what RAM locations are used for register R0 - R7?

SUMMARY

This chapter began with an exploration of the major registers of the 8051, including A, B, R0, R1, R2, R3, R4, R5, R6, R7, DPTR, and PC. The use of these registers was demonstrated in the context of programming examples. The process of creating an Assembly language program was described from writing the source file, to assembling it, linking, and executing the program. The PC (program counter) register always points to the next instruction to be executed. The way the 8051 uses program ROM space was explored because 8051 Assembly language programmers must be aware of where programs are placed in ROM, and how much memory is available.

An Assembly language program is composed of a series of statements that are either instructions or pseudo-instructions, also called *directives*. Instructions are translated by the assembler into machine code. Pseudo-instructions are not translated into machine code: They direct the assembler in how to translate instructions into machine code. Some pseudo-instructions, called *data directives*, are used to define data. Data is allocated in byte-size increments. The data can be in binary, hex, decimal, or ASCII formats.

Flags are useful to programmers since they indicate certain conditions, such as carry or overflow, that result from execution of instructions. The stack is used to store data temporarily during execution of a program. The stack resides in the RAM space of the 8051, which was diagrammed and explained. Manipulation of the stack via POP and PUSH instructions was also explored.

PROBLEMS

SECTION 2.1: INSIDE THE 8051

1. Most registers in the 8051 are _____ bits wide.
2. Registers R0 - R7 are all _____ bits wide
3. Registers ACC and B are _____ bits wide.
4. Name a 16-bit register in the 8051.
5. To load R4 with the value 65H, the pound sign is _____ (necessary, optional) in the instruction "MOV R4,#65H".
6. What is the result of the following code and where is it kept?

```
        MOV   A,#15H
        MOV   R2,#13H
        ADD   A,R2
```

7. Which of the following is (are) illegal?

    (a) MOV R3,#500   (b) MOV R1,#50   (c) MOV R7,#00
    (d) MOV A,#255H   (e) MOV A,#50H   (f) MOV A,#F5H
    (g) MOV R9,#50H

8. Which of the following is (are) illegal?

    (a) ADD R3,#50H   (b) ADD A,#50H   (c) ADD R7,R4
    (d) ADD A,#255H   (e) ADD A,R5   (f) ADD A,#F5H
    (g) ADD R3,A

9. What is the result of the following code and where is it kept?

```
        MOV   R4,#25H
        MOV   A,#1FH
        ADD   A,R4
```

10. What is the result of the following code and where is it kept?

```
        MOV   A,#15
        MOV   R5,#15
        ADD   A,R5
```

SECTION 2.2: INTRODUCTION TO 8051 ASSEMBLY PROGRAMMING and

SECTION 2.3: ASSEMBLING AND RUNNING AN 8051 PROGRAM

11. Assembly language is a_____ (low, high) level language while C is a _____ (low, high) level language.
12. Of C and Assembly language, which is more efficient in terms of code generation (i.e., the amount of ROM space it uses)?
13. Which program produces the "obj" file?
14. True or false. The source file has the extension "src" or "asm".
15. Which file provides the listing of error messages?
16. True or false. The source code file can be a non-ASCII file.
17. True or false. Every source file must have ORG and END directives.
18. Do the ORG and END directives produce opcodes?
19. Why are the ORG and END directives also called pseudocode?
20. True or false. The ORG and END directives appear in the ".lst" file.

21. Every 8051 family member wakes up at address _____ when it is powered up.
22. A programmer puts the first opcode at address 100H. What happens when the microcontroller is powered up?
23. Find the number of bytes each of the following instructions take.
    - (a) `MOV A,#55H`   (b) `MOV R3,#3`   (c) `INC R2`
    - (d) `ADD A,#0`   (e) `MOV A,R1`   (f) `MOV R3,A`
    - (g) `ADD A,R2`
24. Pick up a program listing of your choice, and show the ROM memory addresses and their contents.
25. Find the address of the last location of on-chip ROM for each of the following.
    - (a) DS5000-16   (b) DS5000-8   (c) DS5000-32
    - (d) AT89C52   (e) 8751   (f) AT89C51
    - (g) DS5000-64
26. Show the lowest and highest values (in hex) that the 8051 program counter can take.
27. A given 8051 has 7FFFH as the address of its last location of on-chip ROM. What is the size of on-chip ROM for this 8051?
28. Repeat Question 27 for 3FFH.

SECTION 2.5: 8051 DATA TYPES AND DIRECTIVES

29. Compile and state the contents of each ROM location for the following data.
```
          ORG   200H
MYDAT_1:  DB    "Earth"
MYDAT_2:  DB    "987-65"
MYDAT_3:  DB    "GABEH 98"
```
30. Compile and state the contents of each ROM location for the following data.
```
        ORG   340H
DAT_1:  DB    22,56H,10011001B,32,0F6H,11111011B
```

SECTION 2.6: 8051 FLAG BITS AND THE PSW REGISTER

31. The PSW is a(n) _____ -bit register.
32. Which bits of PSW are used for the CY and AC flag bits, respectively?
33. Which bits of PSW are used for the OV and P flag bits, respectively?
34. In the ADD instruction, when is CY raised?
35. In the ADD instruction, when is AC raised?
36. What is the value of the CY flag after the following code?
```
CLR   C          ;CY = 0
CPL   C          ;complement carry
```
37. Find the CY flag value after each of the following codes.
    - (a) `MOV A,#54H`   (b) `MOV A,#00`   (c) `MOV A,#250`
          `ADD A,#0C4H`   `ADD A,#0FFH`   `ADD A,#05`
38. Write a simple program in which the value 55H is added 5 times.

39. Which bits of the PSW are responsible for selection of the register banks?
40. On power up, what is the location of the first stack?
41. In the 8051, which register bank conflicts with the stack?
42. In the 8051, what is the size of the stack pointer (SP) register?
43. On power up, which of the register banks is used ?
44. Give the address locations of RAM assigned to various banks.
45. Assuming the use of bank 0, find at what RAM location each of the following lines stored the data.

    (a) MOV R4,#32H   (b) MOV R0,#12H

    (c) MOV R7,#3FH   (d) MOV R5,#55H

46. Repeat Problem 45 for bank 2.
47. After power up, show how to select bank 2 with a single instruction.
48. Show the stack and stack pointer for each line of the following program.

```
        ORG   0
MOV   R0,#66H
MOV   R3,#7FH
MOV   R7,#5DH
PUSH  0
PUSH  3
PUSH  7
CLR   A
MOV   R3,A
MOV   R7,A
POP   3
POP   7
POP   0
```

49. In Problem 48, does the sequence of POP instructions restore the original values of registers R0, R3, and R7? If not, show the correct sequence of instructions.
50. Show the stack and stack pointer for each line of the following program.

```
        ORG   0
MOV   SP,#70H
MOV   R5,#66H
MOV   R2,#7FH
MOV   R7,#5DH
PUSH  5
PUSH  2
PUSH  7
CLR   A
MOV   R2,A
MOV   R7,A
POP   7
POP   2
POP   5
```

ANSWERS TO REVIEW QUESTIONS

SECTION 2.1: INSIDE THE 8051

1. MOV A,#34H
   MOV B,#3FH
   ADD A,B
2. MOV A,#16H
   ADD A,#0CDH
   MOV R2,A
3. False
4. FF hex and 255 in decimal
5. 8

SECTION 2.2: INTRODUCTION TO 8051 ASSEMBLY PROGRAMMING

1. The real work is performed by instructions such as MOV and ADD. Pseudo-instructions, also called assembly directives, instruct the assembler in doing its job.
2. The instruction mnemonics, pseudo-instructions
3. False
4. All except (c)
5. Assembler directive
6. True
7. (c)

SECTION 2.3: ASSEMBLING AND RUNNING AN 8051 PROGRAM

1. True   2. True   3. (a)   4. (b) and (d)   5. (d)

SECTION 2.4: THE PROGRAM COUNTER AND ROM SPACE IN THE 8051

1. 16   2. True   3. 0000H   4. 2
5. With 8K bytes, we have 8192 ($8 \times 1024 = 8192$) bytes, and the ROM space is 0000 to 1FFFH.

SECTION 2.5: 8051 DATA TYPES AND DIRECTIVES

1. DB   2. 7
3. If the value is to be changed later, it can be done once in one place instead of at every occurrence.
4. (a) 4 bytes   (b) 7 bytes
5. This places the ASCII values for each character in memory locations starting at 200H. Notice that all values are in hex.
   200 = (41)
   201 = (42)
   202 = (43)
   203 = (31)
   204 = (32)
   205 = (33)

SECTION 2.6: 8051 FLAG BITS AND THE PSW REGISTER

1. PSW (program status register)   2. 8 bits
3. D1 and D5 which are referred to as PSW.1 and PSW.5, respectively.
4.

```
   Hex            binary
    FF            1111  1111
 +  1          +          1
   100            10000 0000    This leads to CY=1 and AC=1
```

5.

```
  Hex       binary
  C2        1100 0010
+ 3D      + 0011 1101
  FF        1111 1111
```

This leads to CY = 0 and AC = 0.

SECTION 2.7: 8051 REGISTER BANKS AND STACK

1. 8-bit   2. Incremented   3. Decremented
4. 08   5. 0   6. 0 - 7
7. Register bank 3
8. RAM locations 18H to 1FH

CHAPTER 3

JUMP, LOOP, AND CALL INSTRUCTIONS

OBJECTIVES

Upon completion of this chapter, you will be able to:

≫≫   **Code 8051 Assembly language instructions using loops**
≫≫   **Code 8051 Assembly language conditional jump instructions**
≫≫   **Explain conditions that determine each conditional jump instruction**
≫≫   **Code long jump instructions for unconditional jumps**
≫≫   **Code short jump instructions for unconditional short jumps**
≫≫   **Calculate target addresses for jump instructions**
≫≫   **Code 8051 subroutines**
≫≫   **Describe precautions in using the stack in subroutines**
≫≫   **Discuss crystal frequency versus machine cycle**
≫≫   **Code 8051 programs to generate a time delay**

In the sequence of instructions to be executed, it is often necessary to transfer program control to a different location. There are many instructions in the 8051 to achieve this. This chapter covers the control transfer instructions available in 8051 Assembly language. In the first section we discuss instructions used for looping, as well as instructions for conditional and unconditional jumps. In the second section we examine CALL instructions and their uses. In the third section, time delay subroutines are described.

SECTION 3.1: LOOP AND JUMP INSTRUCTIONS

In this section we first discuss how to perform a looping action in the 8051 and then talk about jump instructions, both conditional and unconditional.

Looping in the 8051

Repeating a sequence of instructions a certain number of times is called a *loop*. The loop is one of most widely used actions that any microprocessor performs. In the 8051, the loop action is performed by the instruction "DJNZ reg, label". In this instruction, the register is decremented; if it is not zero, it jumps to the target address referred to by the label. Prior to the start of the loop the register is loaded with the counter for the number of repetitions. Notice that in this instruction both the register decrement and the decision to jump are combined into a single instruction.

Example 3-1

Write a program to
(a) clear ACC, then
(b) add 3 to the accumulator ten times.

Solution:

```
;This program adds value 3 to the ACC ten times

          MOV   A,#0       ;A=0, clear ACC
          MOV   R2,#10     ;load counter R2=10
AGAIN:    ADD   A,#03      ;add 03 to ACC
          DJNZ  R2,AGAIN   ;repeat until R2=0(10 times)
          MOV   R5,A       ;save A in R5
```

In the program in Example 3-1, the R2 register is used as a counter. The counter is first set to 10. In each iteration the instruction DJNZ decrements R2 and checks its value. If R2 is not zero, it jumps to the target address associated with label "AGAIN". This looping action continues until R2 becomes zero. After R2 becomes zero, it falls through the loop and executes the instruction immediately below it, in this case the "MOV R5,A" instruction.

Notice in the DJNZ instruction that the registers can be any of R0 - R7. The counter can also be a RAM location as we will see in Chapter 5.

Example 3-2

What is the maximum number of times that the loop in Example 3-1 can be repeated?

Solution:

Since R2 holds the count and R2 is an 8-bit register, it can hold a maximum of FFH (255 decimal); therefore, the loop can be repeated a maximum of 256 times.

Loop inside a loop

As shown in Example 3-2, the maximum count is 256. What happens if we want to repeat an action more times than 256? To do that, we use a loop inside a loop, which is called a *nested loop*. In a nested loop, we use two registers to hold the count. See Example 3-3.

Example 3-3

Write a program to (a) load the accumulator with the value 55H, and (b) complement the ACC 700 times.

Solution:

Since 700 is larger than 255 (the maximum capacity of any register), we use two registers to hold the count. The following code shows how to use R2 and R3 for the count.

```
        MOV  A,#55H      ;A=55H
        MOV  R3,#10      ;R3=10, the outer loop count
NEXT:   MOV  R2,#70      ;R2=70, the inner loop count
AGAIN:  CPL  A           ;complement A register
        DJNZ R2,AGAIN    ;repeat it 70 times (inner loop)
        DJNZ R3,NEXT
```

In this program, R2 is used to keep the inner loop count. In the instruction "DJNZ R2,AGAIN", whenever R2 becomes 0 it falls through and "DJNZ R3,NEXT" is executed. This instruction forces the CPU to load R2 with the count 70 and the inner loop starts again. This process will continue until R3 becomes zero and the outer loop is finished.

Other conditional jumps

Conditional jumps for the 8051 are summarized in Table 3-1. More details of each instruction are provided in Appendix A. In Table 3-1, notice that some of the instructions, such as JZ (jump if A = zero) and JC (jump if carry), jump only if a certain condition is met. Next we examine some conditional jump instructions with examples.

CHAPTER 3: JUMP, LOOP, AND CALL INSTRUCTIONS   **67**

JZ (jump if A = 0)

In this instruction the content of register A is checked. If it is zero, it jumps to the target address. For example, look at the following code.

```
MOV   A,R0          ;A=R0
JZ    OVER          ;jump if A = 0
MOV   A,R1          ;A=R1
JZ    OVER          ;jump if A = 0
      . . .
OVER:
```

In this program, if either R0 or R1 is zero, it jumps to the label OVER. Notice that the JZ instruction can be used only for register A. It can only check to see whether the accumulator is zero, and it does not apply to any other register. More importantly, you don't have to perform an arithmetic instruction such as decrement to use the JNZ instruction. See Example 3-4.

Table 3-1: 8051 Conditional Jump Instructions

Instruction	Action
JZ	Jump if A = 0
JNZ	Jump if A ≠ 0
DJNZ	Decrement and jump if A ≠ 0
CJNE A,byte	Jump if A ≠ byte
CJNE reg,#data	Jump if byte ≠ #data
JC	Jump if CY = 1
JNC	Jump if CY = 0
JB	Jump if bit = 1
JNB	Jump if bit = 0
JBC	Jump if bit = 1 and clear bit

Example 3-4

Write a program to determine if R5 contains the value 0. If so, put 55H in it.

Solution:

```
      MOV   A,R5        ;copy R5 to A
      JNZ   NEXT        ;jump if A is not zero
      MOV   R5,#55H
NEXT:     . . .
```

JNC (jump if no carry, jumps if CY = 0)

In this instruction, the carry flag bit in the flag (PSW) register is used to make the decision whether to jump. In executing "JNC label", the processor looks at the carry flag to see if it is raised (CY = 1). If it is not, the CPU starts to fetch and execute instructions from the address of the label. If CY = 1, it will not jump but will execute the next instruction below JNC.

It needs to be noted that there is also a "JC label" instruction. In the JC instruction, if CY = 1 it jumps to the target address. We will give more examples of these instructions in the context of applications in future chapters.

There is also a JB (jump if bit is high) and JNB (jump if bit is low). These are discussed in Chapters 4 and 8 when bit manipulation instructions are discussed.

Example 3-5

Find the sum of the values 79H, F5H, and E2H. Put the sum in registers R0 (low byte) and R5 (high byte).

Solution:

```
        MOV   A,#0        ;clear A(A=0)
        MOV   R5,A        ;clear R5
        ADD   A,#79H      ;A=0+79H=79H
        JNC   N_1         ;if no carry, add next number
        INC   R5          ;if CY=1, increment R5
N_1:    ADD   A,#0F5H     ;A=79+F5=6E and CY=1
        JNC   N_2         ;jump if CY=0
        INC   R5          ;If CY=1 then increment R5(R5=1)
N_2:    ADD   A,#0E2H     ;A=6E+E2=50 and CY=1
        JNC   OVER        ;jump if CY=0
        INC   R5          ;if CY=1, increment 5
OVER:MOV     R0,A         ;Now R0=50H, and R5=02
```

All conditional jumps are short jumps

It must be noted that all conditional jumps are short jumps, meaning that the address of the target must be within -128 to $+127$ bytes of the contents of the program counter (PC). This very important concept is discussed at the end of this section.

Unconditional jump instructions

The unconditional jump is a jump in which control is transferred unconditionally to the target location. In the 8051 there are two unconditional jumps: LJMP (long jump) and SJMP (short jump). Each is discussed below.

LJMP (long jump)

LJMP is an unconditional long jump. It is a 3-byte instruction in which the first byte is the opcode, and the second and third bytes represent the 16-bit address of the target location. The 2-byte target address allows a jump to any memory location from 0000 to FFFFH.

Remember that although the program counter in the 8051 is 16-bit, thereby giving a ROM address space of 64K bytes, not all 8051 family members have that much on-chip program ROM. The original 8051 had only 4K bytes of on-chip ROM for program space; consequently, every byte was precious. For this reason there is also a SJMP (short jump) instruction which is a 2-byte instruction as opposed to the 3-byte LJMP instruction. This can save some bytes of memory in many applications where memory space is in short supply. SJMP is discussed next.

SJMP (short jump)

In this 2-byte instruction, the first byte is the opcode and the second byte is the relative address of the target location. The relative address range of 00 - FFH is divided into forward and backward jumps; that is, within –128 to +127 bytes of memory relative to the address of the current PC (program counter). If the jump is forward, the target address can be within a space of 127 bytes from the current PC. If the target address is backward, the target address can be within –128 bytes from the current PC. This is explained in detail next.

Calculating the short jump address

In addition to the SJMP instruction, all conditional jumps such as JNC, JZ, and DJNZ are also short jumps due to the fact that they are all two-byte instructions. In these instructions the first byte is the opcode and the second byte is the relative address. The target address is relative to the value of the program counter. To calculate the target address, the second byte is added to the PC of the instruction immediately below the jump. To understand this, look at Example 3-6.

Example 3-6

Using the following list file, verify the jump forward address calculation.

Line	PC	Opcode	Mnemonic	Operand
01	0000		ORG	0000
02	0000	7800	MOV	R0,#0
03	0002	7455	MOV	A,#55H
04	0004	6003	JZ	NEXT
05	0006	08	INC	R0
06	0007	04	AGAIN: INC	A
07	0008	04	INC	A
08	0009	2477	NEXT: ADD	A,#77h
09	000B	5005	JNC	OVER
10	000D	E4	CLR	A
11	000E	F8	MOV	R0,A
12	000F	F9	MOV	R1,A
13	0010	FA	MOV	R2,A
14	0011	FB	MOV	R3,A
15	0012	2B	OVER: ADD	A,R3
16	0013	50F2	JNC	AGAIN
17	0015	80FE	HERE: SJMP	HERE
18	0017		END	

Solution:

First notice that the JZ and JNC instructions both jump forward. The target address for a forward jump is calculated by adding the PC of the following instruction to the second byte of the short jump instruction, which is called the relative address. In line 4 the instruction "JZ NEXT" has opcode of 60 and operand of 03 at the addresses of 0004 and 0005. The 03 is the relative address, relative to the address of the next instruction INC R0, which is 0006. By adding 0006 to 3, the target address of the label NEXT, which is 0009, is generated. In the same way for line 9, the "JNC OVER" instruction has opcode and operand of 50 and 05 where 50 is the opcode and 05 the relative address. Therefore, 05 is added to 000D, the address of instruction "CLR A", giving 12H, the address of label OVER.

Example 3-7

Verify the calculation of backward jumps in Example 3-6.

Solution:

In that program list, "JNC AGAIN" has opcode 50 and relative address F2H. When the relative address of F2H is added to 15H, the address of the instruction below the jump, we have 15H + F2H = 07 (the carry is dropped). Notice that 07 is the address of label AGAIN. Look also at "SJMP HERE", which has 80 and FE for the opcode and relative address, respectively. The PC of the following instruction, 0017H, is added to FEH, the relative address, to get 0015H, address of the HERE label (17H + FEH = 15H). Notice that FEH is –2 and 17H + (–2) = 15H. For further discussion of the addition of negative numbers, see Chapter 6.

Jump backward target address calculation

While in the case of a forward jump, the displacement value is a positive number between 0 to 127 (00 to 7F in hex), for the backward jump the displacement is a negative value of 0 to –128 as explained in Example 3-7.

It must be emphasized that regardless of whether the SJMP is a forward or backward jump, for any short jump the address of the target address can never be more than –128 to +127 bytes from the address associated with the instruction below the SJMP. If any attempt is made to violate this rule, the assembler will generate an error stating the jump is out of range.

Review Questions

1. The mnemonic DJNZ stands for _____.
2. True or false. "DJNZ R5,BACK" combines a decrement and a jump in a single instruction.
3. "JNC HERE" is a ___-byte instruction.
4. In "JZ NEXT", which register's content is checked to see if it is zero?
5. LJMP is a ___-byte instruction.

SECTION 3.2: CALL INSTRUCTIONS

Another control transfer instruction is the CALL instruction, which is used to call a subroutine. Subroutines are often used to perform tasks that need to be performed frequently. This makes a program more structured in addition to saving memory space. In the 8051 there are two instructions for call: LCALL (long call) and ACALL (absolute call). Deciding which one to use depends on the target address. Each instruction is explained next.

LCALL (long call)

In this 3-byte instruction, the first byte is the opcode and the second and third bytes are used for the address of the target subroutine. Therefore, LCALL can be used to call subroutines located anywhere within the 64K byte address space of the 8051. To make sure that after execution of the called subroutine the 8051 knows where to come back to, it automatically saves on the stack the address of the instruction immediately below the LCALL. When a subroutine is called, control is transferred to that subroutine, and the processor saves the PC (program counter) on the stack and begins to fetch instructions from the new location. After finishing execution of the subroutine, the instruction RET (return) transfers control back to the caller. Every subroutine needs RET as the last instruction. See Example 3-8.

The following points should be noted for the program in Example 3-8.

1. Notice the DELAY subroutine. Upon executing the first "LCALL DELAY", the address of the instruction right below it, "MOV A,#0AAH", is pushed onto the stack, and the 8051 starts to execute instructions at address 300H.

2. In the DELAY subroutine, first the counter R5 is set to 255 (R5 = FFH); therefore, the loop is repeated 256 times. When R5 becomes 0, control falls to the RET instruction which pops the address from the stack into the program counter and resumes executing the instructions after the CALL.

Example 3-8

Write a program to toggle all the bits of port 1 by sending to it the values 55H and AAH continuously. Put a time delay in between each issuing of data to port 1. This program will be used to test the ports of the 8051 in the next chapter.

Solution:

```
            ORG    0
BACK:       MOV    A,#55H      ;load A with 55H
            MOV    P1,A        ;send 55H to port 1
            LCALL  DELAY       ;time delay
            MOV    A,#0AAH     ;load A with AA (in hex)
            MOV    P1,A        ;send AAH to port 1
            LCALL  DELAY
            SJMP   BACK        ;keep doing this indefinitely
;────── this is the delay subroutine
            ORG    300H        ;put time delay at address 300H
DELAY:      MOV    R5,#0FFH    ;R5=255(FF in hex),the counter
AGAIN:      DJNZ   R5,AGAIN    ;stay here until R5 becomes 0
            RET                ;return to caller (when R5 = 0)
            END                ;end of asm file
```

The amount of time delay in Example 3-8 depends on the frequency of the 8051. How to calculate the exact time will be explained in detail in Chapter 4. However you can increase the time delay by using a nested loop as shown below.

```
DELAY:                    ;nested loop delay
          MOV R4,#255     ;R4=255(FF in hex)
NEXT:     MOV R5,#255     ;R5=255(FF in hex)
AGAIN:    DJNZ R5,AGAIN   ;stay here until R5 becomes 0
          DJNZ R4,NEXT    ;decrement R4
                          ;keep loading R5 until R4=0
          RET             ;return (when R4 = 0)
```

CALL instruction and the role of the stack

The stack and stack pointer were covered in the last chapter. To understand the importance of the stack in microcontrollers, we now examine the contents of the stack and stack pointer for Example 3-8. This is shown in Example 3-9.

Example 3-9

Analyze the stack contents after the execution of the first LCALL in the following.

Solution:

```
001  0000                        ORG   0
002  0000  7455   BACK:   MOV   A,#55H   ;load A with 55H
003  0002  F590            MOV   P1,A     ;send 55H to port 1
004  0004  120300          LCALL DELAY    ;time delay
005  0007  74AA            MOV   A,#0AAH  ;load A with AAH
006  0009  F590            MOV   P1,A     ;send AAH to port 1
007  000B  120300          LCALL DELAY
008  000E  80F0            SJMP  BACK     ;keep doing this
009  0010
010  0010  ;————————this is the delay subroutine
011  0300                        ORG 300H
012  0300           DELAY:
013  0300  7DFF            MOV   R5,#0FFH ;R5=255
014  0302  DDFE   AGAIN:   DJNZ  R5,AGAIN ;stay here
015  0304  22              RET            ;return to caller
016  0305                  END            ;end of asm file
```

When the first LCALL is executed, the address of the instruction "MOV A,#0AAH" is saved on the stack. Notice that the low byte goes first and the high byte is last. The last instruction of the called subroutine must be a RET instruction which directs the CPU to POP the top bytes of the stack into the PC and resume executing at address 07. The diagram shows the stack frame after the first LCALL.

0A	
09	00
08	07

SP = 09

Use of PUSH and POP instructions in subroutines

Upon calling a subroutine, the stack keeps track of where the CPU should return after completing the subroutine. For this reason, we must be very careful in any manipulation of stack contents. The rule is that the number of PUSH and POP instructions must always match in any called subroutine. In other words, for every PUSH there must be a POP. See Example 3-10.

Calling subroutines

In Assembly language programming it is common to have one main program and many subroutines that are called from the main program. This allows you to make each subroutine into a separate module. Each module can be tested separately and then brought together with the main program. More importantly, in a large program the modules can be assigned to different programmers in order to shorten development time.

Example 3-10

Analyze the stack for the first LCALL instruction in the following program.

```
01  0000                  ORG   0
02  0000  7455  BACK:     MOV   A,#55H      ;load A with 55H
03  0002  F590            MOV   P1,A        ;send 55H to port 1
04  0004  7C99            MOV   R4,#99H
05  0006  7D67            MOV   R5,#67H
06  0008  120300          LCALL DELAY       ;time delay
07  000B  74AA            MOV   A,#0AAH     ;Load A with AA
08  000D  F590            MOV   P1,A        ;send AAH to port 1
09  000F  120300          LCALL DELAY
10  0012  80EC            SJMP  BACK        ;keep doing this
11  0014                  ;————this is the delay subroutine
12  0300                  ORG 300H
13  0300  C004  DELAY:PUSH 4                ;PUSH R4
14  0302  C005        PUSH 5                ;PUSH R5
15  0304  7CFF        MOV   R4,#0FFH        ;R4=FFH
16  0306  7DFF  NEXT: MOV   R5,#0FFH        ;R5=255
17  0308  DDFE  AGAIN:DJNZ R5,AGAIN
18  030A  DCFA        DJNZ R4,NEXT
19  030C  D005        POP   5               ;POP INTO R5
20  030E  D004        POP   4               ;POP INTO R4
21  0310  22          RET                   ;return to caller
22  0311              END                   ;end of asm file
```

Solution:

First notice that for the PUSH and POP instructions we must specify the direct address of the register being pushed or popped. Here is the stack frame.

After the first LCALL		After PUSH 4			After PUSH 5		
0B		0B			0B	67	R5
0A		0A	99	R4	0A	99	R4
09	00	PCH	09	00	PCH	09	00	PCH
08	0B	PCL	08	0B	PCL	08	0B	PCL

It needs to be emphasized that in using LCALL, the target address of the subroutine can be anywhere within the 64K bytes memory space of the 8051. This is not the case for the other call instruction, ACALL, which is explained next.

```
;MAIN program calling subroutines
          ORG   0
MAIN:     LCALL      SUBR_1
          LCALL      SUBR_2
          LCALL      SUBR_3

HERE:     SJMP       HERE
;———————end of MAIN
;
SUBR_1:   ....
          ....
          RET
;———————end of subroutine 1
;
SUBR_2:   ....
          ....
          RET
;———————end of subroutine 2

SUBR_3:   ....
          ....
          RET
;———————end of subroutine 3
          END        ;end of the asm file
```

Figure 3-1. 8051 Assembly Main Program That Calls Subroutines

ACALL (absolute call)

ACALL is a 2-byte instruction in contrast to LCALL, which is 3 bytes. Since ACALL is a 2-byte instruction, the target address of the subroutine must be within 2K bytes address because only 11 bits of the 2 bytes are used for the address. There is no difference between ACALL and LCALL in terms of saving the program counter on the stack or the function of the RET instruction. The only difference is that the target address for LCALL can be anywhere within the 64K byte address space of the 8051 while the target address of ACALL must be within a 2K-byte range. In many variations of the 8051 marketed by different companies, on-chip ROM is as low as 1K bytes. In such cases, the use of ACALL instead of LCALL can save a number of bytes of program ROM space.

Example 3-11
A developer is using the Atmel AT89C1051 microcontroller chip for a product. This chip has only 1K bytes of on-chip flash ROM. Which of the instructions LCALL and ACALL is most useful in programming this chip? **Solution:** The ACALL instruction is more useful since it is a 2-byte instruction. It saves one byte each time the call instruction is used.

Of course in addition to using compact instructions, we can program efficiently by having a detailed knowledge of all the instructions supported by a given microprocessor, and using them wisely. Look at Example 3-12.

Example 3-12

Rewrite Example 3-8 as efficiently as you can.

Solution:

```
        ORG  0
        MOV  A,#55H    ;load A with 55H
BACK:   MOV  P1,A      ;issue value in reg A to port 1
        ACALL DELAY    ;time delay
        CPL  A         ;complement reg A
        SJMP BACK      ;keep doing this indefinitely

;————this is the delay subroutine
DELAY:
        MOV  R5,#0FFH  ;R5=255(FF in hex),the counter
AGAIN:  DJNZ R5,AGAIN  ;stay here until R5 becomes 0
        RET            ;return to caller
        END            ;end of asm file
```

Notice in this program that register A is set to 55H. By complementing 55H, we have AAH; and by complementing AAH we have 55H. Why? "01010101" in binary (55H) becomes "10101010" in binary (AAH) when it is complemented; and "10101010" becomes "01010101" if it is complemented.

Review Questions

1. What do the mnemonics "LCALL" and "ACALL" stand for?
2. True or false. In the 8051, control can be transferred anywhere within the 64K bytes of code space if using the LCALL instruction.
3. How does the CPU know where to return to after executing the RET instruction?
4. Describe briefly the function of the RET instruction.
5. The LCALL instruction is a ___-byte instruction.

SECTION 3.3: TIME DELAY GENERATION AND CALCULATION

In the last section we used the DELAY subroutine. How to generate various time delays and calculate exact delays is discussed in this section.

Machine cycle

For the CPU to execute an instruction takes a certain number of clock cycles. In the 8051 family, these clock cycles are referred to as *machine cycles*. Appendix A.2 provides the list of 8051 instructions and their machine cycles. To calculate a time delay, we use this list. In the 8051 family, the length of the machine cycle depends on the frequency of the crystal oscillator connected to the

8051 system. The crystal oscillator, along with on-chip circuitry, provide the clock source for the 8051 CPU (see Chapter 4). The frequency of the crystal connected to the 8051 family can vary from 4 MHz to 30 MHz, depending on the chip rating and manufacturer. Very often the 11.0592 MHz crystal oscillator is used to make the 8051-based system compatible with the serial port of the IBM PC (see Chapter 10). In the 8051, one machine cycle lasts 12 oscillator periods. Therefore, to calculate the machine cycle, we take 1/12 of the crystal frequency, then take its inverse, as shown in Example 3-13.

Example 3-13

The following shows crystal frequency for three different 8051-based systems. Find the period of the machine cycle in each case.
(a) 11.0592 MHz   (b) 16 MHz   (c) 20 MHz

Solution:

(a) 11.0592/12 = 921.6 kHz; machine cycle is 1/921.6 kHz = 1.085 μs (microsecond)
(b) 16 MHz/12 = 1.333 MHz; machine cycle (MC) = 1/.1.333 MHz = 0.75 μs
(c) 20 MHz/12 = 1.66 MHz; MC = 1/1.66 MHz = 0.60 μs

Example 3-14

For an 8051 system of 11.0592 MHz, find how long it takes to execute each of the following instructions.

(a) MOV R3,#55   (b) DEC R3   (c) DJNZ R2,target
(d) LJMP   (e) SJMP   (f) NOP (no operation)
(g) MUL AB

Solution:

The machine cycle for a system of 11.0592 MHz is 1.085 μs as shown in Example 3-13. Table A-1 in Appendix A shows machine cycles for each of the above instructions. Therefore, we have:

Instruction	Machine cycles	Time to execute
(a) MOV R3,#55	1	1×1.085 μs = 1.085 μs
(b) DEC R3	1	1×1.085 μs = 1.085 μs
(c) DJNZ R2,target	2	2×1.085 μs = 2.17 μs
(d) LJMP	2	2×1.085 μs = 2.17 μs
(e) SJMP	2	2×1.085 μs = 2.17 μs
(f) NOP	1	1×1.085 μs = 1.085 μs
(g) MUL AB	4	4×1.085 μs = 4.34 μs

Delay calculation

As seen in the last section, a delay subroutine consists of two parts: (1) setting a counter, and (2) a loop. Most of the time delay is performed by the body of the loop, as shown in Example 3-15.

Example 3-15

Find the size of the delay in the following program, if the crystal frequency is 11.0592 MHz.

```
            MOV   A,#55H
AGAIN:      MOV   P1,A
            ACALL DELAY
            CPL   A
            SJMP  AGAIN
;----Time delay
DELAY:      MOV   R3,#200
HERE:       DJNZ  R3,HERE
            RET
```

Solution:

From Table A-1 in Appendix A, we have the following machine cycles for each instruction of the DELAY subroutine.

		Machine Cycle
DELAY:	MOV R3,#200	1
HERE:	DJNZ R3,HERE	2
	RET	1

Therefore, we have a time delay of $[(200 \times 2) + 1 + 1] \times 1.085$ μs = 436.17 μs.

Very often we calculate the time delay based on the instructions inside the loop and ignore the clock cycles associated with the instructions outside the loop.

In Example 3-15, the largest value the R3 register can take is 255; therefore, one way to increase the delay is to use NOP instructions in the loop. NOP, which stands for "no operation," simply wastes time. This is shown in Example 3-16.

Loop inside loop delay

Another way to get a large delay is to use a loop inside a loop, which is also called a *nested loop*. See Example 3-17.

Example 3-16

Find the time delay for the following subroutine, assuming a crystal frequency of 11.0592 MHz.

```
                                   Machine Cycle
DELAY:      MOV   R3,#250               1

HERE:       NOP                         1
            NOP                         1
            NOP                         1
            NOP                         1
            DJNZ R3,HERE                2

            RET                         1
```

Solution:

The time delay inside the HERE loop is [250 (1+1+1+1+2)] × 1.085 μs = 1500 × 1.085 μs = 1627.5 μs. Adding the two instructions outside the loop we have 1627.5 μs + 2 × 1.085 μ s = 1629.67 μs.

Example 3-17

For a machine cycle of 1.085 μs, find the time delay in the following subroutine.

```
DELAY:                              Machine Cycle
            MOV   R2,#200               1
AGAIN:      MOV   R3,#250               1
HERE:       NOP                         1
            NOP                         1
            DJNZ R3,HERE                2
            DJNZ R2,AGAIN               2
            RET                         1
```

For the HERE loop, we have (4 × 250) 1.085 μs = 1085 μs. The AGAIN loop repeats the HERE loop 200 times; therefore, we have 200 × 1085 μs = 217000, if we do not include the overhead. However, the instructions "MOV R3,#250" and "DJNZ R2,AGAIN" at the beginning and end of the AGAIN loop add (3 × 200 × 1.085 μs) = 651 μs to the time delay. As a result we have 217000 + 651 = 217651 μs = 217.651 milliseconds for total time delay associated with the above DELAY subroutine. Notice that in the case of a nested loop, as in all other time delay loops, the time is approximate since we have ignored the first and last instructions in the subroutine.

Review Questions

1. True or false. In the 8051, the machine cycle lasts 12 clock cycles of the crystal frequency.
2. The minimum number of machine cycles needed to execute an 8051 instruction is _____.
3. For Question 2, what is the maximum number of cycles needed, and for which instructions?
4. Find the machine cycle for a crystal frequency of 12 MHz.
5. Assuming a crystal frequency of 12 MHz, find the time delay associated with the loop section of the following DELAY subroutine.

```
DELAY:
            MOV   R3,#100
HERE:       NOP
            NOP
            NOP
            DJNZ  R3,HERE
            RET
```

SUMMARY

The flow of a program proceeds sequentially, from instruction to instruction, unless a control transfer instruction is executed. The various types of control transfer instructions in Assembly language include conditional and unconditional jumps, and call instructions.

The looping action in 8051 Assembly language is performed using a special instruction which decrements a counter and jumps to the top of the loop if the counter is not zero. Other jump instructions jump conditionally, based on the value of the carry flag, the accumulator, or bits of the I/O port. Unconditional jumps can be long or short, depending on the relative value of the target address. Special attention must be given to the effect of LCALL and ACALL instructions on the stack.

PROBLEMS

SECTION 3.1: LOOP AND JUMP INSTRUCTIONS

1. In the 8051, looping action with instruction "DJNZ Rx, rel address" is limited to ____ iterations.
2. If a conditional jump is not taken, what is the next instruction to be executed?
3. In calculating the target address for a jump, a displacement is added to the contents of register _____.
4. The mnemonic SJMP stands for _____ and it is a ___-byte instruction.
5. The mnemonic LJMP stands for _____ and it is a ___-byte instruction.
6. What is the advantage of using SJMP over LJMP?
7. True or false. The target of a short jump is within –128 to +127 bytes of the current PC.
8. True or false. All 8051 jumps are short jumps.

9. Which of the following instructions is (are) not a short jump?
   (a) JZ   (b) JNC   (c) LJMP   (d) DJNZ
10. A short jump is a ___-byte instruction. Why?
11. True or false. All conditional jumps are short jumps.
12. Show code for a nested loop to perform an action 1000 times.
13. Show code for a nested loop to perform an action 100,000 times.
14. Find the number of times the following loop is performed.

```
        MOV   R6,#200
BACK:   MOV   R5,#100
HERE:   DJNZ  R5,HERE
        DJNZ  R6,BACK
```

15. The target address of a jump backward is a maximum of _____ bytes from the current PC.
16. The target address of a jump forward is a maximum of _____ bytes from the current PC.

SECTION 3.2: CALL INSTRUCTIONS

17. LCALL is a ___-byte instruction.
18. ACALL is a ___-byte instruction.
19. The ACALL target address is limited to ____ bytes from the present PC.
20. The LCALL target address is limited to ____ bytes from the present PC.
21. When LCALL is executed, how many bytes of the stack are used?
22. When ACALL is executed, how many bytes of the stack are used?
23. Why do the number of PUSH and POP instructions in a subroutine need to be equal?
24. Describe the action associated with the POP instruction.
25. Show the stack for the following code.

```
000B  120300            LCALL DELAY
000E  80F0              SJMP  BACK      ;keep doing this
0010
0010  ;————————this is the delay subroutine
0300                    ORG 300H
0300          DELAY:
0300  7DFF              MOV   R5,#0FFH  ;R5=255
0302  DDFE  AGAIN:      DJNZ  R5,AGAIN  ;stay here
0304  22                RET             ;return
```

26. Reassemble Example 3-10 at ORG 200 (instead of ORG 0) and show the stack frame for the first LCALL instruction.

SECTION 3.3:TIME DELAY GENERATION AND CALCULATION

27. Find the system frequency if the machine cycle = 1.2 μs.
28. Find the machine cycle if crystal frequency is 18 MHz.
29. Find the machine cycle if crystal frequency is 12 MHz.
30. Find the machine cycle if crystal frequency is 25 MHz.

31. True or false. LJMP and SJMP instructions take the same amount of time to execute even though one is a 3-byte instruction and the other one is a 2-byte instruction.

32. Find the time delay for the delay subroutine shown to the right, if the system frequency is 11.0592 MHz.

```
DELAY:    MOV   R3,#150
HERE:     NOP
          NOP
          NOP
          DJNZ  R3,HERE
          RET
```

33. Find the time delay for the delay subroutine shown to the right, if the system frequency is 16 MHz.

```
DELAY:    MOV   R3,#200
HERE:     NOP
          NOP
          NOP
          DJNZ  R3,HERE
          RET
```

34. Find the time delay for the delay subroutine shown to the right, if the system frequency is 11.0592 MHz.

```
DELAY:    MOV   R5,#100
BACK:     MOV   R2,#200
AGAIN:    MOV   R3,#250
HERE:     NOP
          NOP
          DJNZ  R3,HERE
          DJNZ  R2,AGAIN
          DJNZ  R5,BACK
          RET
```

35. Find the time delay for the delay subroutine shown to the right, if the system frequency is 16 MHz.

```
DELAY:    MOV   R2,#150
AGAIN:    MOV   R3,#250
HERE:     NOP
          NOP
          NOP
          DJNZ  R3,HERE
          DJNZ  R2,AGAIN
          RET
```

ANSWERS TO REVIEW QUESTIONS

SECTION 3.1: LOOP AND JUMP INSTRUCTIONS
1. Decrement and jump if not zero   2. True   3. 2   4. A   5. 3

SECTION 3.2: CALL INSTRUCTIONS
1. Long CALL and Absolute CALL   2. True
3. The address of where to return is in the stack.
4. Upon executing the RET instruction, the CPU pops off the top two bytes of the stack into the program counter (PC) register and starts to execute from this new location.
5. 3

SECTION 3.3: TIME DELAY GENERATION AND CALCULATION
1. True   2. 1   3. MUL and DIV each take 4 machine cycles.
4. 12 MHz / 12 = 1 MHz, and MC = 1/1 MHz = 1 μs.
5. [100 (1+1+1+2)] $\times$ 1 μs = 500 μs = 0.5 milliseconds.

CHAPTER 4

I/O PORT PROGRAMMING

OBJECTIVES

Upon completion of this chapter, you will be able to:

≫≫ Explain the purpose of each pin of the 8051 microcontroller
≫≫ List the 4 ports of the 8051
≫≫ Describe the dual role of port 0 in providing both data and addresses
≫≫ Code Assembly language to use the ports for input or output
≫≫ Explain the use of port 3 for interrupt signals
≫≫ Code 8051 instructions for I/O handling
≫≫ Code bit-manipulation instructions in the 8051

This chapter describes the 8051 pins and then shows I/O port programming of the 8051 with many examples.

SECTION 4.1: PIN DESCRIPTION OF THE 8051

Although 8051 family members (e.g., 8751, 89C51, DS5000) come in different packages, such as DIP (dual in-line package), QFP (quad flat package), and LLC (leadless chip carrier), they all have 40 pins that are dedicated for various functions such as I/O, $\overline{RD}$, $\overline{WR}$, address, data, and interrupts. It must be noted that some companies provide a 20-pin version of the 8051 with a reduced number of I/O ports for less demanding applications. However, since the vast majority of developers use the 40-pin DIP package chip, we will concentrate on that.

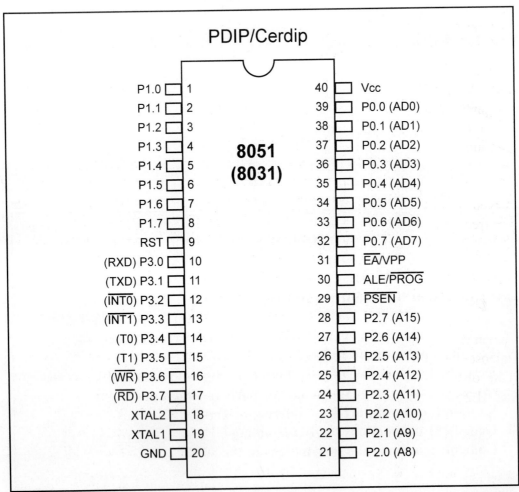

Figure 4-1. 8051 Pin Diagram

Examining Figure 4-1, note that of the 40 pins, a total of 32 pins are set aside for the four ports P0, P1, P2, and P3, where each port takes 8 pins. The rest of the pins are designated as V_{CC}, GND, XTAL1, XTAL2, RST, $\overline{EA}$, $\overline{PSEN}$. Of these 8 pins, six of them (V_{CC}, GND, XTAL1, XTAL2, RST, and $\overline{EA}$) are used by all members of the 8051 and 8031 families. In other words, they must be connected in order for the system to work, regardless of whether the microcontroller is of the 8051 or 8031 family. The other two pins, $\overline{\text{PSEN}}$ and ALE, are used mainly in 8031-based systems. We first describe the function of each pin. Ports are discussed separately.

V_{CC}

Pin 40 provides supply voltage to the chip. The voltage source is +5V.

GND

Pin 20 is the ground.

XTAL1 and XTAL2

The 8051 has an on-chip oscillator but requires an external clock to run it. Most often a quartz crystal oscillator is connected to inputs XTAL1 (pin 19) and XTAL2 (pin 18). The quartz crystal oscillator connected to XTAL1 and XTAL2 also needs two capacitors of 30 pF value. One side of each capacitor is connected to the ground as shown in Figure 4-2 (a).

It must be noted that there are various speeds of the 8051 family. Speed refers to the maximum oscillator frequency connected to XTAL. For example, a 12-MHz chip must be connected to a crystal with 12 MHz frequency or less. Likewise, a 20-MHz microcontroller requires a crystal frequency of no more than 20 MHz. When the 8051 is connected to a crystal oscillator and is powered up, we can observe the frequency on the XTAL2 pin using the oscilloscope.

If you decide to use a frequency source other than a crystal oscillator, such as a TTL oscillator, it will be connected to XTAL1; XTAL2 is left unconnected, as shown in Figure 4-2 (b).

RST

Pin 9 is the RESET pin. It is an input and is active high (normally low). Upon applying a high pulse to this pin, the microcontroller will reset and terminate all activities. This is often referred to

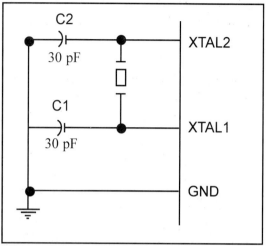

Figure 4-2 (a). XTAL Connection to 8051

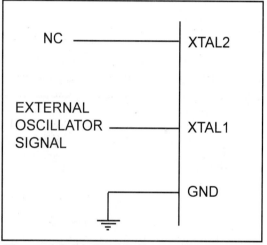

Figure 4-2 (b). XTAL Connection to an External Clock Source

Table 4-1: RESET Value of Some 8051 Registers

Register	Reset Value
PC	0000
ACC	0000
B	0000
PSW	0000
SP	0007
DPTR	0000

as a *power-on reset*. Activating a power-on reset will cause all values in the registers to be lost. Table 4-1 provides a partial list of 8051 registers and their values after power-on reset.

Notice that the value of the PC (program counter) is 0 upon reset, forcing the CPU to fetch the first opcode from ROM memory location 0000. This means that we must place the first line of source code in ROM location 0 because that is where the CPU wakes up and expects to find the first instruction. Figure 4-3 shows two ways of connecting the RST pin to the power-on reset circuitry.

In order for the RESET input to be effective, it must have a minimum duration of 2 machine cycles. In other words, the high pulse must be high for a minimum of 2 machine cycles before it is allowed to go low.

In the 8051, a machine cycle is defined as 12 oscillator periods, as discussed in Chapter 3, as shown again in Example 4-1.

$\overline{EA}$

The 8051 family members, such as the 8751, 89C51, or DS5000, all come with on-chip ROM to store programs. In such cases, the $\overline{EA}$ pin is connected to V_{CC}. For family members such as the 8031 and 8032 in which there is no on-chip ROM, code is stored on an external ROM and is fetched by the 8031/32. Therefore, for the 8031 the $\overline{EA}$ pin must be connected to GND to indicate that the code is stored externally. $\overline{EA}$, which stands for "external access," is pin number 31 in the DIP packages. It is an input pin and must be connected to either V_{CC} or GND. In other words, it cannot be left unconnected.

In Chapter 14, we will show how the 8031 uses this pin along with $\overline{PSEN}$ to access programs stored in ROM memory located outside the 8031. In 8051 chips with on-chip ROM, such as the 8751, 89C51, or DS5000, $\overline{EA}$ is connected to V_{CC}, as we will see in the next section.

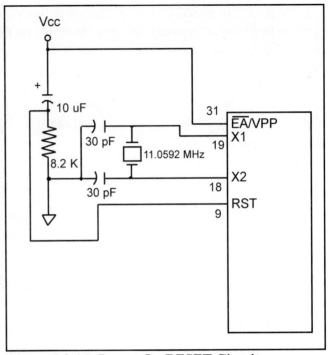

Figure 4-3 (a). Power-On RESET Circuit

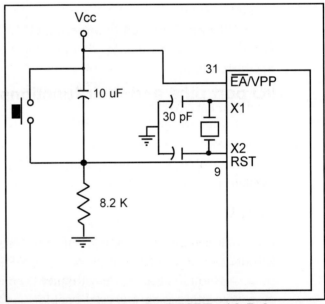

Figure 4-3 (b). Power-On RESET with Debounce

Example 4-1

Find the machine cycle for (a) XTAL = 11.0592 MHz (b) XTAL = 16 MHz.

Solution:

(a) 11.0592 MHz / 12 = 921.6 kHz;
    machine cycle = 1 / 921.6 kHz = 1.085 μs

(b) 16 MHz / 12 = 1.333 MHz;
    machine cycle = 1 / 1.333 MHz = 0.75 μs

The pins discussed so far must be connected no matter which family member is used. The next two pins are used mainly in 8031-based systems and are discussed in more detail in Chapter 14. The following is a brief description of each.

$\overline{PSEN}$

This is an output pin. $\overline{PSEN}$ stands for "program store enable." In an 8031-based system in which an external ROM holds the program code, this pin is connected to the OE pin of the ROM. See Chapter 14 to see how this is used.

ALE

ALE (address latch enable) is an output pin and is active high. When connecting an 8031 to external memory, port 0 provides both address and data. In other words, the 8031 multiplexes address and data through port 0 to save pins. The ALE pin is used for demultiplexing the address and data by connecting to the G pin of the 74LS373 chip. This is discussed in detail in Chapter 14.

I/O port pins and their functions

The four ports P0, P1, P2, and P3 each use 8 pins, making them 8-bit ports. All the ports upon RESET are configured as output, ready to be used as output ports. To use any of these ports as an input port, it must be programmed, as we will explain throughout this section. First, we describe each port.

Port 0

Port 0 occupies a total of 8 pins (pins 32 - 39). It can be used for input or output. To use the pins of port 0 as both input and output ports, each pin must be connected externally to a 10K ohm pull-up resistor. This is due to the fact that P0 is an open drain, unlike P1, P2, and P3, as we will soon see. *Open drain* is a term used for MOS chips in the same way that *open collector* is used for TTL chips. In any system using the 8751, 89C51, or DS5000 chips, we normally connect P0 to pull-up resistors. See Figure 4-4. In this way we take advantage of port 0 for both input and output. With external pull-up resistors connected upon reset, port 0 is configured as an output port. For example, the following code will continuously send out to port 0 the alternating values 55H and AAH.

```
            MOV     A,#55H
BACK:       MOV     P0,A
            ACALL   DELAY
            CPL     A
            SJMP    BACK
```

Port 0 as input

With resistors connected to port 0, in order to make it an input, the port must be programmed by writing 1 to all the bits. In the following code, port 0 is configured first as an input port by writing 1s to it, and then data is received from that port and sent to P1.

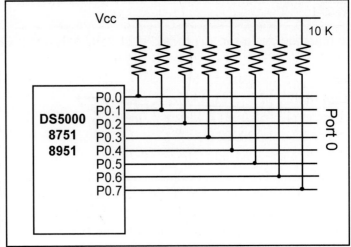

Figure 4-4. Port 0 with Pull-Up Resistors

```
            MOV  A,#0FFH    ;A = FF hex
            MOV  P0,A       ;make P0 an input port
                            ;by writing all 1s to it
BACK:       MOV  A,P0       ;get data from P0
            MOV  P1,A       ;send it to port 1
            SJMP BACK       ;keep doing it
```

Dual role of port 0

As shown in Figure 4-1, port 0 is also designated as AD0 - AD7, allowing it to be used for both address and data. When connecting an 8051/31 to an external memory, port 0 provides both address and data. The 8051 multiplexes address and data through port 0 to save pins. ALE indicates if P0 has address or data. When ALE = 0, it provides data D0 - D7, but when ALE = 1 it has address A0 - A7. Therefore, ALE is used for demuliplexing address and data with the help of a 74LS373 latch, as we will see in Chapter 14.

Port 1

Port 1 occupies a total of 8 pins (pins 1 through 8). It can be used as input or output. In contrast to port 0, this port does not need any pull-up resistors since it already has pull-up resistors internally. Upon reset, port 1 is configured as an output port. For example, the following code will continuously send out to port 1 the alternating values 55H and AAH.

```
            MOV  A,#55H
BACK:       MOV  P1,A
            ACALL DELAY
            CPL  A
            SJMP BACK
```

Port 1 as input

To make port 1 an input port, it must programmed as such by writing 1 to all its bits. The reason for this is discussed in Appendix C.2. In the following code, port 1 is configured first as an input port by writing 1s to it, then data is received from that port and saved in R7, R6, and R5.

```
MOV   A,#0FFH    ;A=FF hex
MOV   P1,A       ;make P1 an input port
                 ;by writing all 1s to it
MOV   A,P1       ;get data from P1
MOV   R7,A       ;save it in reg R7
ACALL DEALY      ;wait
MOV   A,P1       ;get another data from P1
MOV   R6,A       ;save it in reg R6
ACALL DELAY      :wait
MOV   A,P1       ;get another data from P1
MOV   R5,A       ;save it in reg R5
```

Port 2

Port 2 occupies a total of 8 pins (pins 21 through 28). It can be used as input or output. Just like P1, port 2 does not need any pull-up resistors since it already has pull-up resistors internally. Upon reset, port 2 is configured as an output port. For example, the following code will send out continuously to port 2 the alternating values 55H and AAH. That is, all the bits of P2 toggle continuously.

```
        MOV   A,#55H
BACK:   MOV   P2,A
        ACALL DELAY
        CPL   A
        SJMP  BACK
```

Port 2 as input

To make port 2 an input, it must programmed as such by writing 1 to all its bits. In the following code, port 2 is configured first as an input port by writing 1s to it. Then data is received from that port and is sent to P1 continuously.

```
        MOV   A,#0FFH    ;A=FF hex
        MOV   P2,A       ;make P2 an input port by
                         ;writing all 1s to it
BACK:   MOV   A,P2       ;get data from P2
        MOV   P1,A       ;send it to Port 1
        SJMP  BACK       ;keep doing that
```

Dual role of port 2

In systems based on the 8751, 89C51, and DS5000, P2 is used as simple I/O. However, in 8031-based systems, port 2 must be used along with P0 to provide the 16-bit address for the external memory. As shown in Figure 4-1, Port 2 is also designated as A8 - A15, indicating its dual function. Since an 8031 is capable of accessing 64K bytes of external memory, it needs a path for the 16 bits of the address. While P0 provides the lower 8 bits via A0 - A7, it is the job of P2 to provide bits A8 - A15 of the address. In other words, when the 8031 is connected to external memory, P2 is used for the upper 8 bits of the 16-bit address, and it cannot be used for I/O. This is discussed in detail in Chapter 14.

From the discussion so far, we conclude that in systems based on 8751, 89C51, or DS5000 microcontrollers, we have three ports, P0, P1, and P2, for I/O operations. This should be enough for most microcontroller applications. That leaves port 3 for interrupts as well as other signals, as we will see next.

Port 3

Port 3 occupies a total of 8 pins, pins 10 through 17. It can be used as input or output. P3 does not need any pull-up resistors, the same as P1 and P2 did not. Although port 3 is configured as an output port upon reset, this is not the way it is most commonly used. Port 3 has the additional function of providing some extremely important signals such as interrupts. Table 4-2 provides these alternate functions of P3. This information applies to both 8051 and 8031 chips.

Table 4-2: Port 3 Alternate Functions

P3 Bit	Function	Pin
P3.0	RxD	10
P3.1	TxD	11
P3.2	$\overline{\text{INT0}}$	12
P3.3	$\overline{\text{INT1}}$	13
P3.4	T0	14
P3.5	T1	15
P3.6	$\overline{\text{WR}}$	16
P3.7	$\overline{\text{RD}}$	17

P3.0 and P3.1 are used for the RxD and TxD serial communications signals. See Chapter 10 to see how they are connected. Bits P3.2 and P3.3 are set aside for external interrupts, and are discussed in Chapter 11. Bits P3.4 and P3.5 are used for timers 0 and 1, and are discussed in Chapter 9 where timers are discussed. Finally, P3.6 and P3.7 are used to provide the $\overline{\text{WR}}$ and $\overline{\text{RD}}$ signals of external memories connected in 8031-based systems. Chapter 14 discusses how they are used in 8031-based systems. In systems based on the 8751, 89C51, or DS5000, pins 3.6 and 3.7 are used for I/O while the rest of the pins in Port 3 are normally used in the alternate function role.

Review Questions

1. A given 8051 chip has a speed of 16 MHz. What is the range of frequency that can be applied to the XTAL1 and XTAL2 pins?
2. A 16-MHz 8051 system has a machine cycle of _____.
3. Which pin is used to inform the 8051 that the on-chip ROM contains the program?
4. There are total of _____ ports in the 8051 and each has _____ bits.
5. True or false. All of the 8051 ports can be used for both input and output.
6. Upon power up, the program counter (PC) has a value of _____.
7. Upon power up, the 8051 fetches the first opcode from ROM address location _____.
8. Which of the 8051 ports need pull-up resistors to function as an I/O port?

SECTION 4.2: I/O PROGRAMMING; BIT MANIPULATION

In this section we further examine 8051 I/O instructions. We pay special attention to I/O bit manipulation since it is a powerful and widely used 8051 feature. A detailed discussion of I/O ports of the 8051 is given in Appendix C.2.

Different ways of accessing the entire 8 bits

In the following code, as in many previous I/O examples, the entire 8 bits of Port 1 are accessed.

```
BACK:       MOV    A,#55H
            MOV    P1,A
            ACALL  DELAY
            MOV    A,#0AAH
            MOV    P1,A
            ACALL  DELAY
            SJMP   BACK
```

The above code toggles every bit of P1 continuously. We have seen a variation of the above program before. Now we can rewrite the above code in a more efficient manner by accessing the port directly without going through the accumulator. This is shown next.

```
BACK:       MOV    P1,#55H
            ACALL  DELAY
            MOV    P1,#0AAH
            ACALL  DELAY
            SJMP   BACK
```

We can write another variation of the above code by using a technique called *read-modify-write*. This is shown next.

Read-modify-write feature

The ports in the 8051 can be accessed by the read-modify-write technique. This feature saves many lines of code by combining in a single instruction all three actions of (1) reading the port, (2) modifying it, and (3) writing to the port. The following code first places 01010101 (binary) into port 1. Next, the instruction "XLR P1,#0FFH" performs an XOR logic operation on P1 with 1111 1111 (binary), and then writes the result back into P1.

```
            MOV  P1,#55H      ;P1=01010101
AGAIN:      XLR  P1,#0FFH     ;EX-OR P1 with 1111 1111
            ACALL DELAY
            SJMP AGAIN
```

Notice that the XOR of 55H and FFH gives AAH. Likewise, the XOR of AAH and FFH gives 55H. Logic instructions are discussed in Chapter 7.

Single-bit addressability of ports

There are times that we need to access only 1 or 2 bits of the port instead of the entire 8 bits. A powerful feature of 8051 I/O ports is their capability to access individual bits of the port without altering the rest of the bits in that port. For example, the following code toggles the bit P1.2 continuously.

```
BACK:       CPL P1.2        ;complement P1.2 only
            ACALL DELAY
            SJMP BACK


;another variation of the above program follows
AGAIN:      SETB P1.2       ;change only P1.2=high
            ACALL DELAY
            CLR  P1.2       ;change only P1.2=low
            ACALL DELAY
            SJMP AGAIN
```

Notice that P1.2 is the third bit of P1, since the first bit is P1.0, the second bit is P1.1, and so on. Table 4-3 shows the bits of 8051 I/O ports. See Example 4-2 for an example of bit manipulation of I/O bits. Notice in Example 4-2 that unused portions of Ports 1 and 2 are undisturbed. This single-bit addressability of I/O ports is one of most powerful features of the 8051 microcontroller.

Table 4-3: Single-Bit Addressability of Ports

P0	P1	P2	P3	Port Bit
P0.0	P1.0	P2.0	P3.0	D0
P0.1	P1.1	P2.1	P3.1	D1
P0.2	P1.2	P2.2	P3.2	D2
P0.3	P1.3	P2.3	P3.3	D3
P0.4	P1.4	P2.4	P3.4	D4
P0.5	P1.5	P2.5	P3.5	D5
P0.6	P1.6	P2.6	P3.6	D6
P0.7	P1.7	P2.7	P3.7	D7

Example 4-2

Write a program to perform the following.
(a) Keep monitoring the P1.2 bit until it becomes high,
(b) When P1.2 becomes high, write value 45H to port 0, and
(c) Send a high-to-low (H-to-L) pulse to P2.3.

Solution:

```
            SETB P1.2           ;make P1.2 an input
            MOV  A,#45H         ;A=45H
AGAIN:      JNB  P1.2,AGAIN     ;get out when P1.2=1
            MOV  P0,A           ;issue A to P0
            SETB P2.3           ;make P2.3 high
            CLR  P2.3           ;make P2.3 low for H-to-L
```

In this program, instruction "JNB P1.2,AGAIN" (JNB means jump if no bit) stays in the loop as long as P1.2 is low. When P1.2 becomes high, it gets out of the loop, writes the value 45H to port 0, and creates a H-to-L pulse by the sequence of instructions SETB and CLR.

Review Questions

1. Upon reset, the 8051 ports are configured as
   (a) input (b) output (c) both input and output.
2. True or false. The instruction "SETB P2.1" makes pin P2.1 high while leaving other bits of P2 unchanged.
3. Why do we use 55H and AAH to test the bits of the port?
4. Is the following a valid instruction: "MOV P1,#99H"? Explain your answer.
5. Using the instruction "JNB P2.5,HERE" assumes that bit P2.5 is an _____ (input, output).

SUMMARY

This chapter began by describing the function of each pin of the 8051. The four ports of the 8051, P0, P1, P2, and P3, each use 8 pins, making them 8-bit ports. These ports can be used for input or output. Port 0 can be used for either address or data. Port 3 can be used to provide interrupt and serial communication signals. Then I/O instructions of the 8051 were explained, and numerous examples were given.

PROBLEMS

SECTION 4.1: PIN DESCRIPTION OF THE 8051

1. The 8051 DIP package is a _____-pin package.
2. Which pins are assigned to V_{CC} and GND?
3. In the 8051, how many pins are designated as I/O port pins?
4. The crystal oscillator is connected to pins ____ and ____ .
5. If an 8051 is rated as 25 MHz, what is the maximum frequency that can be connected to it?
6. Indicate the pin number assigned to RST in the DIP package.
7. RST is an _____ (input, output) pin.
8. The RST pin is normally _____ (low, high) and needs a _____ (low, high) signal to be activated.
9. What are the contents of the PC (program counter) upon RESET of the 8051?
10. What are the contents of the SP register upon RESET of the 8051?
11. What are the contents of the A register upon RESET of the 8051?
12. Find the machine cycle for the following crystal frequencies connected to X1 and X2.
    (a) 12 MHz (b) 20 MHz (c) 25 MHz (d) 30 MHz
13. $\overline{EA}$ stands for _____ and is an_____ (input, output) pin.
14. For 8051 family members with on-chip ROM such as the 8751 and the 89C51, pin $\overline{EA}$ is connected to _____ (V_{CC}, GND).
15. $\overline{PSEN}$ is an _____ (input, output) pin.
16. ALE is an _____ (input, output) pin.
17. ALE is used mainly in systems based on the _____ (8051, 8031).
18. How many pins are designated as P0 and what are those in the DIP package?

19. How many pins are designated as P1 and what are those in the DIP package?
20. How many pins are designated as P2 and what are those in the DIP package?
21. How many pins are designated as P3 and what are those in the DIP package?
22. Upon RESET, all the bits of ports are configured as _____ (input, output).
23. In the 8051, which port needs a pull-up resistor to be used as I/O?
24. Which port of the 8051 does not have any alternate function and can be used solely for I/O?
25. Write a program to get 8-bit data from P1 and send it to ports P0, P2, and P3.
26. Write a program to get an 8-bit data from P2 and send it to ports P0 and P1.
27. In P3, which pins are for RxD and TxD?
28. At what memory location does the 8051 wake up upon RESET? What is the implication of that?
29. Write a program to toggle all the bits of P1 and P2 continuously
    (a) using AAH and 55H (b) using the CPL instruction.
30. What is the address of the last location of on-chip ROM for the 8751?

SECTION 4.2: I/O PROGRAMMING; BIT MANIPULATION

31. Which ports of the 8051 are bit-addressable?
32. What is the advantage of bit-addressability for 8051 ports?
33. When P1 is accessed as a single bit port, it is designated as _____.
34. Is the instruction "CPL P1" a valid instruction?
35. Write a program to toggle P1.2 and P1.5 continuously without disturbing the rest of the bits.
36. Write a program to toggle P1.3, P1.7, and P2.5 continuously without disturbing the rest of the bits.
37. Write a program to monitor bit P1.3. When it is high, send 55H to P2.
38. Write a program to monitor the P2.7 bit. When it is low, send 55H and AAH to P0 continuously.
39. Write a program to monitor the P2.0 bit. When it is high, send 99H to P1.
40. Write a program to monitor the P1.5 bit. When it is high, make a low-to-high-to-low pulse on P1.3.

ANSWERS TO REVIEW QUESTIONS

SECTION 4.1: PIN DESCRIPTION OF THE 8051
1. From 0 to 16 MHz, but no more than 16 MHz.
2. 1/12th of 16 MHz is 1.33 MHz. and the machine cycle is = 0.75 μs
3. EA   4. 4, 8   5. True
6. PC = 0000   7. 0000   8. Port 0

SECTION 4.2: I/O PROGRAMMING; BIT MANIPULATION
1. (b)   2. True   3. They are the complement of each other.
4. Yes. This is called immediate addressing mode (discussed in Chapter 5).
5. input

CHAPTER 5

8051 ADDRESSING MODES

OBJECTIVES

Upon completion of this chapter, you will be able to:

>> List the five addressing modes of the 8051 microcontroller
>> Contrast and compare the addressing modes
>> Code 8051 Assembly language instructions using each addressing mode
>> Access RAM using various addressing modes
>> List the SFR (special function registers) addresses
>> Discuss how to access the SFR
>> Manipulate the stack using direct addressing mode
>> Code 8051 instructions to manipulate a look-up table

The CPU can access data in various ways. The data could be in a register, or in memory, or be provided as an immediate value. These various ways of accessing data are called *addressing modes*. In this chapter we discuss 8051 addressing modes in the context of some examples.

The various addressing modes of a microprocessor are determined when it is designed and therefore cannot be changed by the programmer. The 8051 provides a total of five distinct addressing modes. They are as follows:

(1) immediate
(2) register
(3) direct
(4) register indirect
(5) indexed

In the first section of this chapter we look at immediate and register addressing modes. In the second section we cover accessing memory using the direct, register indirect, and indexed addressing modes. In subsequent chapters we will use these addressing modes in many different applications. Since the reader is now familiar with ADD and MOV, these two instructions are used to explain addressing modes.

SECTION 5.1: IMMEDIATE AND REGISTER ADDRESSING MODES

In this section, first we examine immediate addressing mode and then register addressing mode.

Immediate addressing mode

In this addressing mode, the source operand is a constant. In immediate addressing mode, as the name implies, when the instruction is assembled, the operand comes immediately after the opcode. Notice that the immediate data must be preceded by the pound sign, "#". This addressing mode can be used to load information into any of the registers, including the DPTR register. Examples follow.

```
MOV A,#25H          ;load 25H into A
MOV R4,#62          ;load the decimal value 62 into R4
MOV B,#40H          ;load 40H into B
MOV DPTR,#4521H     ;DPTR=4512H
```

Although the DPTR register is 16-bit, it can also be accessed as two 8-bit registers, DPH and DPL, where DPH is the high byte and DPL is the low byte. Look at the following code.

```
MOV DPTR,#2550H
    is the same as:
MOV DPL,#50H
MOV DPH,#25H
```

Also notice that the following would produce an error since the value is larger than 16-bit.

```
MOV DPTR,#68975 ;illegal!! value > 65535 (FFFFFH)
```

We can use the EQU directive to access immediate data as shown below.

```
COUNT       EQU  30
...          ..
MOV         R4,#COUNT       ;R4=1E(30=1EH)
MOV         DPTR,#MYDATA    ;DPTR=200H

            ORG  200H
MYDATA:     DB   "America"
```

Notice that we can also use immediate addressing mode to send data to 8051 ports. For example, "MOV P1,#55H" is a valid instruction.

Register addressing mode

Register addressing mode involves the use of registers to hold the data to be manipulated. Examples of register addressing mode follow.

```
MOV A,R0  ;copy the contents of R0 into A
MOV R2,A  ;copy the contents of A into R2
ADD A,R5  ;add the contents of R5 to contents of A
ADD A,R7  ;add the contents of R7 to contents of A
MOV R6,A  ;save accumulator in R6
```

It should be noted that the source and destination registers must match in size. In other words, coding "MOV DPTR,A" will give an error, since the source is an 8-bit register and the destination is a 16-bit register. See the following.

```
MOV DPTR,#25F5H
MOV R7,DPL
MOV R6,DPH
```

Notice that we can move data between the accumulator and Rn (for n = 0 to 7) but movement of data between Rn registers is not allowed. For example, the instruction "MOV R4,R7" is invalid.

In the first two addressing modes, the operands are either inside one of the registers or tagged along with the instruction itself. In most programs, the data to be processed is often in some memory location of RAM or in the code space of ROM. There are many ways to access this data. The next section describes these different methods.

Review Problems

1. Can the programmer of a microcontroller make up new addressing modes?
2. Show the instruction to load 1000 0000 (binary) into R3.
3. Why is the following invalid? "MOV R2,DPTR"
4. True or false. DPTR is a 16-bit register that is also accessible in low-byte and high-byte formats.
5. Is the PC (program counter) also available in low-byte and high-byte formats?

SECTION 5.2: ACCESSING MEMORY USING VARIOUS ADDRESSING MODES

We can use direct or register indirect addressing modes to access data stored either in RAM or registers of the 8051. This topic will be discussed thoroughly in this section. We will also show how to access on-chip ROM containing data using indexed addressing mode.

Direct addressing mode

As mentioned in Chapter 2, there are 128 bytes of RAM in the 8051. The RAM has been assigned addresses 00 to 7FH. The following is summary of the allocation of these 128 bytes.
1. RAM locations 00 - 1FH are assigned to the register banks and stack.
2. RAM locations 20 - 2FH are set aside as bit-addressable space to save single-bit data. This is discussed in Chapter 8.
3. RAM locations 30 - 7FH are available as a place to save byte- sized data.

Although the entire 128 bytes of RAM can be accessed using direct addressing mode, it is most often used to access RAM locations 30 - 7FH. This is due to the fact that register bank locations are accessed by the register names of R0 - R7, but there is no such name for other RAM locations. In direct addressing mode, the data is in a RAM memory location whose address is known, and this address is given as a part of the instruction. Contrast this with immediate addressing mode in which the operand itself is provided with the instruction. The "#" sign distinguishes between the two modes. See the examples below, and note the absence of the "#" sign.

```
MOV R0,40H    ;save content of RAM location 40H in R0
MOV 56H,A     ;save content of A in RAM location 56H
MOV R4,7FH    ;move contents of RAM location 7FH to R4
```

As discussed earlier, RAM locations 0 to 7 are allocated to bank 0 registers R0 - R7. These registers can be accessed in two ways, as shown below.

```
MOV A,4       ;is same as
MOV A,R4      ;which means copy R4 into A

MOV A,7       ;is same as
MOV A,R7      ;which means copy R7 into A
```

```
MOV A,2        ;is the same as
MOV A,R2       ;which means copy R2 into A

MOV A,0        ;is the same as
MOV A,R0       ;which means copy R0 into A
```

The above examples should reinforce the importance of the "#" sign in 8051 instructions. See the following code.

```
MOV R2,#5 ;R2=05
MOV A,2   ;copy R2 to A (A=R2=05)
MOV B,2   ;copy R2 to B (B=R2=05)
MOV 7,2   ;copy R2 to R7
          ;since "MOV R7,R2" is invalid
```

Although it is easier to use the names R0 - R7 than their memory addresses, RAM locations 30H to 7FH cannot be accessed in any way other than their addresses since they have no names.

SFR registers and their addresses

Among the registers we have discussed so far, we have seen that R0 - R7 are part of the 128 bytes of RAM memory. What about registers A, B, PSW, and DPTR? Do they also have addresses? The answer is yes. In the 8051, registers A, B, PSW, and DPTR are part of the group of registers commonly referred to as SFR (special function registers). There are many special function registers and they are widely used, as we will discuss in future chapters. The SFR can be accessed by their names (which is much easier) or by their addresses. For example, register A has address E0H, and register B has been designated the address F0H, as shown in Table 5-1. Notice how the following pairs of instructions mean the same thing.

```
MOV 0E0H,#55H  ;is the same as
MOV A,#55H     ;which means load 55H into A (A=55H)

MOV 0F0H,#25H  ;is the same as
MOV B,#25H     ;which means load 25H into B (B=25H)

MOV 0E0H,R2    ;is the same as
MOV A,R2       ;which means copy R2 into A

MOV 0F0H,R0    ;is the same as
MOV B,R0       ;which means copy R0 into B
```

Table 5-1 lists the 8051 special function registers (SFR) and their addresses. The following two points should be noted about the SFR addresses.

1. The special function registers have addresses between 80H and FFH. These addresses are above 80H, since the addresses 00 to 7FH are addresses of RAM memory inside the 8051.
2. Not all the address space of 80 to FF is used by the SFR. The unused locations 80H to FFH are reserved and must not be used by the 8051 programmer.

Table 5-1: Special Function Register (SFR) Addresses

Symbol	Name	Address
ACC*	Accumulator	0E0H
B*	B register	0F0H
PSW*	Program status word	0D0H
SP	Stack pointer	81H
DPTR	Data pointer 2 bytes	
DPL	Low byte	82H
DPH	High byte	83H
P0*	Port 0	80H
P1*	Port 1	90H
P2*	Port 2	0A0H
P3*	Port 3	0B0H
IP*	Interrupt priority control	0B8H
IE*	Interrupt enable control	0A8H
TMOD	Timer/counter mode control	89H
TCON*	Timer/counter control	88H
T2CON*	Timer/counter 2 control	0C8H
T2MOD	Timer/counter mode control	0C9H
TH0	Timer/counter 0 high byte	8CH
TL0	Timer/counter 0 low byte	8AH
TH1	Timer/counter 1 high byte	8DH
TL1	Timer/counter 1 low byte	8BH
TH2	Timer/counter 2 high byte	0CDH
TL2	Timer/counter 2 low byte	0CCH
RCAP2H	T/C 2 capture register high byte	0CBH
RCAP2L	T/C 2 capture register low byte	0CAH
SCON*	Serial control	98H
SBUF	Serial data buffer	99H
PCON	Power control	87H

* Bit addressable (discussed further in Chapter 8)

Regarding direct addressing mode, it must be noted that the address value is limited to one byte, 00 - FFH. This means that the use of this addressing mode is limited to accessing RAM locations and registers with addresses located inside the 8051.

Example 5-1

Write code to send 55H to ports P1 and P2, using (a) their names (b) their addresses.

Solution:

(a)
```
    MOV A,#55H        ;A=55H
    MOV P1,A          ;P1=55H
    MOV P2,A          ;P2=55H
```

(b)
```
    From Table 5-1, P1 address = 80H; P2 address = A0H
    MOV A,#55H        ;A=55H
    MOV 80H,A         ;P1=55H
    MOV 0A0H,A        ;P2=55H
```

Stack and direct addressing mode

Another major use of direct addressing mode is the stack. In the 8051 family, only direct addressing mode is allowed for pushing onto the stack. Therefore, an instruction such as "PUSH A" is invalid. Pushing the accumulator onto the stack must be coded as "PUSH 0E0H" where 0E0H is the address of register A. Similarly, pushing R3 of bank 0 is coded as "PUSH 03". Direct addressing mode must be used for the POP instruction as well. For example, "POP 04" will pop the top of the stack into R4 of bank 0.

Example 5-2

Show the code to push R5, R6, and A onto the stack and then pop them back them into R2, R3, and B, where register B = register A, R2 = R6, and R3 = R5.

Solution:

```
    PUSH 05           ;push R5 onto stack
    PUSH 06           ;push R6 onto stack
    PUSH 0E0H         ;push register A onto stack
    POP  0F0H         ;pop top of stack into register B
                      ;now register B = register A
    POP  02           ;pop top of stack into R2
                      ;now R2 = R6
    POP  03           ;pop top of stack into R3
                      ;now R3 = R5
```

Register indirect addressing mode

In the register indirect addressing mode, a register is used as a pointer to the data. If the data is inside the CPU, only registers R0 and R1 are used for this purpose. In other words, R2 - R7 cannot be used to hold the address of an operand located in RAM when using this addressing mode. When R0 and R1 are used as pointers, that is, when they hold the addresses of RAM locations, they must be pre-ceded by the "@" sign, as shown below.

```
MOV A,@R0 ;move contents of RAM location whose
          ;address is held by R0 into A
MOV @R1,B ;move contents of B into RAM location
          ;whose address is held by R1
```

Notice that R0 (as well as R1) is preceded by the "@" sign. In the absence of the "@" sign, MOV will be interpreted as an instruction moving the contents of register R0 to A, instead of the contents of the memory location pointed to by R0.

Example 5-3

Write a program to copy the value 55H into RAM memory locations 40H to 45H using
(a) direct addressing mode,
(b) register indirect addressing mode without a loop, and
(c) with a loop.

Solution:

(a)
```
        MOV A,#55H       ;load A with value 55H
        MOV 40H,A        ;copy A to RAM location 40H
        MOV 41H,A        ;copy A to RAM location 41H
        MOV 42H,A        ;copy A to RAM location 42H
        MOV 43H,A        ;copy A to RAM location 43H
        MOV 44H,A        ;copy A to RAM location 44H
```

(b)
```
        MOV A,#55H       ;load A with value 55H
        MOV R0,#40H      ;load the pointer. R0=40H
        MOV @R0,A        ;copy A to RAM location R0 points to
        INC R0           ;increment pointer. Now R0=41H
        MOV @R0,A        ;copy A to RAM location R0 points to
        INC R0           ;increment pointer. Now R0=42H
        MOV @R0,A        ;copy A to RAM location R0 points to
        INC R0           ;increment pointer. Now R0=43H
        MOV @R0,A        ;copy A to RAM location R0 points to
        INC R0           ;increment pointer. Now R0=44H
        MOV @R0,A
```

(c)
```
        MOV   A,#55    ;A=55H
        MOV   R0,#40H  ;load pointer. R0=40H, RAM address
        MOV   R2,#05   ;load counter, R2=5
AGAIN:  MOV   @R0,A    ;copy 55 to RAM location R0 points to
        INC   R0       ;increment R0 pointer
        DJNZ  R2,AGAIN      ;loop until counter = zero
```

Advantage of register indirect addressing mode

One of the advantages of register indirect addressing mode is that it makes accessing data dynamic rather than static as in the case of direct addressing mode. Example 5-3 shows two cases of copying 55H into RAM locations 40H to 45H. Notice in solution (b) that there are two instructions that are repeated numerous times. We can create a loop with those two instructions as shown in solution (c). Solution (c) is the most efficient and is possible only because of register indirect addressing mode. Looping is not possible in direct addressing mode. This is the main difference between the direct and register indirect addressing modes.

Example 5-4

Write a program to clear 16 RAM locations starting at RAM address 60H.

Solution:
```
        CLR     A            ;A=0
        MOV     R1,#60H      ;load pointer. R1=60H
        MOV     R7,#16       ;load counter, R7=16 (10 in hex)
AGAIN:  MOV     @R1,A        ;clear RAM location R1 points to
        INC     R1           ;increment R1 pointer
        DJNZ    R7,AGAIN     ;loop until counter = zero
```

An example of how to use both R0 and R1 in the register indirect addressing mode in a block transfer is given in Example 5-5.

Example 5-5

Write a program to copy a block of 10 bytes of data from RAM locations starting at 35H to RAM locations starting at 60H.

Solution:
```
        MOV   R0,#35H    ;source pointer
        MOV   R1,#60H    ;destination pointer
        MOV   R3,#10     ;counter
BACK:   MOV   A,@R0      ;get a byte from source
        MOV   @R1,A      ;copy it to destination
        INC   R0         ;increment source pointer
        INC   R1         ;increment destination pointer
        DJNZ  R3,BACK    ;keep doing it for all ten bytes
```

Limitation of register indirect addressing mode in the 8051

As stated earlier, R0 and R1 are the only registers that can be used for pointers in register indirect addressing mode. Since R0 and R1 are 8 bits wide, their use is limited to accessing any information in the internal RAM (scratch pad memory of 30H - 7FH, or SFR). However, there are times when we need to access data stored in external RAM or in the code space of on-chip ROM. Whether accessing externally connected RAM or on-chip ROM, we need a 16-bit pointer. In such cases, the DPTR register is used, as shown next.

Indexed addressing mode and on-chip ROM access

Indexed addressing mode is widely used in accessing data elements of look-up table entries located in the program ROM space of the 8051. The instruction used for this purpose is "MOVC A, @A+DPTR". The 16-bit register DPTR and register A are used to form the address of the data element stored in on-chip ROM. Because the data elements are stored in the program (code) space ROM of the 8051, it uses the instruction MOVC instead of MOV. The "C" means code. In this instruction the contents of A are added to the 16-bit register DPTR to form the 16-bit address of the needed data. See Example 5-6.

Example 5-6

In this program, assume that the word "USA" is burned into ROM locations starting at 200H, and that the program is burned into ROM locations starting at 0. Analyze how the program works and state where "USA" is stored after this program is run.

Solution:

```
        ORG   0000H          ;burn into ROM starting at 0
        MOV   DPTR,#200H      ;DPTR=200H look-up table address
        CLR   A               ;clear A(A=0)
        MOVC  A,@A+DPTR        ;get the char from code space
        MOV   R0,A             ;save it in R0
        INC   DPTR             ;DPTR=201 pointing to next char
        CLR   A               ;clear A(A=0)
        MOVC  A,@A+DPTR        ;get the next char
        MOV   R1,A             ;save it in R1
        INC   DPTR             ;DPTR=202 pointing to next char
        CLR   A               ;clear A(A=0)
        MOVC  A,@A+DPTR        ;get the next char
        MOV   R2,A             ;save it in R2
HERE:SJMP HERE                ;stay here

;Data is burned into code space starting at 200H
        ORG 200H
MYDATA:   DB "USA"
        END                   ;end of program
```

In the above program ROM locations 200H - 202H have the following contents.
200=('U') 201=('S') 202=('A')
We start with DPTR = 200H, and A = 0. The instruction "MOVC A, @A+DPTR" moves the contents of ROM location 200H (200H + 0 = 200H) to register A. Register A contains 55H, the ASCII value for "U". This is moved to R0. Next, DPTR is incremented to make DPTR = 201H. A is set to 0 again to get the contents of the next ROM location 201H, which holds character "S". After this program is run, we have R0 = 55H, R1 = 53H, and R2 = 41H, the ASCII values for the characters "U", "S" and "A".

Example 5-7

Assuming that ROM space starting at 250H contains "America", write a program to transfer the bytes into RAM locations starting at 40H.

Solution:

```
;(a) This method uses a counter
        ORG   0000
        MOV   DPTR,#MYDATA   ;LOAD ROM POINTER
        MOV   R0,#40H        ;LOAD RAM POINTER
        MOV   R2,#7          ;LOAD COUNTER
BACK:   CLR   A              ;A=0
        MOVC  A,@A+DPTR      ;MOVE DATA FROM CODE SPACE
        MOV   @R0,A          ;SAVE IT IN RAM
        INC   DPTR           ;INCREMENT ROM POINTER
        INC   R0             ;INCREMENT RAM POINTER
        DJNZ  R2,BACK        ;LOOP UNTIL COUNTER=0
HERE:   SJMP  HERE
;----------ON-CHIP CODE SPACE USED FOR STORING DATA
        ORG   250H
MYDATA: DB    "AMERICA"
        END

;(b)This method uses null char for end of string
        ORG   0000
        MOV   DPTR,#MYDATA   ;LOAD ROM POINTER
        MOV   R0,#40H        ;LOAD RAM POINTER
BACK:   CLR   A              ;A=0
        MOVC  A,@A+DPTR      ;MOVE DATA FROM CODE SPACE
        JZ    HERE           ;EXIT IF NULL CHAR
        MOV   @R0,A          ;SAVE IT IN RAM
        INC   DPTR           ;INCREMENT ROM POINTER
        INC   R0             ;INCREMENT RAM POINTER
        SJMP  BACK           ;LOOP
HERE:   SJMP  HERE

;----------ON-CHIP CODE SPACE USED FOR STORING DATA
        ORG   250H
MYDATA: DB    "AMERICA",0    ;notice null char for
                             ;end of string
        END
```

Notice the null character, 0, as end of string and how we use the JZ instruction to detect that.

Look-up table and the use of indexed addressing mode

      The look-up table is a widely used concept in microprocessor programming. It allows access to elements of a frequently used table with minimum operations. As an example, assume that for a certain application we need x^2 values in the range of 0 to 9. We can use a look-up table instead of calculating it. This is shown in Example 5-8.

Example 5-8

Write a program to get the x value from P1 and send x^2 to P2, continuously.

Solution:

```
        ORG  0
        MOV  DPTR,#300H      ;LOAD LOOK-UP TABLE ADDRESS
        MOV  A,#0FFH         ;A=FF
        MOV  P1,A            ;CONFIGURE P1 AS INPUT PORT
BACK:   MOV  A,P1            ;GET X
        MOVC A,@A+DPTR       ;GET X SQAURE FROM TABLE
        MOV  P2,A            ;ISSUE IT TO P2
        SJMP BACK            ;KEEP DOING IT


        ORG  300H
XSQR_TABLE:
        DB   0,1,4,9,16,25,36,49,64,81
        END
```

Notice that the first instruction could be replaced with "MOV DPTR,#XSQR_TABLE".

Example 5-9

Answer the following questions for Example 5-8.
(a) Indicate the content of ROM locations 300 - 309H.
(b) At what ROM location is the square of 6, and what value should be there?
(c) Assume that P1 has a value of 9: what value is at P2 (in binary)?

Solution:

(a) All values are in hex.
```
     300 = (00)      301 = (01)   302 = (04)   303 = (09)
     304 = (10)      4 × 4 = 16 = 10 in hex
     305 = (19)      5 × 5 = 25 = 19 in hex
     306 = (24)      6 × 6 = 36 = 24H
     307 = (31)      308 = (40)   309 = (51)
```

(b) 306H; it is 24H

(c) 01010001B which is 51H and 81 in decimal ($9^2 = 81$).

In addition to the use of DPTR to access program ROM, it can be used to access memory externally connected to the 8051. This is discussed in Chapter 14.

Another register used in indexed addressing mode is the program counter. This is discussed in Appendix A.

In many of the examples above, the MOV instruction was used for the sake of clarity, even though one can use any instruction as long as that instruction supports the addressing mode. For example, the instruction "ADD A, @R0" would add the contents of the memory location pointed to by R0 to the contents of register A. We will see more examples of using addressing modes with various instructions in the next few chapters.

Review Questions

1. The instruction "MOV A, 40H" uses _____ addressing mode. Why?
2. What address is assigned to register R2 of bank 0?
3. What address is assigned to register R2 of bank 2?
4. What address is assigned to register A?
5. Which registers are allowed to be used for register indirect addressing mode if the data is in on-chip RAM?

SUMMARY

This chapter described the five addressing modes of the 8051. Immediate addressing mode uses a constant for the source operand. Register addressing mode involves the use of registers to hold data to be manipulated. Direct or register indirect addressing modes can be used to access data stored either in RAM or registers of the 8051. Direct addressing mode is also used to manipulate the stack. Register indirect addressing mode uses a register as a pointer to the data. The advantage of this is that it makes addressing dynamic rather than static. Indexed addressing mode is widely used in accessing data elements of look-up table entries located in the program ROM space of the 8051.

A group of registers called the SFR (special function registers) can be accessed by their names or their addresses.

PROBLEMS

1. Which of the following are invalid uses of immediate addressing mode?
    (a) MOV A,#24H   (b) MOV R1,30H   (c) MOV R4,#60H
2. Identify the addressing mode for each of the following.
    (a) MOV B,#34H   (b) MOV A,50H   (c) MOV R2,07
    (d) MOV R3,#0   (e) MOV R7,0   (f) MOV R6,#7FH
    (g) MOV R0,A   (h) MOV B,A   (i) MOV A,@R0
    (j) MOV R7,A   (k) MOV A,@R1
3. Indicate the address assigned to each of the following.
    (a) R0 of bank 0   (b) ACC   (c) R7 of bank 0
    (d) R3 of bank 2   (e) B   (f) R7 of bank 3
    (g) R4 of bank 1   (h) DPL   (i) R6 of bank 1
    (j) R0 of bank 3   (k) DPH   (l) P0

4. Which register bank shares the same space with the stack?
5. In accessing the stack, we must use _____ addressing mode.
6. What does the following instruction do? "MOV A, 0F0H"
7. What does the following instruction do? "MOV A, 1FH"
8. Write code to push R0, R1, and R3 of bank 0 onto the stack and pop them back into R5, R6, and R7 of bank 3.
9. Which registers are allowed to be used for register indirect addressing mode when accessing data in RAM?
10. Write a program to copy FFH into RAM locations 50H to 6FH
11. Write a program to copy 10 bytes of data starting at ROM address 400H to RAM locations starting at 30H.
12. Write a program to find y where $y = x^2 + 2x + 5$, and x is between 0 and 9.
13. Write a program to add the following data and store the result in RAM location 30H.

```
            ORG   200H
MYDATA:     DB    06,09,02,05,07
```

ANSWERS TO REVIEW QUESTIONS

SECTION 5.1: IMMEDIATE AND REGISTER ADDRESSING MODES
1. No
2. MOV R3,#10000000B
3. Source and destination register's sizes do not match
4. True   5. No

SECTION 5.2: ACCESSING MEMORY USING VARIOUS ADDRESSING MODES
1. Direct; because there is no "#" sign
2. 02   3. 12H
4. E0H   5. R0 and R1

CHAPTER 6

ARITHMETIC INSTRUCTIONS AND PROGRAMS

OBJECTIVES

Upon completion of this chapter, you will be able to:

>> Define the range of numbers possible in 8051 unsigned data
>> Code addition and subtraction instructions for unsigned data
>> Explain the BCD (binary coded decimal) system of data representation
>> Contrast and compare packed and unpacked BCD data
>> Perform addition and subtraction on BCD data
>> Code 8051 unsigned data multiplication and division instructions
>> Define the range of numbers possible in 8051 signed data
>> Code 8051 signed data arithmetic instructions
>> Explain carry and overflow problems and their corrections

This chapter describes all 8051 arithmetic instructions for both signed and unsigned numbers. Program examples are given to illustrate the application of these instructions. In Section 6.1 we discuss instructions and programs related to addition and subtraction of unsigned numbers, including BCD (binary coded decimal) data. Multiplication and division are explored in Section 6.2. Signed numbers are discussed in Section 6.3.

SECTION 6.1: UNSIGNED ADDITION AND SUBTRACTION

Unsigned numbers are defined as data in which all the bits are used to represent data, and no bits are set aside for the positive or negative sign. This means that the operand can be between 00 and FFH (0 to 255 decimal) for 8-bit data. The topic of signed numbers is discussed separately in Section 6.3.

Addition of unsigned numbers

In the 8051, in order to add numbers together, the accumulator register (A) must be involved. The form of the ADD instruction is

```
ADD  A,source   ;A = A + source
```

The instruction ADD is used to add two operands. The destination operand is always in register A while the source operand can be a register, immediate data, or in memory. Remember that memory-to-memory arithmetic operations are never allowed in 8051 Assembly language. The instruction could change any of the AF, CF, or PF bits of the flag register, depending on the operands involved. The effect of the ADD instruction on the overflow flag is discussed in Section 6.3 since it is used mainly in signed number operations. Look at Example 6-1.

Example 6-1

Show how the flag register is affected by the following instructions.

```
    MOV A,#0F5H         ;A=F5 hex
    ADD A,#0BH          ;A=F5+0B=00
```

Solution:

```
      F5H       1111 0101
  +   0BH    +  0000 1011
      100H      0000 0000
```

After the addition, register A (destination) contains 00 and the flags are as follows:

CY = 1 since there is a carry out from D7.
PF = 1 because the number of 1s is zero (an even number), PF is set to 1.
AC = 1 since there is a carry from D3 to D4.

Addition of individual bytes

Chapter 2 contained a program that added 5 bytes of data. The sum was purposely kept less than FFH, the maximum value an 8-bit register can hold. To calculate the sum of any number of operands, the carry flag should be checked after the addition of each operand. Example 6-2 uses R7 to accumulate carries as the operands are added to A.

Example 6-2

Assume that RAM locations 40 - 44 have the following values. Write a program to find the sum of the values. At the end of the program, register A should contain the low byte and R7 the high byte. All values are in hex.

```
40=(7D)
41=(EB)
42=(C5)
43=(5B)
44=(30)
```

Solution:

```
            MOV   R0,#40H     ;load pointer
            MOV   R2,#5       ;load counter
            CLR   A           ;A=0
            MOV   R7,A        ;clear R7
AGAIN:      ADD   A,@R0       ;add the byte pointer to by R0
            JNC   NEXT        ;if CY=0 don't accumulate carry
            INC   R7          ;keep track of carries
NEXT:       INC   R0          ;increment pointer
            DJNZ  R2,AGAIN    ;repeat until R2 is zero
```

Analysis of Example 6-2

Three iterations of the loop are shown below. Tracing of the program is left to the reader as an exercise.

1. In the first iteration of the loop, 7DH is added to A with CY = 0 and R7= 00, and the counter R2 = 04.
2. In the second iteration of the loop, EBH is added to A, which results in A = 68H and CY=1. Since a carry occurred, R7 is incremented. Now the counter R2 = 03.
3. In the third iteration, C5H is added to A, which makes A = 2DH. Again a carry occurred, so R7 is incremented again. Now counter R2 = 02.

At the end when the loop is finished, the sum is held by registers A and R7, where A has the low byte and R7 has the high byte.

ADDC and addition of 16-bit numbers

When adding two 16-bit data operands, we need to be concerned with the propagation of a carry from the lower byte to the higher byte. The instruction ADDC (add with carry) is used on such occasions. For example, look at the addition of 3CE7H + 3B8DH, as shown below.

```
        1
      3C  E7
  +   3B  8D
      78  74
```

When the first byte is added (E7 + 8D = 74, CY = 1). The carry is propagated to the higher byte, which results in 3C + 3B + 1 = 78 (all in hex). Example 6-3 shows the above steps in an 8051 program.

Example 6-3

Write a program to add two 16-bit numbers. The numbers are 3CE7H and 3B8DH. Place the sum in R7 and R6; R6 should have the lower byte.

Solution:

```
CLR   C           ;make CY=0
MOV   A,#0E7H      ;load the low byte now A=E7H
ADD   A,#8DH       ;add the low byte now A=74H and CY=1
MOV   R6,A         ;save the low byte of the sum in R6
MOV   A,#3CH       ;load the high byte
ADDC  A,#3BH       ;add with the carry
                   ;3B + 3C + 1 = 78(all in hex)
MOV   R7,A         ;save the high byte of the sum
```

BCD (binary coded decimal) number system

BCD stands for binary coded decimal. BCD is needed because in everyday life we use the digits 0 to 9 for numbers, not binary or hex numbers. Binary representation of 0 to 9 is called BCD (see Figure 6-1). In computer literature one encounters two terms for BCD numbers, (1) unpacked BCD, and (2) packed BCD. We describe each one next.

Unpacked BCD

In unpacked BCD, the lower 4 bits of the number represent the BCD number, and the rest of the bits are 0. Example: "0000 1001" and "0000 0101" are unpacked BCD for 9 and 5, respectively. Unpacked BCD requires 1 byte of memory or an 8-bit register to contain it.

Digit	BCD
0	0000
1	0001
2	0010
3	0011
4	0100
5	0101
6	0110
7	0111
8	1000
9	1001

Figure 6.1. BCD Code

112

Packed BCD

In packed BCD, a single byte has two BCD numbers in it, one in the lower 4 bits, and one in the upper 4 bits. For example, "0101 1001" is packed BCD for 59H. It takes only 1 byte of memory to store the packed BCD operands. This is one reason to use packed BCD since it is twice as efficient in storing data.

There is a problem with adding BCD numbers, which must be corrected. The problem is that after adding packed BCD numbers, the result is no longer BCD. Look at the following.

```
MOV A,#17H
ADD A,#28H
```

Adding these two numbers gives 0011 1111B (3FH), which is not BCD! A BCD number can only have digits from 0000 to 1001 (or 0 to 9). In other words, adding two BCD numbers must give a BCD result. The result above should have been 17 + 28 = 45 (0100 0101). To correct this problem, the programmer must add 6 (0110) to the low digit: 3F + 06 = 45H. The same problem could have happened in the upper digit (for example, in 52H + 87H = D9H). Again to solve this problem, 6 must be added to the upper digit (D9H + 60H = 139H) to ensure that the result is BCD (52 + 87 = 139). This problem is so pervasive that most microprocessors such as the 8051 have an instruction to deal with it. In the 8051 the instruction "DA   A" is designed to correct the BCD addition problem. This is discussed next.

DA instruction

The DA (decimal adjust for addition) instruction in the 8051 is provided to correct the aforementioned problem associated with BCD addition. The mnemonic "DA" has as its only operand the accumulator "A". The DA instruction will add 6 to the lower nibble or higher nibble if needed; otherwise, it will leave the result alone. The following example will clarify these points.

```
MOV   A,#47H      ;A=47H first BCD operand
MOV   B,#25H      ;B=25 second BCD operand
ADD   A,B         ;hex(binary) addition (A=6CH)
DA    A           ;adjust for BCD addition (A=72H)
```

After the program is executed, register A will contain 72H (47 + 25 = 72). The "DA" instruction works only on A. In other words, while the source can be an operand of any addressing mode, the destination must be in register A in order for DA to work. It also needs to be emphasized that DA must be used after the addition of BCD operands and that BCD operands can never have any digit greater than 9. In other words, no A - F digits are allowed. It is also important to note that DA works only after an ADD instruction; it will not work after the INC instruction.

Summary of DA action

After an ADD or ADDC instruction,

1. If the lower nibble (4 bits) is greater than 9, or if AC = 1, add 0110 to the lower 4 bits.
2. If the upper nibble is greater than 9, or if CY = 1, add 0110 to the upper 4 bits.

In reality there is no other use for the AC (auxiliary carry) flag bit except for BCD addition and correction. For example, adding 29H and 18H will result in 41H, which is incorrect as far as BCD is concerned.

```
    Hex              BCD
    29               0010 1001
+   18          +    0001 1000
    41               0100 0001    AC=1
+    6          +         0110
    47               0100 0111
```

Since AC = 1 after the addition, "DA   A" will add 6 to the lower nibble. The final result is in BCD format.

Example 6-4

Assume that 5 BCD data items are stored in RAM locations starting at 40H, as shown below. Write a program to find the sum of all the numbers. The result must be in BCD.

```
40=(71)
41=(11)
42=(65)
43=(59)
44=(37)
```

Solution:

```
        MOV  R0,#40H     ;load pointer
        MOV  R2,#5       ;load counter
        CLR  A           ;A=0
        MOV  R7,A        ;clear R7
AGAIN:  ADD  A,@R0       ;add the byte pointer to by R0
        DA   A           ;adjust for BCD
        JNC  NEXT        ;if CY=0 don't accumulate carry
        INC  R7          ;keep track of carries
NEXT:   INC  R0          ;increment pointer
        DJNZ R2,AGAIN    ;repeat until R2 is zero
```

Subtraction of unsigned numbers

```
SUBB   A,source ;A = A - source - CY
```

In many microprocessors there are two different instructions for subtraction: SUB and SUBB (subtract with borrow). In the 8051 we have only SUBB. To make SUB out of SUBB, we have to make CY = 0 prior to the execution of the instruction. Therefore, there are two cases for the SUBB instruction: (1) with CY = 0, and (2) with CY = 1. First we examine the case where CY = 0 prior to the execution of SUBB. Notice that we use the CY flag for the borrow.

SUBB (subtract with borrow) when CY = 0

In subtraction, the 8051 microprocessors (indeed, all modern CPUs) use the 2's complement method. Although every CPU contains adder circuitry, it would be too cumbersome (and take too many transistors) to design separate subtracter circuitry. For this reason, the 8051 uses adder circuitry to perform the subtraction command. Assuming that the 8051 is executing a simple subtract instruction and that CY = 0 prior to the execution of the instruction, one can summarize the steps of the hardware of the CPU in executing the SUBB instruction for unsigned numbers, as follows.

1. Take the 2's complement of the subtrahend (source operand).
2. Add it to the minuend (A).
3. Invert the carry.

These three steps are performed for every SUBB instruction by the internal hardware of the 8051 CPU, regardless of the source of the operands, provided that the addressing mode is supported. It is after these three steps that the result is obtained and the flags are set. Example 6-5 illustrates the three steps.

Example 6-5

Show the steps involved in the following.

```
CLR   C           ;make  CY=0
MOV   A,#3FH      ;load 3FH into A (A=3FH)
MOV   R3,#23H     ;load 23H into R3(R3=23H)
SUBB  A,R3        ;subtract A - R3, place result in A
```

Solution:

```
A =   3F    0011 1111        0011 1111
R3 =  23    0010 0011      + 1101 1101  (2's complement)
      1C                   1 0001 1100
                           0  CF=0 (step 3)
```

The flags would be set as follows: CY = 0, AC = 0, and the programmer must look at the carry flag to determine if the result is positive or negative.

If after the execution of SUBB the CY = 0, the result is positive; if CY = 1, the result is negative and the destination has the 2's complement of the result. Normally, the result is left in 2's complement, but the CPL (complement) and INC instructions can be used to change it. The CPL instruction performs the 1's complement of the operand; then the operand is incremented (INC) to get the 2's complement. See Example 6-6.

Example 6-6

Analyze the following program:

```
      CLR   C
      MOV   A,#4C       ;load A with value 4CH (A=4CH)
      SUBB  A,#6EH      ;subtract 6E from A
      JNC   NEXT        ;if CY=0 jump to NEXT target
      CPL   A           ;if CY=1 then take 1's complement
      INC   A           ;and increment to get 2's complement
NEXT:MOV    R1,A        ;save A in R1
```

Solution:

Following are the steps for "SUBB A,6EH":

```
    4C    0100 1100                         0100 1100
 -  6E    0110 1110        2's comp =       1001 0010
    -22                                    01101 1110
CY=1, the result is negative, in 2's complement.
```

SUBB (subtract with borrow) when CY = 1

This instruction is used for multibyte numbers and will take care of the borrow of the lower operand. If CY = 1 prior to executing the SUBB instruction, it also subtracts 1 from the result. See Example 6-7.

Example 6-7

Analyze the following program:

```
      CLR   C            ;CY=0
      MOV   A,#62H       ;A=62H
      SUBB  A,#96H       ;62H-96H=CCH with CY=1
      MOV   R7,A         ;save the result
      MOV   A,#27H       ;A=27H
      SUBB  A,#12H       ;27H-12H-1=14H
      MOV   R6,A         ;save the result
```

Solution:

After the SUBB, A = 62H – 96H = CCH and the carry flag is set high indicating there is a borrow. Since CY = 1, when SUBB is executed the second time A = 27H – 12H – 1 =14H. Therefore, we have 2762H – 1296H = 14CCH.

Review Questions

1. The instruction "ADD A, source" places the sum in _____.
2. Why is the following ADD instruction illegal?
   ADD   R1,R2
3. Rewrite the instruction above in correct form.
4. The instruction "ADDC A, source" places the sum in _____.
5. Find the value of the A and CY flags in each of the following.
   (a)   MOV A,#4FH   (b)   MOV A,#9CH
         ADD A,#0B1H   ADD A,#63H
6. Show how the CPU would subtract 05H from 43H.
7. If CY = 1, A = 95H, and B= 4FH prior to the execution of "SUBB A,B", what will be the contents of A after the subtraction?

SECTION 6.2: UNSIGNED MULTIPLICATION AND DIVISION

In multiplying or dividing two numbers in the 8051, the use of registers A and B is required since the multiplication and division instructions work only with these two registers. We first discuss multiplication.

Multiplication of unsigned numbers

The 8051 supports byte by byte multiplication only. The bytes are assumed to be unsigned data. The syntax is as follows:

```
MUL AB     ;A x B, place 16-bit result in B and A
```

In byte by byte multiplication, one of the operands must be in register A, and the second operand must be in register B. After multiplication, the result is in the A and B registers; the lower byte is in A, and the upper byte is in B. The following example multiplies 25H by 65H. The result is a 16-bit data that is held by the A and B registers.

```
MOV   A,#25H    ;load 25H to reg. A
MOV   B,#65H    ;load 65H in reg. B
MUL   AB        ;25H * 65H = E99 where
                ;B = 0EH and A = 99H
```

Table 6-1: Unsigned Multiplication Summary (MUL AB)

Multiplication	Operand 1	Operand 2	Result
byte × byte	A	B	A = low byte, B = high byte

Note: Multiplication of operands larger than 8-bit takes some manipulation. It is left to the reader to experiment with.

CHAPTER 6: ARITHMETIC INSTRUCTIONS AND PROGRAMS

Division of unsigned numbers

In the division of unsigned numbers, the 8051 supports byte over byte only. The syntax is as follows.

```
DIV AB     ;divide A by B
```

In dividing a byte by a byte, the numerator must be in register A and the denominator must be in B. After the DIV instruction is performed, the quotient is in A and the remainder is in B. See the following example.

```
MOV  A,#95     ;load 95 into A
MOV  B,#10     ;load 10 into B
DIV  AB        ;now  A = 09(quotient) and
               ;B=05(remainder)
```

Notice the following points for instruction "DIV AB".

1. This instruction always makes CY = 0 and OV = 0 if the numerator is not 0.
2. If the numerator is 0 (B = 0), OV = 1 indicates an error, and CY = 0. The standard practice in all microprocessors when dividing a number by 0 is to indicate in some way the invalid result of infinity. In the 8051, the OV flag is set to 1.

Table 6-2: Unsigned Division Summary (DIV AB)

Division	Numerator	Denominator	Quotient	Remainder
byte / byte	A	B	A	B

An application for DIV instructions

There are times when an ADC (analog-to-digital converter) is connected to a port and the ADC represents some quantity such as temperature or pressure. The 8-bit ADC provides data in hex in the range of 00 - FFH. This hex data must be converted to decimal. We do that by dividing it by 10 repeatedly, saving the remainders as shown in Example 6-8.

Review Questions

1. In multiplication of two bytes in the 8051, we must place one byte in register _____ and the other one in register _____.
2. In unsigned byte by byte multiplication, the product will be placed in register(s) _____ .
3. Is this a valid 8051 instruction? Explain your answer. "MUL A,R1".
4. In byte/byte division, the numerator must be placed in register_____ and the denominator in register_____.
5. In unsigned byte/byte division, the quotient will be placed in register_____ and the remainder in register_____.
6. Is this a valid 8051 instruction? Explain your answer. "DIV A,R1"

Example 6-8

(a) Write a program to get hex data in the range of 00 - FFH from port 1 and convert it to decimal. Save the digits in R7, R6, and R5, where the least significant digit is in R7.
(b) Analyze the program, assuming that P1 has a value of FDH for data.

Solution:

(a)

```
            MOV   A,#0FFH
            MOV   P1,A          ;make P1 an input port
            MOV   A,P1          ;read data from P1
            MOV   B,#10         ;B=0A hex (10 dec)
            DIV   AB            ;divide by 10
            MOV   R7,B          ;save lower digit
            MOV   B,#10         ;
            DIV   AB            ;divide by 10 once more
            MOV   R6,B          ;save the next digit
            MOV   R5,A          ;save the last digit
```

(b)
To convert a binary (hex) value to decimal, we divide it by 10 repeatedly until the quotient is less than 10. After each division the remainder is saved. In the case of an 8-bit binary such as FDH we have 253 decimal as shown below (all in hex).

	Q	*R*	
FD/0A=	19	3	(low digit)
19/0A=	2	5	(middle digit)
		2	(high digit)

Therefore, we have FDH = 253. In order to display this data it must be converted to ASCII, which is described in the next chapter.

SECTION 6.3: SIGNED NUMBER CONCEPTS AND ARITHMETIC OPERATIONS

All data items used so far have been unsigned numbers, meaning that the entire 8-bit operand was used for the magnitude. Many applications require signed data. In this section the concept of signed numbers is discussed along with related instructions. If your applications do not involve signed numbers, you can bypass this section.

Concept of signed numbers in computers

In everyday life, numbers are used that could be positive or negative. For example, a temperature of 5 degrees below zero can be represented as −5, and 20 degrees above zero as +20. Computers must be able to accommodate such numbers. To do that, computer scientists have devised the following arrangement for

---

the representation of signed positive and negative numbers: The most significant bit (MSB) is set aside for the sign (+ or –), while the rest of the bits are used for the magnitude. The sign is represented by 0 for positive (+) numbers and 1 for negative (–) numbers. Signed byte representation is discussed below.

Signed 8-bit operands

In signed byte operands, D7 (MSB) is the sign and D0 to D6 are set aside for the magnitude of the number. If D7 = 0, the operand is positive, and if D7 = 1, it is negative.

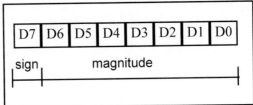

Figure 6-2. 8-Bit Signed Operand

Positive numbers

The range of positive numbers that can be represented by the format shown in Figure 6-2 is 0 to +127. If a positive number is larger than +127, a 16-bit size operand must be used. Since the 8051 does not support 16-bit data, we will not discuss it.

```
   0      0000 0000
  +1      0000 0001
 . . .    . . .
  +5      0000 0101
 . . .    . . .
+127      0111 1111
```

Negative numbers

For negative numbers, D7 is 1; however, the magnitude is represented in its 2's complement. Although the assembler does the conversion, it is still important to understand how the conversion works. To convert to negative number representation (2's complement), follow these steps.

1. Write the magnitude of the number in 8-bit binary (no sign).
2. Invert each bit.
3. Add 1 to it.

Examples 6-9, 6-10, and 6-11 demonstrate these three steps.

Example 6-9

Show how the 8051 would represent –5.

Solution:

Observe the following steps.

```
1.   0000 0101      5 in 8-bit binary
2.   1111 1010      invert each bit
3    1111 1011      add 1 (which becomes FB in hex)
```

Therefore –5 = FBH, the signed number representation in 2's complement for –5.

Example 6-10

Show how the 8051 would represent –34H.

Solution:

Observe the following steps.

```
1.    0011 0100      34H given in binary
2.    1100 1011      invert each bit
3     1100 1100      add 1 (which is CC in hex)
```

Therefore –34 = CCH, the signed number representation in 2's complement for –34H.

Example 6-11

Show how the 8051 would represent –128.

Solution:

Observe the following steps.

```
1.    1000 0000      128 in 8-bit binary
2.    0111 1111      invert each bit
3     1000 0000      add 1 (which becomes 80 in hex)
```

Therefore –128 = 80H, the signed number representation in 2's complement for –128.

From the examples above it is clear that the range of byte-sized negative numbers is –1 to –128. The following lists byte-sized signed numbers ranges:

Decimal	Binary	Hex
-128	1000 0000	80
-127	1000 0001	81
-126	1000 0010	82
...	....	..
-2	1111 1110	FE
-1	1111 1111	FF
0	0000 0000	00
+1	0000 0001	01
+2	0000 0010	02
..	....	..
+127	0111 1111	7F

The above explains the mystery behind the relative address of –128 to +127 in the short jump discussed in Chapter 3.

Overflow problem in signed number operations

When using signed numbers, a serious problem arises that must be dealt with. This is the overflow problem. The 8051 indicates the existence of an error by raising the OV (overflow) flag, but it is up to the programmer to take care of the erroneous result. The CPU understands only 0s and 1s and ignores the human convention of positive and negative numbers. What is an overflow? If the result of an operation on signed numbers is too large for the register, an overflow has occurred and the programmer must be notified. Look at Example 6-12.

Example 6-12

Examine the following code and analyze the result.

```
    MOV   A,#+96      ;A=0110 0000(A=60H)
    MOV   R1,#+70     ;R1=0100 0110(R1=46H)
    ADD   A,R1        ;A=1010 0110
                      ;A=A6H= -90 decimal, INVALID!!)
```

Solution:

```
  +96  0110 0000
+ +70  0100 0110
+ 166  1010 0110   and OV=1
```

According to the CPU, the result is –90, which is wrong. The CPU sets OV = 1 to indicate the overflow.

In Example 6-12, +96 is added to +70 and the result according to the CPU was –90. Why? The reason is that the result was larger than what A could contain. Like all other 8-bit registers, A could only contain up to +127. The designers of the CPU created the overflow flag specifically for the purpose of informing the programmer that the result of the signed number operation is erroneous.

When is the OV flag set?

In 8-bit signed number operations, OV is set to 1 if either of the following two conditions occurs:

1. There is a carry from D6 to D7 but no carry out of D7 (CY = 0).
2. There is a carry from D7 out (CY=1) but no carry from D6 to D7.

In other words, the overflow flag is set to 1 if there is a carry from D6 to D7 or from D7 out, but not both. This means that if there is a carry both from D6 to D7 and from D7 out, OV = 0. In Example 6-12, since there is only a carry from D6 to D7 and no carry from D7 out, OV = 1. Examples 6-13, 6-14, and 6-15 give further illustrations of the use of the overflow flag in signed arithmetic.

Example 6-13

Observe the following, noting the role of the OV flag.

```
        MOV   A,#-128    ;A=1000 0000(A=80H)
        MOV   R4,#-2     ;R4=1111 1110(R4=FEH)
        ADD   A,R4       ;A=0111 1110 (A=7EH=+126, invalid)
```

Solution:

```
   -128        1000 0000
 + - 2         1111 1110
 - 130         0111 1110   and OV=1
```

According to the CPU, the result is +126, which is wrong (OV = 1).

Example 6-14

Observe the following, noting the OV flag.

```
MOV   A,#-2            ;A=1111 1110 (A=FEH)
MOV   R1,#-5           ;R1=1111 1011(R1=FBH)
ADD   A,R1            ;A=1111 1001 (A=F9H=-7,correct,OV=0)
```

Solution:

```
     -2   1111 1110
  + -5    1111 1011
   - 7    1111 1001   and OV = 0
```

According to the CPU, the result is –7, which is correct (OV = 0).

Example 6-15

Examine the following, noting the role of OV.

```
MOV   A,#+7           ;A=0000 0111 (A=07H)
MOV   R1,#+18         ;R1=0001 0010(R1=12H)
ADD   A,R1            ;A=0001 1001 (A=19H=+25, correct,OV=0)
```

Solution:

```
     7 0000 0111
  +  18 0001 0010
     25 0001 1001   and OV = 0
```

According to the CPU, this is +25, which is correct (OV = 0)

From the above examples we conclude that in any signed number addition, OV indicates whether the result is valid or not. If OV = 1, the result is erroneous; if OV = 0, the result is valid. We can state emphatically that in unsigned number addition we must monitor the status of CY (carry flag), and in signed number addition, the OV (overflow) flag must be monitored by the programmer. In the 8051, instructions such as JNC and JC allow the program to branch right after the addition of unsigned numbers, as we saw in Section 6.1. There is no such instruction for the OV flag. However, this can be achieved by "JB PSW.2" or "JNB PSW.2" since PSW, the flag register, is a bit-addressable register. This is discussed in Chapter 8.

Review Questions

1. In an 8-bit operand, bit _____ is used for the sign bit.
2. Convert –16H to its 2's complement representation.
3. The range of byte-sized signed operands is – _____ to + _____.
4. Show +9 and –9 in binary.
5. Explain the difference between a carry and an overflow.

SUMMARY

This chapter discussed arithmetic instructions for both signed and unsigned data in the 8051. Unsigned data uses all 8 bits of the byte for data, making a range of 0 to 255 decimal. Signed data uses 7 bits for data and 1 for the sign bit, making a range of –128 to + 127 decimal.

Binary coded decimal, or BCD, data represents the digits 0 through 9. Both packed and unpacked BCD formats were discussed. The 8051 contains special instructions for arithmetic operations on BCD data.

In coding arithmetic instructions for the 8051, special attention has to be given to the possibility of a carry or overflow condition.

PROBLEMS

SECTION 6.1: UNSIGNED ADDITION AND SUBTRACTION

1. Find the CY and AC flags for each of the following.

(a)
```
MOV A,#3FH
ADD A,#45H
```
(b)
```
MOV A,#99H
ADD A,#58H
```
(c)
```
MOV A,#0FFH
SETB C
ADDC A,#00
```
(d)
```
MOV A,#0FFH
ADD A,#1
```
(e)
```
MOV A,#0FEH
SETB C
ADDC A,#01
```
(f)
```
CLR C
MOV A,#0FFH
ADDC A,#01
ADDC A,#0
```

2. Write a program to add all the digits of your ID number and save the result in R3. The result must be in BCD.

3. Write a program to add the following numbers and save the result in R2, R3. The data is stored in on-chip ROM.

```
          ORG  250H
MYDATA:   DB   53,94,56,92,74,65,43,23,83
```

4. Modify Problem 3 to make the result in BCD.

5. Write a program to (a) write the value 55H to RAM locations 40H - 4FH, and (b) add all these RAM locations contents together, and save the result in RAM locations 60H and 61H.

6. State the steps that the SUBB instruction will go through and for each of the following.
   (a) 23H – 12H   (b) 43H – 53H   (c) 99 – 99

7. For Problem 6, write a program to perform each operation.

8. True or false. The "DA A" instruction works on register A and it must be used after the ADD and ADDC instructions.

9. Write a program to add 897F9AH to 34BC48H and save the result in RAM memory locations starting at 40H.

10. Write a program to subtract 197F9AH from 34BC48H, and save the result in RAM memory locations starting at 40H.

11. Write a program to add BCD 197795H to 344548H and save the BCD result in RAM memory locations starting at 40H.

SECTION 6.2: UNSIGNED MULTIPLICATION AND DIVISION

12. Show how to perform 77×34 in the 8051.

13. Show how to perform 77/3 in the 8051.

14. True or false. The MUL and DIV instructions work on any register of the 8051.

15. Write a program with three subroutines to (a) transfer the following data from on-chip ROM to RAM locations starting at 30H, (b) add them and save the result in 70H, (c) find the average of the data and store it in R7. Notice that the data is stored in a code space of on-chip ROM.

```
          ORG  250H
MYDATA:   DB   3,9,6,9,7,6,4,2,8
```

SECTION 6.3: SIGNED NUMBER CONCEPTS AND ARITHMETIC OPERATIONS

16. Show how the following are represented by the assembler.
    (a) –23   (b) +12   (c) –28
    (d) +6FH   (e) –128   (f) +127

17. The memory addresses in computers are _____ (signed, unsigned) numbers.

18. Write a program for each of the following and indicate the status of the OV flag for each.
    (a) (+15) + (–12)   (b) (–123) + (–127)
    (c) (+25H) + (+34H)   (d) (–127) + (+127)

---

19. Explain the difference between the CY and OV flags and where each one is used.
20. Explain when the OV flag is raised.
21. Which register holds the OV flag?
22. How do you detect the OV flag in the 8051? How do you detect the CY flag?

ANSWERS TO REVIEW QUESTIONS

SECTION 6.1: UNSIGNED ADDITION AND SUBTRACTION

1. A, the accumulator   2. Because the accumulator must be involved
3.
```
   MOV        A,R1
   ADD        A,R2
```
4. Accumulator
5. (a) A = 00 and CY = 1 (b) A = FF and CY = 0
6.
```
     43H   0100  0011                        0100  0011
   - 05H   0000  0101  2's complement    +   1111  1011
     3EH                                     0011  1110
```
7. A = 95H – 4FH – 1 = 45H

SECTION 6.2: UNSIGNED MULTIPLICATION AND DIVISION
1. A, B
2. A and B, where A has the low byte and B has the high byte
3. No. We must use registers A and B for this operation.
4. A, B
5. A, B
6. No. We must use registers A and B for this operation.

SECTION 6.3: SIGNED NUMBER CONCEPTS AND ARITHMETIC OPERATIONS
1. D7
2. 16H is 00010110 in binary and its 2's complement is 1110 1010 or
    −16H=EA in hex.
3. −128 to +127
4. +9=00001001 and −9=11110111 or F7 in hex.
5. An overflow is a carry into the sign bit (D7) but the carry is a carry out of register (D7).

CHAPTER 7

LOGIC INSTRUCTIONS AND PROGRAMS

OBJECTIVES

Upon completion of this chapter, you will be able to:

>> Define the truth tables for logic functions AND, OR, XOR
>> Code 8051 Assembly language logic function instructions
>> Use 8051 logic instructions for bit manipulation
>> Use compare and jump instructions for program control
>> Code 8051 rotate and swap instructions
>> Code 8051 programs for ASCII and BCD data conversion

The study of logic instructions is the main topic of this chapter. In Section 7.1, we discuss the logic instructions AND, OR, and XOR, as well as COMPARE instructions for the 8051. The ROTATE and SWAP instructions are discussed in Section 7.2. Finally, in Section 7.3 we use these instructions to provide some real-world applications such as BCD-to-ASCII conversion.

SECTION 7.1: LOGIC AND COMPARE INSTRUCTIONS

In this section we cover boolean logic instructions such as AND, OR, exclusive-or (XOR), and complement. We will also study the compare instruction.

AND

```
ANL destination,source ;dest = dest AND source
```

This instruction will perform a logical AND on the two operands and place the result in the destination. The destination is normally the accumulator. The source operand can be a register, in memory, or immediate. See Appendix A.1 for more on the addressing modes for this instruction. The ANL instruction for byte-size operands has no effect on any of the flags. The ANL instruction is often used to mask (set to 0) certain bits of an operand. See Example 7-1.

X	Y	X AND Y
0	0	0
0	1	0
1	0	0
1	1	1

Example 7-1

Show the results of the following.

```
MOV   A,#35H   ;A = 35H
ANL   A,#0FH   ;A = A AND 0FH (now A = 05)
```

Solution:

```
35H   0 0 1 1 0 1 0 1
0FH   0 0 0 0 1 1 1 1
05H   0 0 0 0 0 1 0 1        35H AND 0FH = 05H
```

OR

```
ORL destination,source    ;dest = dest OR source
```

The destination and source operands are ORed, and the result is placed in the destination. The ORL instruction can be used to set certain bits of an operand to 1. The destination is normally the accumulator. The source operand can be a register, in memory, or immediate. See Appendix A for more on the addressing modes supported by this instruction. The ORL instruction for byte-size operands has no effect on any of the flags. See Example 7-2.

X	Y	X OR Y
0	0	0
0	1	1
1	0	1
1	1	1

Example 7-2

Show the results of the following.

```
MOV  A,#04            ;A = 04
ORL  A,#68H           ;A = 6C
```

Solution:

```
04H        0000 0100
68H        0110 1000
6CH        0110 1100      04 OR 68 = 6CH
```

XOR

```
XRL destination,source   ;dest = dest XOR source
```

X	Y	X XOR Y
0	0	0
0	1	1
1	0	1
1	1	0

This instruction will perform the XOR operation on the two operands, and place the result in the destination. The destination is normally the accumulator. The source operand can be a register, in memory, or immediate. See Appendix A.1 for the addressing modes of this instruction. The XRL instruction for byte-size operands has no effect on any of the flags. See Examples 7-3 and 7-4.

Example 7-3

Show the results of the following.

```
MOV   A,#54H
XRL   A,#78H
```

Solution:

```
54H        0 1 0 1 0 1 0 0
78H        0 1 1 1 1 0 0 0
2CH        0 0 1 0 1 1 0 0    54H XOR 78H = 2CH
```

Example 7-4

The XRL instruction can be used to clear the contents of a register by XORing it with itself. Show how "XRL A,A" clears A, assuming that AH = 45H.

Solution:

```
45H        01000101
45H        01000101
00         00000000 XOR a number with itself = 0
```

XRL can also be used to see if two registers have the same value. "XRL A,R1" will exclusive-or register A and register R1, and put the result in A. If both registers have the same value, 00 is placed in A. Then we can use the JZ instruction to make a decision based on the result. See Example 7-5.

CHAPTER 7: LOGIC INSTRUCTIONS AND PROGRAMS 129

Example 7-5

Read and test P1 to see whether it has the value 45H. If it does, send 99H to P2; otherwise, it stays cleared.

Solution:

```
      MOV   P2,#00     ;clear P2
      MOV   P1,#0FFH   ;make P1 an input port
      MOV   R3,#45H    ;R3 =45H
      MOV   A,P1       ;read P1
      XRL   A,R3
      JNZ   EXIT       ;jump if A has value other than 0
      MOV   P2,#99H
EXIT:...
```

In the program in Example 7-5 notice the use of the JNZ instruction. JNZ and JZ test the contents of the accumulator only. In other words, there is no such thing as a zero flag in the 8051.

Another widely used application of XRL is to toggle bits of an operand. For example, to toggle bit 2 of register A, we could use the following code. This code causes D2 of register A to change to the opposite value, while all the other bits remain unchanged.

```
      XRL   A,#04H     ;EX-OR  A with 0000 0100
```

CPL A (complement accumulator)

This instruction complements the contents of register A. The complement action changes the 0s to 1s and the 1s to 0s. This is also called *1's complement*.

```
      MOV   A,#55H
      CPL   A     ;now A=AAH
                  ;0101 0101(55H)becomes 1010 1010 (AAH)
```

To get the 2's complement, all we have to do is to add 1 to the 1's complement. There is no 2's complement instruction in the 8051. Notice that in complementing a byte, the data must be in register A. The CPL instruction does not support any other addressing mode. See Example 7-6.

Example 7-6

Find the 2's complement of the value 85H.

Solution:

```
      MOV   A,#85H                    85H = 1000 0101
      CPL   A      ;1's comp.  1'S = 0111 1010
      ADD   A,#1   ;2's comp.             + 1
                                     0111 1011  = 7BH
```

Compare instruction

The 8051 has an instruction for the compare operation. It has the following syntax.

```
CJNE destination,source,relative address
```

In the 8051, the actions of comparing and jumping are combined into a single instruction called CJNE (compare and jump if not equal). The CJNE instruction compares two operands, and jumps if they are not equal. In addition, it changes the CY flag to indicate if the destination operand is larger or smaller. It is important to notice that the operands themselves remain unchanged. For example, after the execution of the instruction "CJNE A, #67H, NEXT", register A still has its original value. This instruction compares register A with value 67H and jumps to the target address NEXT only if register A has a value other than 67H.

Example 7-7

Examine the following code, then answer the following questions.
(a) Will it jump to NEXT?
(b) What is in A after the CJNE instruction is executed?

```
        MOV   A,#55H
        CJNE  A,#99H,NEXT
        ...
NEXT:   ...
```

Solution:

(a) Yes, it jumps because 55H and 99H are not equal.
(b) A = 55H, its original value before the comparison.

In CJNE, the destination operand can be in the accumulator or in one of the Rn registers. The source operand can be in a register, in memory, or immediate. See Appendix A for the addressing modes of this instruction. This instruction affects the carry flag only. CY is changed as shown in Table 7-1. The following shows how the comparison works for all possible conditions.

Table 7-1: Carry Flag Setting For CJNE Instruction

Compare	Carry Flag
destination > source	CY = 0
destination < source	CY = 1

```
            CJNE R5,#80,NOT_EQUAL    ;check R5 for 80
            ...                      ;R5=80
NOT_EQUAL:  JNC   NEXT               ;jump if R5>80
            ...                      ;R5<80
NEXT:       ...
```

Notice in the CJNE instruction that any Rn register can be compared with an immediate value. There is no need for register A to be involved. Also notice that CY is always checked for cases of greater or less than, but only after it is determined that they are not equal. See Examples 7-8 and 7-9.

Example 7-8

Write code to determine if register A contains the value 99H. If so, make R1 = FFH; otherwise, make R1 = 0.

Solution:

```
    MOV  R1,#0        ;clear R1
    CJNE A,#99H,NEXT  ;if A not equal 99, then jump
    MOV  R1,#0FFH     ;they are equal, make R1=FFH
NEXT:...              ;not equal so R1=0
OVER:...
```

Example 7-9

Assume that P1 is an input port connected to a temperature sensor. Write a program to read the temperature and test it for the value 75. According to the test results, place the temperature value into the registers indicated by the following.

If T = 75   then A = 75
If T < 75   then R1 = T
If T > 75   then R2 = T

Solution:

```
        MOV  P1,#0FFH     ;make P1 an input port
        MOV  A,P1         ;read P1 port, temperature
        CJNE A,#75,OVER   ;jump if A not equal to 75
        SJMP EXIT         ;A=75, exit
OVER:   JNC  NEXT         ;if CY=0 then A>75
        MOV  R1,A         ;CY=1, A<75, save in R1
        SJMP EXIT         ; and exit
NEXT:   MOV  R2,A         ;A>75, save it in R2
EXIT:   ...
```

The compare instruction is really a subtraction, except that the values of the operands do not change. Flags are changed according to the execution of the SUBB instruction. It must be emphasized again that in the CJNE instruction, the operands are not affected, regardless of the result of the comparison. Only the CY flag is affected. This is despite the fact that CJNE uses the subtract operation to set or reset the CY flag.

Example 7-10

Write a program to monitor P1 continuously for the value 63H. It should get out of the monitoring only if P1 = 63H.

Solution:

```
    MOV P1,#0FFH        ;make P1 an input port
HERE: MOV A,P1          ;get P1
    CJNE A,#63,HERE     ;keep monitoring unless P1=63H
```

Example 7-11

Assume internal RAM memory locations 40H - 44H contain the daily temperature for five days, as shown below. Search to see if any of the values equals 65. If value 65 does exist in the table, give its location to R4; otherwise, make R4 = 0.

40H=(76)   41H=(79)   42H=(69)   43H=(65)   44H=(62)

Solution:

```
            MOV  R4,#0      ;R4=0
            MOV  R0,#40H    ;load pointer
            MOV  R2,#05     ;load counter
            MOV  A,#65      ;A=65, value searched for
    BACK:   CJNE A,@R0,NEXT ;compare RAM data with 65
            MOV  R4,R0      ;if 65, save address
            SJMP EXIT       ;and exit
    NEXT:   INC  R0         ;otherwise increment pointer
            DJNZ R2,BACK    ;keep checking until count=0
    EXIT    ...
```

Review Questions

1. Find the content of register A after the following code in each case.

   (a) MOV A,#37H   (b) MOV A,#37H   (c) MOV A,#37H
       ANL A,#0CAH   ORL A,#0CAH   XRL A,#0CAH

2. To mask certain bits of the accumulator we must ANL it with _____.
3. To set certain bits of the accumulator to 1 we must ORL it with _____.
4. XRLing an operand with itself results in _____.
5. True or false. The CJNE instruction alters the contents of its operands.
6. What value must R4 have in order for the following instruction not to jump?
   ```
   CJNE R4,#53,OVER
   ```
7. Find the contents of register A after execution of the following code.
   ```
   CLR  A
   ORL  A,#99H
   CPL  A
   ```

CHAPTER 7: LOGIC INSTRUCTIONS AND PROGRAMS   133

SECTION 7.2: ROTATE AND SWAP INSTRUCTIONS

In many applications there is a need to perform a bitwise rotation of an operand. In the 8051 the rotation instructions RL, RR, RLC, and RRC are designed specifically for that purpose. They allow a program to rotate the accumulator right or left. We explore the rotate instructions next since they are widely used in many different applications. In the 8051, to rotate a byte the operand must be in register A. There are two type of rotations. One is a simple rotation of the bits of A, and the other is a rotation through the carry. Each is explained below.

Rotating the bits of A right and left

```
RR    A    ;rotate right A
```

In rotate right, the 8 bits of the accumulator are rotated right one bit, and bit D0 exits from the least significant bit and enters into D7 (most significant bit). See the code and diagram.

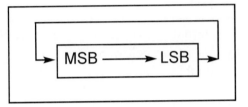

```
MOV   A,#36H   ;A=0011 0110
RR    A         ;A=0001 1011
RR    A         ;A=1000 1101
RR    A         ;A=1100 0110
RR    A         ;A=0110 0011
```

```
RL    A        ;rotate left A
```

In rotate left, the 8 bits of the accumulator are rotated left one bit, and bit D7 exits from the MSB (most significant bit) and enters into D0 (least significant bit). See the code and diagram.

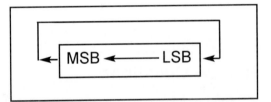

```
MOV   A,#72H   ;A=0111 0010
RL    A         ;A=1110 0100
RL    A         ;A=1100 1001
```

Notice in the RR and RL instructions that no flags are affected.

Rotating through the carry

There are two more rotate instructions in the 8051. They involve the carry flag. Each is shown next.

```
RRC A      ;rotate right through carry
```

In RRC A, as bits are rotated from left to right, they exit the LSB to the carry flag, and the carry flag enters the MSB. In other words, in RRC A the LSB is moved to CY and CY is moved to the MSB. In reality, the carry flag acts as if it is part of register A, making it a 9-bit register.

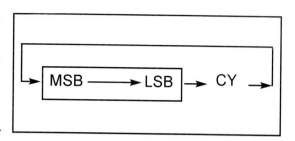

```
CLR  C              ;make CY=0
MOV  A,#26H         ;A=0010 0110
RRC  A              ;A=0001 0011 CY=0
RRC  A              ;A=0000 1001 CY=1
RRC  A              ;A=1000 0100 CY=1
```

```
RLC  A     ;rotate left through carry
```

In RLC A, as bits are shifted from right to left they exit the MSB and enter the carry flag, and the carry flag enters the LSB. In other words, in RCL the MSB is moved to CY (carry flag) and CY is moved to the LSB. See the following code and diagram.

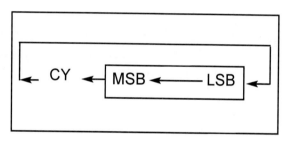

```
SETB C              ;make CY=1
MOV  A,#15H         ;A=0001 0101
RLC  A              ;A=0010 1011 CY=0
RLC  A              ;A=0101 0110 CY=0
RLC  A              ;A=1010 1100 CY=0
RLC  A              ;A=0101 1000 CY=1
```

SWAP A

Another useful instruction is the SWAP instruction. It works only on the accumulator (A). It swaps the lower nibble and the higher nibble. In other words, the lower 4 bits are put into the higher 4 bits and the higher 4 bits are put into the lower 4 bits. See the diagram below and Example 7-12.

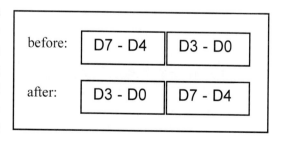

Example 7-12

(a) Find the contents of register A in the following code.
(b) In the absence of a SWAP instruction, how would you exchange the nibbles?
Write a simple program to show the process.

Solution:

(a)
```
MOV   A,#72H      ;A = 72H
SWAP  A           ;A = 27H
```

(b)
```
MOV   A,#72H      ;A=0111 0010
RL    A           ;A=1110 0100
RL    A           ;A=1100 1001
RL    A           ;A=1001 0011
RL    A           ;A=0010 0111
```

Example 7-13

Write a program that finds the number of 1s in a given byte.

Solution:

```
        MOV   R1,#0       ;R1 keeps the number of 1s
        MOV   R7,#8       ;counter = 08 rotate 8 times
        MOV   A,#97H      ;find the number of 1s in 97H
AGAIN:  RLC   A           ;rotate it through the CY once
        JNC   NEXT        ;check for CY
        INC   R1          ;if CY=1 then add one to count
NEXT:   DJNZ  R7,AGAIN    ;go through this 8 times
```

To transfer a byte of data serially, the data can be converted from parallel to serial with the rotate instuctions as shown below:

```
RRC  A           ;first bit to carry
MOV  P1.3,C      ;output carry as data bit
RRC  A           ;second bit to carry
MOV  P1.3,C      ;output carry as data bit
RRC  A           ;third bit to carry
MOV  P1.3,C      ;output carry as data bit
. . .
```

The above is a widely used method in tranferring data to serial memories such as serial EEPROMs.

Review Questions

1. What is the value of register A after each of the following instructions?
   ```
   MOV A,#25H
   RR   A
   RR   A
   RR   A
   RR   A
   ```

2. What is the value of register A after each of the following instructions?
   ```
   MOV   A,#A2H
   RL    A
   RL    A
   RL    A
   RL    A
   ```

3. What is the value of register A after each of the following instructions?
   ```
   CLR  A
   SETB C
   RRC  A
   SETB
   RRC  A
   ```

4. Why does "RLC R1" give an error in the 8051?

5. What is in register A after the execution of the following code?
   ```
   MOV   A,#85H
   SWAP  A
   ANL   A,#0F0H
   ```

SECTION 7.3: BCD AND ASCII APPLICATION PROGRAMS

BCD numbers were discussed in Chapter 6. As stated there, in many newer microcontrollers there is a real time clock (RTC) where the time and date are kept even when the power is off. These microcontrollers provide the time and date in BCD. However, to display them they must be converted to ASCII. In this section we show the application of logic and rotate instructions in the conversion of BCD and ASCII.

Table 7-2. ASCII Code for Digits 0 - 9

Key	ASCII (hex)	Binary	BCD (unpacked)
0	30	011 0000	0000 0000
1	31	011 0001	0000 0001
2	32	011 0010	0000 0010
3	33	011 0011	0000 0011
4	34	011 0100	0000 0100
5	35	011 0101	0000 0101
6	36	011 0110	0000 0110
7	37	011 0111	0000 0111
8	38	011 1000	0000 1000
9	39	011 1001	0000 1001

ASCII numbers

On ASCII keyboards, when the key "0" is activated, "011 0000" (30H) is provided to the computer. Similarly, 31H (011 0001) is provided for the key "1", and so on, as shown in Table 7-2.

It must be noted that although ASCII is standard in the United States (and many other countries), BCD numbers are universal. Since the keyboard, printers, and monitors all use ASCII, how does data get converted from ASCII to BCD, and vice versa? These are the subjects covered next.

Packed BCD to ASCII conversion

The DS5000T microcontrollers have what is called a *real-time clock* (RTC). The RTC provides the time of day (hour, minute, second) and the date (year, month, day) continuously, regardless of whether the power is on or off. However, this data is provided in packed BCD. For this data to be displayed on a device such as an LCD, or to be printed by the printer, it must be in ASCII format.

To convert packed BCD to ASCII, it must first be converted to unpacked BCD. Then the unpacked BCD is tagged with 011 0000 (30H). The following demonstrates converting from packed BCD to ASCII. See also Example 7-14.

Packed BCD	Unpacked BCD	ASCII
29H	02H & 09H	32H & 39H
0010 1001	0000 0010 &	0011 0010 &
	0000 1001	0011 1001

ASCII to packed BCD conversion

To convert ASCII to packed BCD, it is first converted to unpacked BCD (to get rid of the 3), and then combined to make packed BCD. For example, for 4 and 7 the keyboard gives 34 and 37, respectively. The goal is to produce 47H or "0100 0111", which is packed BCD. This process is illustrated in detail below.

Key	ASCII	Unpacked BCD	Packed BCD
4	34	00000100	
7	37	00000111	01000111 or 47H

```
MOV     A,#'4'      ;A=34H, hex for ASCII char 4
MOV     R1,#'7'     ;R1=37H, hex for ASCII char 7
ANL     A,#0FH      ;mask upper nibble (A=04)
ANL     R1,#0FH     ;mask upper nibble (R1=07)
SWAP    A           ;A=40H
ORL     A,R1        ;A=47H, packed BCD
```

After this conversion, the packed BCD numbers are processed and the result will be in packed BCD format. As we saw in Chapter 6, there is a special instruction "DA A", which requires that the data be in packed BCD format.

Example 7-14

Assume that register A has packed BCD, write a program to convert packed BCD to two ASCII numbers and place them in R2 and R6.

Solution:

```
MOV   A,#29H      ;A=29H, packed BCD
MOV   R2,A        ;keep a copy of BCD data in R2
ANL   A,#0FH      ;mask the upper nibble(A=09)
ORL   A,#30H      ;make it an ASCII, A=39H ('9')
MOV   R6,A        ;save it (R6=39H ASCII char)
MOV   A,R2        ;A=29H;get the original data
ANL   A,#0F0H     ;mask the lower nibble(A=20)
RR    A           ;rotate right
RR    A           ;rotate right
RR    A           ;rotate right
RR    A           ;rotate right,(A=02)
ORL   A,#30H      ;A=32H,ASCII char.'2'
MOV   R2,A        ;save ASCII char in R2
```

Of course, in the above code we can replace all the RR instructions with a single "SWAP A" instruction.

Review Questions

1. For the following decimal numbers, give the packed BCD and unpacked BCD representations.
   (a) 15   (b) 99
2. Show the binary and hex formats for "76" and its BCD version.
3. Does the register A have BCD data after the following instruction is executed?
   ```
   MOV A,#54
   ```
4. 67H in BCD when converted to ASCII is ____H and ____H.
5. Does the following convert unpacked BCD in register A to ASCII?
   ```
   MOV   A,#09
   ADD   A,#30H
   ```

SUMMARY

This chapter defined the logic instructions AND, OR, XOR, and complement. In addition, 8051 Assembly language instructions for these functions were described. Compare and jump instructions were described as well. These functions are often used for bit manipulation purposes.

The rotate and swap instructions of the 8051 are used in many applications. This chapter also described BCD and ASCII formats and conversions.

PROBLEMS

SECTION 7.1: LOGIC AND COMPARE INSTRUCTIONS

1. Assume that these registers contain the following: A = F0, B = 56, and R1 = 90. Perform the following operations. Indicate the result and the register where it is stored.
   Note: The operations are independent of each other.
   (a) `ANL A,#45H`
   (b) `ORL A,B`
   (c) `XRL A,#76H`
   (d) `ANL A,R1`
   (e) `XRL A,R1`
   (f) `ORL A,R1`
   (g) `ANL A,#0FFH`
   (h) `ORL A,#99H`
   (i) `XRL A,#0EEH`
   (j) `XRL A,#0AAH`

2. Find the contents of register A after each of the following instructions.
   (a) `MOV A,#65H`
       `ANL A,#76H`
   (b) `MOV A,#70H`
       `ORL A,#6BH`
   (c) `MOV A,#95H`
       `XRL A,#0AAH`
   (d) `MOV A,#5DH`
       `MOV R3,#78H`
       `ANL A,R3`
   (e) `MOV A,#0C5H`
       `MOV R6,#12H`
       `ORL A,R6`
   (f) `MOV A,#6AH`
       `MOV R4,#6EH`
       `XRL A,R4`
   (g) `MOV A,#37H`
       `ORL A,#26H`

3. True or false. In using the CJNE instruction, we must use the accumulator as the destination.

4. Is the following a valid instruction? "`CJNE R4,#67,HERE`"

5. Does the 8051 have a "CJE" (compare and jump if equal) instruction?

6. Indicate the status of CY after CJNE is executed in each of the following cases.
   (a) `MOV A,#25H`
       `CJNE A,#44H,OVER`
   (b) `MOV A,#0FFH`
       `CJNE A,#6FH,NEXT`
   (c) `MOV A,#34`
       `CJNE A,#34,NEXT`
   (d) `MOV R1,#0`
       `CJNE R1,#0,NEXT`
   (e) `MOV R5,#54H`
       `CJNE R5,#0FFH,NEXT`
   (f) `MOV R3,#0AAH`
       `ANL R3,#55H`
       `CJNE R3,#00,NEXT`

7. In Problem 6, indicate whether or not the jump happens for each case.

SECTION 7.2: ROTATE AND SWAP INSTRUCTIONS

8. Find register A contents after each of the following is executed.
   (a) `MOV A,#56H`
       `SWAP A`
       `RR   A`
       `RR   A`
   (b) `MOV A,#39H`
       `CLR  C`
       `RL   A`
       `RL   A`

```
        (c)  CLR  C                    (d)     SETB  C
             MOV  A,#4DH                        MOV  A,#7AH
             SWAP A                             SWAP A
             RRC  A                             RLC  A
             RRC  A                             RLC  A
             RRC  A
```

9. Show the code to replace the SWAP code
   (a) using the rotate right instructions
   (b) using the rotate left instructions
10. Write a program that finds the number of zeros in an 8-bit data item.
11. Write a program that finds the position of the first high in an 8-bit data item. The data is scanned from D0 to D7. Give the result for 68H.
12. Write a program that finds the position of the first high in an 8-bit data item. The data is scanned from D7 to D0. Give the result for 68H.
13. A stepper motor uses the following sequence of binary numbers to move the motor. How would you generate them?

```
        1100,0110,0011,1001
```

SECTION 7.3: BCD AND ASCII APPLICATION PROGRAMS

14. Write a program to convert a series of packed BCD numbers to ASCII. Assume that the packed BCD is located in ROM locations starting at 300H. Place the ASCII codes in RAM locations starting at 40H.

```
                    ORG 300H
        MYDATA:     DB   76H,87H,98H,43H
```

15. Write a program to convert a series of ASCII numbers to packed BCD. Assume that the ASCII data is located in ROM locations starting at 300H. Place the BCD data in RAM locations starting at 60H.

```
                    ORG 300H
        MYDATA:     DB   "87675649"
```

16. Write a program to get an 8-bit binary number from P1, convert it to ASCII, and save the result in RAM locations 40H, 41H, and 42H. What is the result if P1 has 1000 1101 binary as input?
17. Find the result at points (1), (2), and (3) in the following code?

```
                    CJNE A,#50,NOT_EQU
                    ...                 ;point (1)
        NOT_EQU:    JC   NEXT
                    ...                 ;point (2)
        NEXT:       ...                 ;point (3)
```

ANSWERS TO REVIEW QUESTIONS

SECTION 7.1: LOGIC AND COMPARE INSTRUCTIONS
1. (a) 02   (b) FFH   (c) FDH
2. Zeros   3. One
4. All zeros   5. False
6. 54   7. 66H

SECTION 7.2: ROTATE AND SWAP INSTRUCTIONS
1. 52H 2. 2AH 3. C0H
4. Because all the rotate instructions work with the accumulator only
5. 50H

SECTION 7.3: BCD AND ASCII APPLICATION PROGRAMS
1. (a) 15H = 0001 0101 packed BCD, 0000 0001,0000 0101 unpacked BCD
   (b) 99H = 1001 1001 packed BCD, 0000 1001,0000 1001 unpacked BCD
2. 3736H = 00110111 00110110B
   and in BCD we have 76H = 0111 0110B
3. No. We need to write it 54H (with the H) or 01010100B to make it in BCD. The value 54
   without the "H" is interpreted as 36H by the assembler.
4. 36H, 37H
5. Yes, since A = 39H.

CHAPTER 8

SINGLE-BIT INSTRUCTIONS AND PROGRAMMING

One unique and powerful feature of the 8051 is single-bit operation. Single-bit instructions allow the programmer to set, clear, move, and complement individual bits of a port, memory, or register. In this chapter we cover these instructions in the context of some examples and applications. In Section 8.1 we study single-bit general instructions, and in Section 8.2 we cover single-bit instructions related to the carry flag.

SECTION 8.1: SINGLE-BIT INSTRUCTION PROGRAMMING

In most microprocessors, data is accessed in byte-sized chunks. In these byte-addressable microprocessors, the contents of a register, RAM, or port must be accessed by the byte. In other words, the minimum amount of data that can be accessed is one byte. For example, in the Pentium microprocessor the I/O port is byte-oriented, meaning that to change one bit you must access the entire 8 bits. However, in many applications we need to change one bit, for example, to turn on or off a device. The bit-addressability of the 8051 lends itself perfectly to these applications. The ability to access data in a single bit instead of the whole byte makes the 8051 one of the most powerful 8-bit microcontrollers in the market. Which portions of the CPU, RAM, registers, I/O port, or ROM are bit-addressable? Since ROM simply holds program code for execution, there is no need for bit-addressability. All opcodes are byte-oriented. It is the registers, RAM, and I/O ports that need to be bit-addressable. In the 8051 many internal RAM locations, some registers, and all I/O ports are bit-addressable. Each is covered separately.

Single-bit instructions

Instructions that are used for single-bit operations are given in Table 8-1. In this section we discuss these instructions and give many examples of their usage. Other instructions that allow single-bit operations, but only along with the carry flag (CY), are covered separately in the next section.

Table 8-1: Single-Bit Instructions

Instruction		Function
SETB	bit	Set the bit (bit = 1)
CLR	bit	Clear the bit (bit = 0)
CPL	bit	Complement the bit (bit = NOT bit)
JB	bit,target	Jump to target if bit = 1 (jump if bit)
JNB	bit,target	Jump to target if bit = 0 (jump if no bit)
JBC	bit,target	Jump to target if bit = 1, clear bit (jump if bit, then clear)

I/O ports and bit-addressability

The 8051 has four I/O ports, each of which is 8 bits: P0, P1, P2, and P3. We can access either the entire 8 bits or any single bit without altering the rest. When accessing a port in a single-bit manner, we use the syntax "SETB X.Y" where X is the port number 0, 1, 2, or 3, and Y is the desired bit number from 0 to 7 for data bits D0 to D7. For example, "SETB P1.5" sets high bit 5 of port 1. Remember that D0 is the LSB and D7 is the MSB. See Example 8-1.

Example 8-1

Write the following programs.
(a)   Create a square wave of 50% duty cycle on bit 0 of port 1.
(b)   Create a square wave of 66% duty cycle on bit 3 of port 1.

Solution:

(a)   The 50% duty cycle means that the "on" and "off" state (or the high and low portion of the pulse) have the same length. Therefore, we toggle P1.0 with a time delay in between each state.

```
HERE:       SETB  P1.0      ;set to high bit 0 of port 1
            LCALL DELAY      ;call the delay subroutine
            CLR   P1.0       ;P1.0=0
            LCALL DELAY
            SJMP  HERE       ;keep doing it
```

Another way to write the above program is:

```
HERE:       CPL   P1.0       ;complement bit 0 of port 1
            LCALL DELAY      ;call the delay subroutine
            SJMP  HERE       ;keep doing it
```

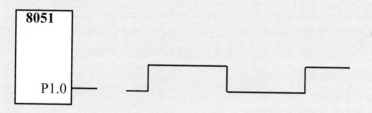

(b)   The 66% duty cycle means the "on" state is twice the "off" state.

```
BACK:       SETB P1.3        ;set port 1 bit 3 high
            LCALL DELAY      ;call the delay subroutine
            LCALL DELAY      ;call the delay subroutine again
            CLR  P1.3        ;clear bit 2 of port 1(P1.3=low)
            LCALL DELAY      ;call the delay subroutine
            SJMP  BACK       ;keep doing it
```

Notice that when code such as "SETB P1.0" is assembled, it becomes "SETB 90H" since P1.0 has the RAM address of 90H. From Figure 8-1 we see that the bit addresses for P0 are 80H to 87H, and for P1 are 90H to 97H, and so on. Figure 8-1 also shows all the registers that are bit-addressable.

Table 8-2: Single-Bit Addressability of Ports

P0	P1	P2	P3	Port's Bit
P0.0	P1.0	P2.0	P3.0	D0
P0.1	P1.1	P2.1	P3.1	D1
P0.2	P1.2	P2.2	P3.2	D2
P0.3	P1.3	P2.3	P3.3	D3
P0.4	P1.4	P2.4	P3.4	D4
P0.5	P1.5	P2.5	P3.5	D5
P0.6	P1.6	P2.6	P3.6	D6
P0.7	P1.7	P2.7	P3.7	D7

Example 8-2

For each of the following instructions, state which bit of which SFR will be affected. Use Figure 8-1.

(a) SETB 86H   (b) CLR 87H   (c) SETB 92H

(d) SETB 0A7H   (e) CLR 0F2H   (f) SETB 0E7H

Solution:

(a) SETB 86H is for SETB P0.6
(b) CLR 87H is for CLR P0.7
(c) SETB 92H is for SETB P1.2
(d) SETB 0A7H is for SETB P2.7
(e) CLR 0F2H is for CLR D2 of register B
(f) SETB 0E7H is for SETB ACC.7 (D7 of register A)

Checking an input bit

The JNB (jump if no bit) and JB (jump if bit = 1) instructions are also widely used single-bit operations. They allow you to monitor a bit and make a decision depending on whether it is 0 or 1. See Example 8-3.

Example 8-3

Assume that bit P2.3 is an input and represents the condition of an oven. If it goes high, it means that the oven is hot. Monitor the bit continuously. Whenever it goes high, send a high-to-low pulse to port P1.5 to turn on a buzzer.

Solution:

```
HERE:JNB  P2.3,HERE      ;keep monitoring for high
     SETB P1.5           ;set bit P1.5=1
     CLR  P1.5           ;make high-to-low
```

146

Instructions JNB and JB can be used for any bits of I/O ports 0, 1, 2, and 3, since all ports are bit-addressable. However, most of port 3 is used for interrupts and serial communication signals, and typically is not used for any I/O, either single-bit or byte-wise. This is discussed in Chapters 10 and 11.

Registers and bit-addressability

While all the I/O ports are bit-addressable, that is not the case with registers, as seen from Figure 8-1. Only registers B, PSW, IP, IE, ACC, SCON, and TCON are bit-addressable. Of the bit-addressable registers, we will concentrate on the familiar registers A, B, and PSW. The rest will be used in future chapters. From Figure 8-1 notice that P0 is assigned the bit addresses 80 - 87H. The bit addresses 88 - 8FH are assigned to TCON. Finally, bit addresses F0 - F7H are assigned to register B. Look at Examples 8-4 and 8-5 using these registers with bit-addressable capability.

Byte adddress	Bit address								
FF									
F0	F7	F6	F5	F4	F3	F2	F1	F0	B
E0	E7	E6	E5	E4	E3	E2	E1	E0	ACC
D0	D7	D6	D5	D4	D3	D2	D1	D0	PSW
B8	--	--	--	BC	BB	BA	B9	B8	IP
B0	B7	B6	B5	B4	B3	B2	B1	B0	P3
A8	AF	--	--	AC	AB	AA	A9	A8	IE
A0	A7	A6	A5	A4	A3	A2	A1	A0	P2
99	not bit addressable								SBUF
98	9F	9E	9D	9C	9B	9A	99	98	SCON
90	97	96	95	94	93	92	91	90	P1
8D	not bit addressable								TH1
8C	not bit addressable								TH0
8B	not bit addressable								TL1
8A	not bit addressable								TL0
89	not bit addressable								TMOD
88	8F	8E	8D	8C	8B	8A	89	88	TCON
87	not bit addressable								PCON
83	not bit addressable								DPH
82	not bit addressable								DPL
81	not bit addressable								SP
80	87	86	85	84	83	82	81	80	P0

Special Function Registers

Figure 8-1. SFR RAM Address (Byte and Bit)

Example 8-4

Write a program to see if the accumulator contains an even number. If so, divide it by 2. If not, make it even and then divide it by 2.

Solution:

```
        MOV   B,#2          ;B=2
        JNB   ACC.0,YES     ; is D0 of reg A 0?, if so jump
        INC   A             ;it is odd, make it even
YES:    DIV   AB            ;A/B
```

Example 8-5

Write a program to see if bits 0 and 5 of register B are 1. If they are not, make them so and save it in R0.

Solution:

```
        JNB  0F0H,NEXT_1    ;jump if B.0 is low
        SETB 0F0H           ;make bit B.0 high
NEXT_1: JNB  0F5H,NEXT_2    ;jump if B.5 is low
        SETB 0F5H           ;make B.5 high
NEXT_2: MOV  R0,B           ;save register B
```

CY	AC	--	RS1	RS0	OV	--	P

RS1	RS0	Register Bank	Address
0	0	0	00H - 07H
0	1	1	08H - 0FH
1	0	2	10H - 17H
1	1	3	18H - 1FH

Figure 8-2. Bits of the PSW Register

As discussed in Chapter 2, in the PSW register two bits are set aside for the selection of the register banks. Upon RESET, bank 0 is selected. We can select any other banks using the bit-addressability of the PSW.

Example 8-6

Write a program to save the accumulator in R7 of bank 2.

Solution:

```
        CLR   PSW.3
        SETB  PSW.4
        MOV   R7,A
```

Example 8-7

While there are instructions such as JNC and JC to check the carry flag bit (CY), there are no such instructions for the overflow flag bit (OV). How would you write code to check OV?

Solution:

The OV flag is PSW.2 of the PSW register. PSW is a bit-addressable register; therefore, we can use the following instruction to check the OV flag.

```
        JB   PSW.2,TARGET    ;jump if OV=1
```

Bit-addressable RAM

Of the 128-byte internal RAM of the 8051, only 16 bytes of it are bit-addressable. The rest must be accessed in byte format. The bit-addressable RAM locations are 20H to 2FH. These 16 bytes provide 128 bits of RAM bit-addressability since $16 \times 8 = 128$. They are addressed as 0 to 127 (in decimal) or 00 to 7FH. Therefore, the bit addresses 0 to 7 are for the first byte of internal RAM location 20H, and 8 to 0FH are the bit addresses of the second byte, RAM location 21H, and so on. The last byte of 2FH has bit addresses of 78H to 7FH. See Figure 8-3. Note that internal RAM locations 20-2FH are both byte-addressable and bit-addressable.

Notice from Figures 8-3 and 8-1 that bit addresses 00 - 7FH belong to RAM byte addresses 20 - 2FH, and bit addresses 80 - F7H belong to SFR P0, P1, etc.

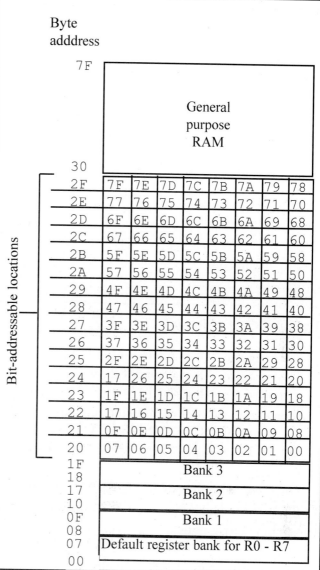

Figure 8-3. 128 Bytes of Internal RAM

Byte address								
7F	General purpose RAM							
30								
2F	7F	7E	7D	7C	7B	7A	79	78
2E	77	76	75	74	73	72	71	70
2D	6F	6E	6D	6C	6B	6A	69	68
2C	67	66	65	64	63	62	61	60
2B	5F	5E	5D	5C	5B	5A	59	58
2A	57	56	55	54	53	52	51	50
29	4F	4E	4D	4C	4B	4A	49	48
28	47	46	45	44	43	42	41	40
27	3F	3E	3D	3C	3B	3A	39	38
26	37	36	35	34	33	32	31	30
25	2F	2E	2D	2C	2B	2A	29	28
24	17	26	25	24	23	22	21	20
23	1F	1E	1D	1C	1B	1A	19	18
22	17	16	15	14	13	12	11	10
21	0F	0E	0D	0C	0B	0A	09	08
20	07	06	05	04	03	02	01	00

(left label: Bit-addressable locations)

1F – 18	Bank 3
17 – 10	Bank 2
0F – 08	Bank 1
07 – 00	Default register bank for R0 - R7

Example 8-8

Find out to which byte each of the following bits belongs. Give the address of the RAM byte in hex.

(a) SETB 42H ;set bit 42H to 1   (d) SETB 28H ;set bit 28H to 1
(b) CLR 67H ;clear bit 67   (e) CLR 12 ;clear bit 12 (decimal)
(c) CLR 0FH ;clear bit 0FH   (f) SETB 05

Solution:

(a) RAM bit address of 42H belongs to D2 of RAM location 28H
(b) RAM bit address of 67H belongs to D7 of RAM location 2CH
(c) RAM bit address of 0FH belongs to D7 of RAM location 21H
(d) RAM bit address of 28H belongs to D0 of RAM location 25H
(e) RAM bit address of 12 belongs to D4 of RAM location 21H
(f) RAM bit address of 05 belongs to D5 of RAM location 20H

One way to use a bit-addressable RAM is to save bits. See Example 8-9.

Example 8-9

The status of bits P1.2 and P1.3 of I/O port P1 must be saved before they are changed. Write a program to save the status of P1.2 in bit location 06 and the status of P1.3 in bit location 07.

Solution:

```
      CLR   06          ;clear bit address 06
      CLR   07          ;clear bit address 07
      JNB   P1.2,OVER   ;check bit P1.2,if 0 then jump
      SETB  06          ;if P1.2=1,set bit location 06 to 1
OVER: JNB   P1.3,NEXT   ;check bit P1.3,if 0 then jump
      SETB  07          ;if P1.3=1, set bit location 07 to 1
NEXT:...
```

Review Questions

1. True or false. All I/O ports of the 8051 are bit-addressable.
2. True or false. All registers of the 8051 are bit-addressable.
3. True or false. All RAM locations of the 8051 are bit-addressable.
4. Indicate which of the following registers are bit-addressable.
   (a) A   (b) B   (c) R4   (d) PSW   (e) R7
5. Of the 128 bytes of RAM in the 8051, how many bytes are bit-addressable? List them.
6. How would you check to see whether bit D0 of R3 is high or low?
7. Find out to which byte each of the following bits belongs. Give the address of the RAM byte in hex.
   (a) SETB 20   (b) CLR 32   (c) SETB 12H
   (d) SETB 95H (e) SETB 0E6H
8. While bit addresses 00 - 7FH belong to _____, bit addresses 80 - F7H belong to _____.
9. True or false. P0, P1, P2, and P3 are part of SFR.
10. True or false. Register TCON is bit-addressable.

SECTION 8.2: SINGLE-BIT OPERATIONS WITH CY

Aside from the fact that the carry flag (CY) is altered by arithmetic and logic instructions, in the 8051 there are also several instructions by which the CY flag can be manipulated directly. These instructions are listed in Table 8-3.

Of the instructions in Table 8-3, we have shown the use of JNC, CLR, and SETB in many examples in the last few chapters. The rest of this section shows some examples of how to use some more instructions from Table 8-3.

Some of the instructions in Table 8-3 deal with the logic operations AND and OR. The examples in this section show how they can be used.

In future chapters, we will show many more examples of the use of single-bit operations in the context of real-world applications.

Table 8-3: Carry Bit-Related Instructions

Instruction	Function
SETB C	make CY = 1
CLR C	clear carry bit (CY = 0)
CPL C	complement carry bit
MOV b,C	copy carry status to bit location (CY = b)
MOV C,b	copy bit location status to carry (b = CY)
JNC target	jump to target if CY = 0
JC target	jump to target if CY = 1
ANL C,bit	AND CY with bit and save it on CY
ANL C,/bit	AND CY with inverted bit and save it on CY
ORL C,bit	OR CY with bit and save it on CY
ORL C,/bit	OR CY with inverted bit and save it on CY

Example 8-10

Write a program to save the status of bits P1.2 and P1.3 on RAM bit locations 6 and 7, respectively.

Solution:

```
MOV C,P1.2      ;save status of P1.2 on CY
MOV 06,C        ;save carry in RAM bit location 06
MOV C,P1.3      ;save status of P1.3 on CY
MOV 07,C        ;save carry in RAM bit location 07
```

Example 8-11

Assume that RAM bit location 12H holds the status of whether there has been a phone call or not. If it is high, it means there has been a new call since it was checked the last time. Write a program to display "New Messages" on an LCD if bit RAM 12H is high. If it is low, the LCD should say "No New Messages".

Solution:

```
        MOV  C,12H          ;copy bit location 12H to carry
        JNC  NO             ;check to see if is high
        MOV  DPTR,#400H     ;yes, load address of message
        LCALL DISPLAY       ;display message (see Chap. 12)
        SJMP EXIT           ;get out
NO:     MOV  DPTR,#420H     ;load the address of No message
        LCALL DISPLAY       ;display it
EXIT:                       ;exit
;————————————data to be displayed on LCD
        ORG  400H
YES_MG:     DB  "New Messages"
        ORG  420H
NO_MG:      DB  "No New Messages"
```

Example 8-12

Assume that the bit P2.2 is used to control the outdoor light and bit P2.5 to control the light inside a building. Show how to turn on the outside light and turn off the inside one.

Solution:

```
STEB  C          ;CY=1
ORL   C,P2.2      ;CY = P2.2 ORed with CY
MOV   P2.2,C      ;turn it "on" if not already "on"
CLR   C          ;CY=0
ANL   C,P2.5      ;CY=P2.5 ANDed with CY
MOV   P2.5,C      ;turn it off if not already off
```

Review Questions

1. Find the status of the CY flag after the following code.
```
        CLR   A
        ADD   A,#0FFH
        JNC   OVER
        CPL   C
OVER:         . . .
```
2. Find the status of the CY flag after the following code.
```
        CLR   C
        JNC   OVER
        SETB  C
OVER:         . . . .
```
3. Find the status of the CY flag after the following code.
```
        CLR   C
        JC    OVER
        CPL   C
OVER:         . . . .
```
4. Show how to save the status of P2.7 in RAM bit location 31.
5. Show how to move the status of RAM bit location 09 to P1.4.

SECTION 8.3: READING INPUT PINS VS. PORT LATCH

In reading a port, some instructions read the status of port pins while others read the status of an internal port latch. Therefore, when reading ports there are two possibilities:
1. Read the status of the input pin.
2. Read the internal latch of the output port.

We must make a distinction between these two categories of instructions since confusion between them is a major source of errors in 8051 programming, especially where external hardware is concerned. In this section we discuss these instructions briefly. However, readers must study and understand the material on this topic and on the internal working of ports that is given in Appendix C.2.

Instructions for reading input port

As stated in Chapter 4, to make any bit of any 8051 port an input port, we must write 1 (logic high) to that bit. After we configure the port bits as input, we can use only certain instructions in order to get the external data present at the pins into the CPU. Table 8-4 shows the list of such instructions.

Table 8-4:Instructions For Reading an Input Port

Mnemonics	Examples	Description
MOV A,PX	MOV A,P2	Bring into A the data at P2 pins
JNB PX.Y,..	JNB P2.1,TARGET	Jump if pin P2.1 is low
JB PX.Y,..	JB P1.3,TARGET	Jump if pin P1.3 is high
MOV C,PX.Y	MOV C,P2.4	Copy status of pin P2.4 to CY

Reading latch for output port

Some instructions read the contents of an internal port latch instead of reading the status of an external pin. Table 8-5 provides the list of these instructions. For example, look at the "ANL P1,A" instruction. The sequence of actions taken when such an instruction is executed is as follows.
1. It reads the internal latch of the port and brings that data into the CPU.
2. This data is ANDed with the contents of register A.
3. The result is rewritten back to the port latch.
4. The port pin data is changed and now has the same value as the port latch.

From the above discussion, we conclude that the instructions which read the port latch normally read a value, perform an operation (and possibly change it), then rewrite it back to the port latch. This is often called "*Read-Modify-Write*". Table 8-5 provides the list of read-modify-write instructions. Notice from Table 8-5 that all the read-modify-write instructions use the port as the destination operand. In other words, they are used only for ports configured as output ports.

Table 8-5:Instructions Reading a Latch (Read-Modify-Write)

Mnemonics	Example
ANL Px	ANL P1,A
ORL Px	ORL P2,A
XRL Px	XRL P0,A
JBC PX.Y,TARGET	JBC P1.1,TARGET
CPL PX.Y	CPL P1.2
INC Px	INC P1
DEC Px	DEC P2
DJNZ PX.Y,TARGET	DJNZ P1,TARGET
MOV PX.Y,C	MOV P1.2,C
CLR PX.Y	CLR P2.3
SETB PX.Y	SETB P2.3

Note: x is 0, 1, 2, or 3 for P0 - P3.

SUMMARY

This chapter described the single-bit operation of the 8051, one of its most powerful features. Single-bit instructions allow the programmer to set, clear, move, and complement individual bits of a port, memory, or register.

In addition, there are several instructions by which the carry flag can be manipulated directly. We also discussed instructions for reading the port pins versus reading the port latch.

PROBLEMS

1. "SETB A" is a(n) _____ (valid, invalid) instruction.
2. "CLR A" is a(n) _____ (valid, invalid) instruction.
3. "CPL A" is a(n) _____ (valid, invalid) instruction.
4. Which of the I/O ports of P0, P1, P2, and P3 are bit-addressable?
5. Which of the registers of the 8051 are bit-addressable?
6. Which of the following instructions are valid? If valid, indicate which bit is altered.

   (a) SETB P1   (b) SETB P2.3   (c) CLR ACC.5
   (d) CLR 90H   (e) SETB B.4   (f) CLR 80H
   (g) CLR PSW.3 (h) CLR 87H

7. Write a program to generate a square wave with 75% duty cycle on bit P1.5.
8. Write a program to generate a square wave with 80% duty cycle on bit P2.7.
9. Write a program to monitor P1.4. When it goes high, the program will generate a sound (square wave of 50% duty cycle) on pin P2.7.
10. Write a program to monitor P2.1. When it goes low, the program will send the value 55H to P0.
11. What bit addresses are assigned to P0?
12. What bit addresses are assigned to P1?
13. What bit addresses are assigned to P2?
14. What bit addresses are assigned to P3?
15. What bit addresses are assigned to the PCON register?
16. What bit addresses are assigned to the TCON register?
17. What bit addresses are assigned to register A?
18. What bit addresses are assigned to register B?
19. What bit addresses are assigned to register PSW?

20. The following are bit addresses. Indicate where each one belongs.
    (a) 85H   (b) 87H   (c) 88H   (d) 8DH   (e) 93H
    (f) A5H   (g) A7H   (h) B3H   (i) D4H   (j) D7H (k) F3H
21. Write a program to save registers A and B on R3 and R5 of bank 2, respectively.
22. Give another instruction for "CLR C".
23. In Problem 22, assemble each instruction and state if there is any difference between them.
24. Show how would you check whether the OV flag is low.
25. Show how would you check whether the CY flag is high.
26. Show how would you check whether the P flag is high.
27. Show how would you check whether the AC flag is high.
28. Give the bit addresses assigned to the flag bit of CY, P, AC, and OV.
29. Of the 128 bytes of RAM locations in the 8051, how many of them are also assigned a bit address as well? Indicate which bytes are those.
30. Indicate the bit addresses assigned to RAM locations 20H to 2FH.
31. The byte addresses assigned to the 128 bytes of RAM are _____ to _____.
32. The byte addresses assigned to the SFR are _____ to _____.
33. Indicate the bit addresses assigned to both of the following. Is there a gap between them?
    (a) RAM locations 20H to 2FH   (b) SFR
34. The following are bit addresses. Indicate where each one belongs.
    (a) 05H   (b) 47H   (c) 18H   (d) 2DH   (e) 53H
    (f) 15H   (g) 67H   (h) 55H   (i) 14H   (j) 37H (k) 7FH
35. True or false. The bit addresses of less than 80H are assigned to RAM locations 20 - 2FH.
36. True or false. The bit addresses of 80H and beyond are assigned to SFR (special function registers).
37. Write instructions to save the CY flag bit in bit location 4.
38. Write instructions to save the AC flag bit in bit location 16H.
39. Write instructions to save the P flag bit in bit location 12H.
40. Write instructions to see whether the D0 and D1 bits of register A are high. If so, divide register A by 4.
41. Write a program to see whether the D7 bit of register A is high. If so, send a message to the LCD stating that ACC has a negative number.
42. Write a program to see whether the D7 bit of register B is low. If so, send a message to the LCD stating that register B has a positive number.
43. Write a program to set high all the bits of RAM locations 20H to 2FH using the following methods:
    (a) using byte addresses   (b) using bit addresses
44. Write a program to see whether the accumulator is divisible by 8.
45. Write a program to find the number of zeros in register R2.

ANSWERS TO REVIEW QUESTIONS

SECTION 8.1: SINGLE-BIT INSTRUCTION PROGRAMMING
1. True 　　2. False 　　3. False 　　4. A,B, and PSW
5. 16 bytes are bit-addressable; they are from byte location 20H to 2FH.
6. MOV　　A,R3
　　JNB　　ACC.0
7. For (a), (b), and (c) use Figure 8-3. 　　(a) RAM byte 22H, bit D4
　　(b) RAM byte 24H, bit D0 　　(c) RAM byte 22H, bit D2
　　For (d) and (e) use Figure 8-1. 　　(d) SETB P1.5 　　(e) SETB ACC.6
8. RAM bytes 00 - 20H, special function registers.
9. True 　　10. True

SECTION 8.2: SINGLE-BIT OPERATIONS WITH CY
1. CY = 0 　　2. CY = 1 　　3. CY = 1
4. MOV C,P2.7 　　;save status of P2.7 on CY
　　MOV 31,C 　　;save carry in RAM bit location 06
5. MOV C,9 　　;save status of RAM bit 09 in CY
　　MOV P1.4,C 　　;save carry in P1.4

CHAPTER 9

COUNTER/TIMER PROGRAMMING IN THE 8051

OBJECTIVES

Upon completion of this chapter, you will be able to:

>> List the timers of the 8051 and their associated registers
>> Describe the various modes of the 8051 timers
>> Program the 8051 timers to generate time delays
>> Program the 8051 counters as event counters

The 8051 has two timers/counters. They can be used either as timers to generate a time delay or as counters to count events happening outside the microcontroller. In this chapter we show how they are programmed and used. In Section 9.1 we see how these timers are used to generate time delays. In Section 9.2 we show how they are used as event counters.

SECTION 9.1: PROGRAMMING 8051 TIMERS

The 8051 has two timers: timer 0 and timer 1. They can be used either as timers or as event counters. In this section we first discuss the timers' registers and then show how to program the timers to generate time delays.

Basic registers of the timer

Both timer 0 and timer 1 are 16 bits wide. Since the 8051 has an 8-bit architecture, each 16-bit timer is accessed as two separate registers of low byte and high byte. Each timer is discussed separately.

Timer 0 registers

The 16-bit register of timer 0 is accessed as low byte and high byte. The low byte register is called TL0 (timer 0 low byte) and the high byte register is referred to as TH0 (timer 0 high byte). These registers can be accessed like any other register, such as A, B, R0, R1, R2, etc. For example, the instruction "MOV TL0,#4FH" moves the value 4FH into TL0, the low byte of timer 0. These registers can also be read like any other register. For example, "MOV R5,TH0" saves TH0 (high byte of timer 0) in R5.

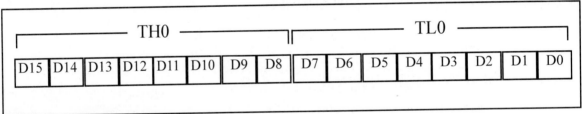

Figure 9-1. Timer 0 Registers

Timer 1 registers

Timer 1 is also 16 bits, and its 16-bit register is split into two bytes, referred to as TL1 (timer 1 low byte) and TH1 (timer 1 high byte). These registers are accessible in the same way as the registers of timer 0.

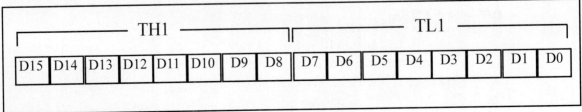

Figure 9-2. Timer 1 Registers

158

TMOD (timer mode) register

Both timers 0 and 1 use the same register, called TMOD, to set the various timer operation modes. TMOD is an 8-bit register in which the lower 4 bits are set aside for timer 0 and the upper 4 bits are set aside for timer 1. In each case, the lower 2 bits are used to set the timer mode and the upper 2 bits to specify the operation. These options are discussed next.

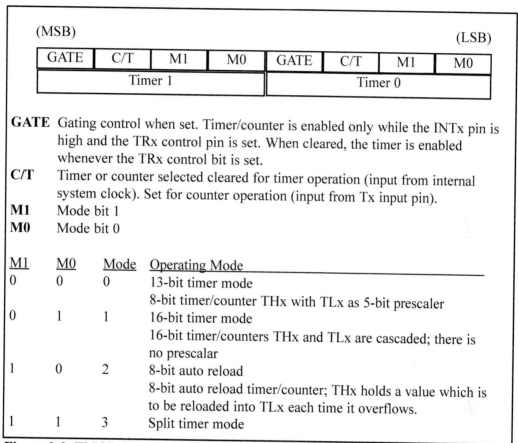

(MSB)   (LSB)

GATE	C/T	M1	M0	GATE	C/T	M1	M0
Timer 1				Timer 0			

GATE Gating control when set. Timer/counter is enabled only while the INTx pin is high and the TRx control pin is set. When cleared, the timer is enabled whenever the TRx control bit is set.

C/T Timer or counter selected cleared for timer operation (input from internal system clock). Set for counter operation (input from Tx input pin).

M1 Mode bit 1

M0 Mode bit 0

M1	M0	Mode	Operating Mode
0	0	0	13-bit timer mode
			8-bit timer/counter THx with TLx as 5-bit prescaler
0	1	1	16-bit timer mode
			16-bit timer/counters THx and TLx are cascaded; there is no prescalar
1	0	2	8-bit auto reload
			8-bit auto reload timer/counter; THx holds a value which is to be reloaded into TLx each time it overflows.
1	1	3	Split timer mode

Figure 9-3. TMOD Register

M1, M0

M0 and M1 select the timer mode. As shown in Figure 9-3, there are three modes: 0, 1, and 2. Mode 0 is a 13-bit timer, mode 1 is a 16-bit timer, and mode 2 is an 8-bit timer. We will mainly concentrate on modes 1 and 2 since they are the ones used most widely. We will soon describe the characteristics of these modes, after describing the rest of the TMOD register.

C/T (clock/timer)

This bit in the TMOD register is used to decide whether the timer is used as a delay generator or an event counter. If C/T = 0, it is used as a timer for time delay generation. The clock source for the time delay is the crystal frequency of the 8051. This section is concerned with this choice. The timer's use as an event counter is discussed in the next section.

Example 9-1

Indicate which mode and which timer are selected for each of the following.
(a) MOV TMOD,#01H (b) MOV TMOD,#20H   (c) MOV TMOD,#12H

Solution:

We convert the values from hex to binary. From Figure 9-3 we have:

(a) TMOD = 00000001, mode 1 of timer 0 is selected.
(b) TMOD = 00100000, mode 2 of timer 1 is selected.
(c) TMOD = 00010010, mode 2 of timer 0, and mode 1 of
    timer 1 are selected.

Clock source for timer

As you know, every timer needs a clock pulse to tick. What is the source of the clock pulse for the 8051 timers? If C/T = 0, the crystal frequency attached to the 8051 is the source of the clock for the timer. This means that the size of the crystal frequency attached to the 8051 also decides the speed at which the 8051 timer ticks. The frequency for the timer is always 1/12th the frequency of the crystal attached to the 8051. See Example 9-2.

Example 9-2

Find the timer's clock frequency and its period for various 8051-based systems, with the following crystal frequencies.
(a)   12 MHz
(b)   16 MHz
(c)   11.0592 MHz

Solution:

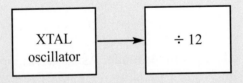

(a) 1/12 × 12 MHz = 1 MHz and   T = 1/1 MHz = 1 μs

(b) 1/12 × 16 MHz = 1.333 MHz and T = 1/1.333 MHz = .75 μs

(c) 1/12 × 11.0592 MHz = 921.6 kHz;
    T = 1/921.6 kHz = 1.085 μs

Although various 8051-based systems have an XTAL frequency of 10 MHz to 40 MHz, we will concentrate on the XTAL frequency of 11.0592 MHz. The reason behind such an odd number has to do with the baud rate for serial communication of the 8051. XTAL = 11.0592 MHz allows the 8051 system to communicate with the IBM PC with no error, as we will see in Chapter 10.

GATE

The other bit of the TMOD register is the GATE bit. Notice in the TMOD register of Figure 9-3 that both timers 0 and 1 have the GATE bit. What is its purpose? Every timer has a means of starting and stopping. Some timers do this by software, some by hardware, and some have both software and hardware controls. The timers in the 8051 have both. The start and stop of the timer are controlled by way of software by the TR (timer start) bits TR0 and TR1. This is achieved by the instructions "SETB TR1" and "CLR TR1" for timer 1, and "SETB TR0" and "CLR TR0" for timer 0. The SETB instruction starts it, and it is stopped by the CLR instruction. These instructions start and stop the timers as long as GATE = 0 in the TMOD register. The hardware way of starting and stopping the timer by an external source is achieved by making GATE = 1 in the TMOD register. However, to avoid further confusion for now, we will make GATE = 0, meaning that no external hardware is needed to start and stop the timers. In using software to start and stop the timer where GATE = 0, all we need are the instructions "SETB TRx" and "CLR TRx". The use of external hardware to stop or start the timer is discussed in Chapter 11 when interrupts are discussed.

Example 9-3

Find the value for TMOD if we want to program timer 0 in mode 2, use 8051 XTAL for the clock source, and use instructions to start and stop the timer.

Solution:

TMOD= 0000 0010 timer 0, mode 2,
$C/T = 0$ to use XTAL clock source, and
gate = 0 to use internal (software)
start and stop method.

Now that we have this basic understanding of the role of the TMOD register, we will look at the timer's modes and how they are programmed to create a time delay. Due to the fact that modes 1 and 2 are so widely used, we describe each of them in detail.

Mode 1 programming

The following are the characteristics and operations of mode 1:
1. It is a 16-bit timer; therefore, it allows values of 0000 to FFFFH to be loaded into the timer's registers TL and TH.
2. After TH and TL are loaded with a 16-bit initial value, the timer must be started. This is done by "SETB TR0" for timer 0 and "SETB TR1" for timer 1.

---

3. After the timer is started , it starts to count up. It counts up until it reaches its limit of FFFFH. When it rolls over from FFFFH to 0000, it sets high a flag bit called TF (timer flag). This timer flag can be monitored. When this timer flag is raised, one option would be to stop the timer with the instructions "CLR TR0" or "CLR TR1", for timer 0 and timer 1, respectively. Again, it must be noted that each timer has its own timer flag: TF0 for timer 0, and TF1 for timer 1.

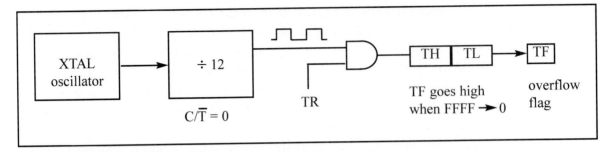

4. After the timer reaches its limit and rolls over, in order to repeat the process the registers TH and TL must be reloaded with the original value, and TF must be reset to 0.

Steps to program in mode 1

To generate a time delay, using the timer's mode 1, the following steps are taken. To clarify these steps, see Example 9-4.
1. Load the TMOD value register indicating which timer (timer 0 or timer 1) is to be used and which timer mode (0 or 1) is selected.
2. Load registers TL and TH with initial count values.
3. Start the timer.
4. Keep monitoring the timer flag (TF) with the "JNB TFx, target" instruction to see if it is raised. Get out of the loop when TF becomes high.
5. Stop the timer.
6. Clear the TF flag for the next round.
7. Go back to Step 2 to load TH and TL again.

To calculate the exact time delay and the square wave frequency generated on pin P1.5, we need to know the XTAL frequency. See Example 9-5.

From Example 9-6 we can develop a formula for delay calculations using mode 1 (16-bit) of the timer for a crystal frequency of XTAL = 11.0592 MHz. This is given in Figure 9-4. The scientific calculator in the Accessory directory of Microsoft Windows can help you to find the TH, TL values. This calculator supports decimal, hex, and binary calculations.

(a) in hex	(b) in decimal
(FFFF - YYXX + 1) × 1.085 µs where YYXX are TH, TL initial values respectively. Notice that values YYXX are in hex.	Convert YYXX values of the TH,TL register to decimal to get a NNNNN decimal number, then (65536 - NNNNN) × 1.085 µs

Figure 9-4. Timer Delay Calculation for XTAL = 11.0592 MHz

Example 9-4

In the following program, we are creating a square wave of 50% duty cycle (with equal portions high and low) on the P1.5 bit. Timer 0 is used to generate the time delay. Analyze the program.

```
              MOV   TMOD,#01       ;Timer 0,mode 1(16-bit mode)
HERE:         MOV   TL0,#0F2H      ;TL0=F2H, the low byte
              MOV   TH0,#0FFH      ;TH0=FFH, the high byte
              CPL   P1.5           ;toggle P1.5
              ACALL DELAY
              SJMP  HERE           ;load TH,TL again
;——————————delay using timer 0
DELAY:
              SETB  TR0            ;start the timer 0
AGAIN:        JNB   TF0,AGAIN      ;monitor timer flag 0 until
                                   ; it rolls over
              CLR   TR0            ;stop timer 0
              CLR   TF0            ;clear timer 0 flag
              RET
```

Solution:

In the above program notice the following steps.

1. TMOD is loaded.
2. FFF2H is loaded into TH0 - TL0.
3. P1.5 is toggled for the high and low portions of the pulse.
4. The DELAY subroutine using the timer is called.
5. In the DELAY subroutine, timer 0 is started by the "SETB TR0" instruction.
6. Timer 0 counts up with the passing of each clock, which is provided by the crystal oscillator. As the timer counts up, it goes through the states of FFF3, FFF4, FFF5, FFF6, FFF7, FFF8, FFF9, FFFA, FFFB, and so on until it reaches FFFFH. One more clock rolls it to 0, raising the timer flag (TF0 = 1). At that point, the JNB instruction falls through.
7. Timer 0 is stopped by the instruction "CLR TR0". The DELAY subroutine ends, and the process is repeated.

Notice that to repeat the process, we must reload the TL and TH registers, and start the timer again.

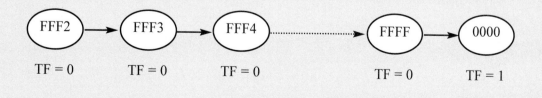

CHAPTER 9: COUNTER/TIMER PROGRAMMING IN THE 8051   163

Example 9-5

In Example 9-4, calculate the amount of time delay in the DELAY subroutine generated by the timer. Assume that XTAL = 11.0592 MHz.

Solution:

The timer works with a clock frequency of 1/12 of the XTAL frequency; therefore, we have 11.0592 MHz / 12 = 921.6 kHz as the timer frequency. As a result, each clock has a period of T = 1 / 921.6 kHz = 1.085 μs. In other words, timer 0 counts up each 1.085 μs resulting in delay = number of counts × 1.085 μs.

The number of counts for the roll over is FFFFH – FFF2H = 0DH (13 decimal). However, we add one to 13 because of the extra clock needed when it rolls over from FFFF to 0 and raises the TF flag. This gives 14 × 1.085 μs = 15.19 μs for half the pulse. For the entire period it is T = 2 × 15.19 μs = 30.38 μs as the time delay generated by the timer.

Example 9-6

In Example 9-5, calculate the frequency of the square wave generated on pin P1.5.

Solution:

In the time delay calculation of Example 9-5, we did not include the overhead due to instructions in the loop. To get a more accurate timing, we need to add clock cycles due to the instructions in the loop. To do that, we use the machine cycles from Table A-1 in Appendix A, as shown below.

```
                                         Cycles
HERE:       MOV   TL0,#0F2H                 2
            MOV   TH0,#0FFH                 2
            CPL   P1.5                      1
            ACALL DELAY                     2
            SJMP  HERE                      2
;-------------delay using timer 0
DELAY:
            SETB  TR0                       1
AGAIN:      JNB   TF0,AGAIN                 14
            CLR   TR0                       1
            CLR   TF0                       1
            RET                             1
                              Total        27
```

T = 2 × 27 × 1.085 μs = 58.59 μs and F = 17067.75 Hz.

Example 9-7

Find the delay generated by timer 0 in the following code, using both of the methods of Figure 9-4. Do not include the overhead due to instructions.

```
            CLR   P2.3              ;clear P2.3
            MOV   TMOD,#01          ;Timer 0,mode 1(16-bit mode)
HERE:       MOV   TL0,#3EH          ;TL0=3EH, the low byte
            MOV   TH0,#0B8H         ;TH0=B8H, the high byte
            SETB  P2.3              ;SET high P2.3
            SETB  TR0               ;start the timer 0
AGAIN:      JNB   TF0,AGAIN         ;monitor timer Flag 0
            CLR   TR0   ;stop the timer 0
            CLR   TF0   ;clear timer 0 flag for next round
            CLR   P2.3
```

Solution:

(a) (FFFF – B83E + 1) = 47C2H = 18370 in decimal and 18370 × 1.085 μs = 19.93145 ms.

(b) Since TH – TL = B83EH = 47166 (in decimal) we have 65536 – 47166 = 18370. This means that the timer counts from B83EH to FFFF. This plus rolling over to 0 goes through a total of 18370 clock cycles, where each clock is 1.085 μs in duration. Therefore, we have 18370 × 1.085 μs = 19.93145 ms as the width of the pulse.

Example 9-8

Modify TL and TH in Example 9-7 to get the largest time delay possible. Find the delay in ms. In your calculation, exclude the overhead due to the instructions in the loop.

Solution:

To get the largest delay we make TL and TH both 0. This will count up from 0000 to FFFFH and then roll over to zero.

```
            CLR   P2.3              ;clear P2.3
            MOV   TMOD,#01          ;Timer 0,mode 1(16-bit mode)
HERE:       MOV   TL0,#0            ;TL0=0, the low byte
            MOV   TH0,#0            ;TH0=0, the high byte
            SETB  P2.3              ;SET high P2.3
            SETB  TR0               ;start timer 0
AGAIN:      JNB   TF0,AGAIN         ;monitor timer Flag 0
            CLR   TR0               ;stop timer 0
            CLR   TF0               ;clear timer 0 flag
            CLR   P2.3
```

Making TH and TL both zero means that the timer will count from 0000 to FFFF, and then roll over to raise the TF flag. As a result, it goes through a total of 65536 states. Therefore, we have delay = (65536 – 0) × 1.085 μs = 71.1065 ms.

CHAPTER 9: COUNTER/TIMER PROGRAMMING IN THE 8051

In Examples 9-7 and 9-8, we did not reload TH and TL since it was a single pulse. Look at Example 9-9 to see how the reloading works in mode 1.

Example 9-9

The following program generates a square wave on pin P1.5 continuously using timer 1 for a time delay. Find the frequency of the square wave if XTAL = 11.0592 MHz. In your calculation do not include the overhead due to instructions in the loop.

```
        MOV   TMOD,#10H      ;timer 1, mode 1(16-bit)
AGAIN:  MOV   TL1,#34H       ;TL1=34H,low byte of timer
        MOV   TH1,#76H       ;TH1=76H,Hi byte
                             ;(7634H=Timer value)
        SETB  TR1            ;start the timer 1
BACK:   JNB   TF1,BACK       ;stay till timer rolls over
        CLR   TR1            ;stop timer 1
        CPL   P1.5           ;comp. p1.5 to get hi, lo
        CLR   TF1            ;clear timer flag 1
        SJMP  AGAIN          ;reload timer since mode 1
                             ;is not auto-reload
```

Solution:

In the above program notice the target of SJMP. In mode 1, the program must reload the TH, TL register every time if we want to have a continuous wave. Now the calculation. Since FFFFH – 7634H = 89CBH + 1 = 89CCH and 89CCH = 35276 clock count and $35276 \times 1.085 \ \mu s = 38.274$ ms and frequency = 26.127 Hz.
Also notice that the high portion and low portion of the square wave pulse are equal. In the above calculation, the overhead due to all the instructions in the loop is not included.

Finding values to be loaded into the timer

Assuming that we know the amount of timer delay we need, the question is how to find the values needed for the TH, TL registers. To calculate the values to be loaded into the TL and TH registers look at the following example where we use crystal frequency of 11.0592 MHz for the 8051 system.

Assuming XTAL = 11.0592 MHz from Example 9-10 we can use the following steps for finding the TH, TL registers' values.

1. Divide the desired time delay by 1.085 μs.
2. Perform 65536 – n, where n is the decimal value we got in Step 1.
3. Convert the result of Step 2 to hex, where $yyxx$ is the initial hex value to be loaded into the timer's registers.
4. Set TL = xx and TH = yy.

Example 9-10

Assume that XTAL = 11.0592 MHz. What value do we need to load into the timer's registers if we want to have a time delay of 5 ms (milliseconds)? Show the program for timer 0 to create a pulse width of 5 ms on P2.3.

Solution:

Since XTAL = 11.0592 MHz, the counter counts up every 1.085 μs. This means that out of many 1.085 μs intervals we must make a 5 ms pulse. To get that, we divide one by the other. We need 5 ms / 1.085 μs = 4608 clocks. To achieve that we need to load into TL and TH the value 65536 − 4608 = 60928 = EE00H. Therefore, we have TH = EE and TL = 00.

```
        CLR   P2.3          ;clear P2.3
        MOV   TMOD,#01      ;Timer 0,mode 1(16-bit mode)
HERE:   MOV   TL0,#0        ;TL0=0, the low byte
        MOV   TH0,#0EEH     ;TH0=EE(hex), the high byte
        SETB  P2.3          ;SET high P2.3
        SETB  TR0           ;start timer 0
AGAIN:  JNB   TF0,AGAIN     ;monitor timer Flag 0
                            ;until it rolls over
        CLR   TR0           ;stop timer 0
        CLR   TF0           ;clear timer 0 flag
```

Example 9-11

Assuming that XTAL = 11.0592 MHz, write a program to generate a square wave of 2 kHz frequency on pin P1.5.

Solution:

This is similar to Example 9-10, except that we must toggle the bit to generate the square wave. Look at the following steps.
(a) T = 1 / f = 1 / 2 kHz = 500 μs the period of square wave.
(b) 1/2 of it for the high and low portion of the pulse is 250 μs.
(c) 250 μs / 1.085 μs = 230 and 65536 − 230 = 65306 which in hex is FF1AH.
(d) TL = 1A and TH = FF, all in hex. The program is as follows.

```
        MOV   TMOD,#10H     ;timer 1, mode 1(16-bit)
AGAIN:  MOV   TL1,#1AH      ;TL1=1A,low byte of timer
        MOV   TH1,#0FFH     ;TH1=FF,Hi byte
        SETB  TR1           ;start the timer 1
BACK:   JNB   TF1,BACK      ;stay until timer rolls over
        CLR   TR1           ;stop timer 1
        CPL   P1.5          ;comp. p1.5 to get hi, lo
        CLR   TF1           ;clear timer flag 1
        SJMP  AGAIN         ;reload timer since mode 1
                            ;is not auto-reload
```

Example 9-12

Assuming XTAL = 11.0592 MHz, write a program to generate a square wave of 50 Hz frequency on pin P2.3.

Solution:

Look at the following steps.
(a) T = 1 / 50 Hz = 20 ms, the period of the square wave.
(b) 1/2 of it for the high and low portion of the pulse = 10 ms
(3) 10 ms / 1.085 μs = 9216 and 65536 − 9216 = 56320 in decimal, and in hex it is DC00H.
(4) TL = 00 and TH = DC (hex)

The program follows.

```
        MOV   TMOD,#10H    ;timer 1, mode 1 (16-bit)
AGAIN:  MOV   TL1,#00      ;TL1=00,low byte of timer
        MOV   TH1,#0DCH    ;TH1=DC,Hi byte
        SETB  TR1          ;start the timer 1
BACK:   JNB   TF1,BACK     ;stay until timer rolls over
        CLR   TR1          ;stop timer 1
        CPL   P2.3         ;comp. p2.3 to get hi, lo
        CLR   TF1          ;clear timer flag 1
        SJMP  AGAIN        ;reload timer since mode 1
                           ;is not auto-reload
```

Generating a large time delay

As we have seen in the examples so far, the size of the time delay depends on two factors, (a) the crystal frequency, and (b) the timer's 16-bit register in mode 1. Both of these factors are beyond the control of the 8051 programmer. We saw earlier that the largest time delay is achieved by making both TH and TL zero. What if that is not enough? Example 9-13 shows how to achieve large time delays.

Using Windows calculator to find TH, TL

The scientific calculator in Microsoft Windows is a handy and easy-to-use tool to find the TH, TL values. Assume that we would like to find the TH, TL values for a time delay where it uses 35,000 clocks of 1.085 μs. The following steps show the calculation.
1. Bring up the scientific calculator in MS Windows and select decimal.
2. Enter 35,000.
3. Select hex. This converts 35,000 to hex, which is 88B8H.
4. Select +/−, to give −35000 decimal (7748H).
5. The lowest 2 digits (48) of this hex value are for TL and the next two (77) are for TH. We ignore all the Fs on the left since our number is 16-bit data.

Example 9-13

Examine the following program and find the time delay in seconds. Exclude the overhead due to the instructions in the loop.

```
          MOV   TMOD,#10H      ;Timer 1,mode 1(16-bit)
          MOV   R3,#200        ;counter for multiple delay
AGAIN:    MOV   TL1,#08        ;TL1=08,low byte of timer
          MOV   TH1,#01        ;TH1=01,Hi byte
          SETB  TR1            ;start the timer 1
BACK:     JNB   TF1,BACK       ;stay until timer rolls over
          CLR   TR1            ;stop timer 1
          CLR   TF1            ;clear timer flag 1
          DJNZ  R3,AGAIN       ;if R3 not zero then
                               ;reload timer
```

Solution:

TH – TL = 0108H = 264 in decimal and 65536 – 264 = 65272. Now 65272 × 1.085 μs = 70.820 ms, and for 200 of them we have 200 × 70.820 ms = 14.164024 seconds.

Mode 0

Mode 0 is exactly like mode 1 except that it is a 13-bit timer instead of 16-bit. The 13-bit counter can hold values between 0000 to 1FFF in TH – TL. Therefore, when the timer reaches its maximum of 1FFH, it rolls over to 0000, and TF is raised.

Mode 2 programming

The following are the characteristics and operations of mode 2.
1. It is an 8-bit timer; therefore, it allows only values of 00 to FFH to be loaded into the timer's register TH.
2. After TH is loaded with the 8-bit value, the 8051 gives a copy of it to TL. Then the timer must be started. This is done by the instruction "SETB TR0" for timer 0 and "SETB TR1" for timer 1. This is just like mode 1.
3. After the timer is started, it starts to count up by incrementing the TL register. It counts up until it reaches its limit of FFH. When it rolls over from FFH to 00, it sets high the TF (timer flag). If we are using timer 0, TF0 goes high; if we are using timer 1, TF1 is raised.

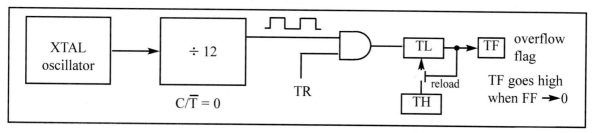

4. When the TL register rolls from FFH to 0 and TF is set to 1, TL is reloaded automatically with the original value kept by the TH register. To repeat the process, we must simply clear TF and let it go without any need by the programmer to reload the original value. This makes mode 2 an auto-reload, in contrast with mode 1 in which the programmer has to reload TH and TL.

It must be emphasized that mode 2 is an 8-bit timer. However, it has an auto-reloading capability. In auto-reload, TH is loaded with the initial count and a copy of it is given to TL. This reloading leaves TH unchanged, still holding a copy of original value. This mode has many applications, including setting the baud rate in serial communication, as we will see in Chapter 10.

Steps to program in mode 2

To generate a time delay using the timer's mode 2, take the following steps.
1. Load the TMOD value register indicating which timer (timer 0 or timer 1) is to be used, and the timer mode (mode 2) is selected.
2. Load the TH registers with the initial count value.
3. Start the timer.
4. Keep monitoring the timer flag (TF) with the "JNB TFx, target" instruction to see whether it is raised. Get out of the loop when TF goes high.
5. Clear the TF flag.
6. Go back to Step 4, since mode 2 is auto-reload.

Example 9-14 illustrates these points. To achieve a larger delay, we can use multiple registers as shown in Example 9-15.

Example 9-14

Assuming that XTAL = 11.0592 MHz, find (a) the frequency of the square wave generated on pin P1.0 in the following program, and (b) the smallest frequency achievable in this program, and the TH value to do that.

(a)
```
          MOV   TMOD,#20H      ;T1/mode 2/8-bit/auto reload
          MOV   TH1,#5         ;TH1=5
          SETB  TR1            ;start the timer 1
BACK:     JNB   TF1,BACK       ;stay till timer rolls over
          CPL   P1.0           ;comp. P1.0 to get hi, lo
          CLR   TF1            ;clear timer flag 1
          SJMP  BACK           ;mode 2 is auto-reload
```

Solution:

(a) First notice the target address of SJMP. In mode 2 we do not need to reload TH since it is auto-reload. Now (256 – 05) × 1.085 μs = 251 × 1.085 μs = 272.33 μs is the high portion of the pulse. Since it is a 50% duty cycle square wave, the period T is twice that; as a result T = 2 × 272.33 μs = 544.67 μs and the frequency = 1.83597 kHz.

(b) To get the smallest frequency, we need the largest T and that is achieved when TH = 00. In that case, we have T = 2 × 256 × 1.085 μs = 555.52 μs and the frequency = 1.8 kHz.

Example 9-15

Find the frequency of a square wave generated on pin P1.0.

Solution:

```
        MOV   TMOD,#2H      ;Timer 0,mode 2
                           ;(8-bit,auto reload)
        MOV   TH0,#0        ;TH0=0
AGAIN:  MOV   R5,#250       ;count for multiple delay
        ACALL DELAY
        CPL   P1.0
        SJMP  AGAIN
DELAY:  SETB  TR0           ;start the timer 0
BACK:   JNB   TF0,BACK      ;stay until timer rolls over
        CLR   TR0           ;stop timer 0
        CLR   TF0           ;clear TF for next round
        DJNZ  R5,DELAY
        RET
```

$T = 2 (250 \times 256 \times 1.085 \ \mu s) = 138.88$ ms, and frequency $= 72$ Hz.

Example 9-16

Assuming that we are programming the timers for mode 2, find the value (in hex) loaded into TH for each of the following cases.

(a) MOV   TH1,#-200   (b) MOV   TH0,#-60
(c) MOV   TH1,#-3   (d) MOV   TH1,#-12
(e) MOV   TH0,#-48

Solution:

You can use the Windows scientific calculator to verify the results provided by the assembler. In Windows calculator, select decimal and enter 200. Then select hex, then +/- to get the TH value. Remember that we only use the right two digits and ignore the rest since our data is an 8-bit data. The following is what we get.

Decimal	*2's complement (TH value)*
–200	38H
– 60	C4H
– 3	FDH
– 12	F4H
– 48	D0H

Assemblers and negative values

Since the timer is 8-bit in mode 2, we can let the assembler calculate the value for TH. For example, in "MOV TH1,#-100", the assembler will calculate the -100 = 9C, and makes THl = 9C in hex. This makes our job easier.

Example 9-17

Find (a) the frequency of the square wave generated in the following code, and (b) the duty cycle of this wave.

```
           MOV    TMOD,#2H        ;Timer 0,mode 2
                                  ;(8-bit,auto reload)
           MOV    TH0,#-150       ;TH0=6AH = 2's comp of -150
AGAIN:     SETB   P1.3            ;P1.3=1
           ACALL DELAY
           ACALL DELAY
           CLR    P1.3            ;P1.3=0
           ACALL DELAY
           SJMP AGAIN


           SETB   TR0             ;start timer 0
BACK:      JNB    TF0,BACK        ;stay until timer rolls over
           CLR    TR0             ;stop timer 0
           CLR    TF0             ;clear TF for next round
           RET
```

Solution:

For the TH value in mode 2, the conversion is done by the assembler as long as we enter a negative number. This also makes the calculation easy. Since we are using 150 clocks, we have time for the DELAY subroutine = 150 × 1.085 μs = 162 μs. The high portion of the pulse is twice that of the low portion (66% duty cycle). Therefore, we have: T = high portion + low portion = 325.5 μs + 162.25 μs = 488.25 μs and frequency = 2.048 kHz.

Notice that in many of the time delay calculations we have ignored the clocks caused by the overhead instructions in the loop. To get a more accurate time delay, and hence frequency, you need to include them. If you use a digital scope and you don't get exactly the same frequency as we have calculated, it is because of the overhead associated with those instructions.

In this section, we used the 8051 timer for time delay generation. However, a more powerful and creative use of these timers is to use them as event counters. We discuss this use of the counter next.

Review Questions

1. How many timers do we have in the 8051?
2. Each timer has _____ registers that are ___ bits wide.
3. TMOD register is a(n) ___-bit register.
4. True or false. The TMOD register is a bit-addressable register.
5. Indicate the selection made in the instruction "MOV TMOD, #20H".
6. In mode 1, the counter rolls over when the counter goes from _____ to _____.
7. In mode 2, the counter rolls over when the counter goes from _____ to _____.
8. In the instruction "MOV TH1, #-200", find the hex value for the TH register.
9. To get a 2-ms delay, what number should be loaded into TH, TL using mode 1? Assume that XTAL = 11.0592 MHz.
10. To get 100 μs delay, what number should be loaded into the TH register using mode 2? Assume XTAL = 11.0592 MHz.

SECTION 9.2: COUNTER PROGRAMMING

In the last section we used the timer/counter of the 8051 to generate time delays. These timers can also be used as counters counting events happening outside the 8051. The use of the timer/counter as an event counter is covered in this section. As far as the use of a timer as an event counter is concerned, everything that we have talked about in programming the timer in the last section also applies to programming it as a counter, except the source of the frequency. For the timer/counter when it is used as a timer, the 8051's crystal is used as the source of the frequency. However, when it is used as a counter, it is a pulse outside of the 8051 that increments the TH, TL registers. In counter mode, notice that the TMOD and TH, TL registers are the same as for the timer discussed in the last section; they even have the same names. The timer's modes are the same as well.

C/T bit in TMOD register

Recall from the last section that the C/T bit in the TMOD register decides the source of the clock for the timer. If C/T = 0, the timer gets pulses from the crystal. In contrast, when C/T = 1, the timer is used as a counter and gets its pulses from outside the 8051. Therefore, when C/T = 1, the counter counts up as pulses are fed from pins 14 and 15. These pins are called T0 (timer 0 input) and T1 (timer 1 input). Notice that these two pins belong to port 3. In the case of timer 0, when C/T = 1, pin P3.4 provides the clock pulse and the counter counts up for each clock pulse coming from that pin. Similarly, for timer 1, when C/T = 1 each clock pulse coming in from pin P3.5 makes the counter count up.

Table 9-1: Port 3 Pins Used For Timers 0 and 1

Pin	Port Pin	Function	Description
14	P3.4	T0	Timer/Counter 0 external input
15	P3.5	T1	Timer/Counter 1 external input

(MSB)							(LSB)
GATE	C/T	M1	M0	GATE	C/T	M1	M0
Timer 1				Timer 0			

Example 9-18

Assuming that clock pulses are fed into pin T1, write a program for counter 1 in mode 2 to count the pulses and display the state of the TL1 count on P2.

Solution:

```
            MOV   TMOD,#01100000B    ;counter 1,mode 2,C/T=1
                                     ;external pulses
            MOV   TH1,#0             ;clear TH1
            SETB  P3.5              ;make T1 input
AGAIN:      SETB  TR1               ;start the counter
BACK:       MOV   A,TL1             ;get copy of count TL1
            MOV   P2,A              ;display it on port 2
            JNB   TF1,Back          ;keep doing it if TF=0
            CLR   TR1               ;stop the counter 1
            CLR   TF1               ;make TF=0
            SJMP  AGAIN             ;keep doing it
```

Notice in the above program the role of the instruction "SETB P3.5". Since ports are set up for output when the 8051 is powered up, we make P3.5 an input port by making it high. In other words, we must configure (set high) the T1 pin (pin P3.5) to allow pulses to be fed into it.

P2 is connected to 8 LEDs and input T0 to pulse.

8051

P2 → to LEDs

T1 — P3.5

In Example 9-18, we are using timer 1 as an event counter where it counts up as clock pulses are fed into pin 3.5. These clock pulses could represent the number of people passing through an entrance, or the number of wheel rotations, or any other event that can be converted to pulses.

In Example 9-18, the TL data was displayed in binary. In Example 9-19, the TL registers are converted to ASCII to be displayed on an LCD.

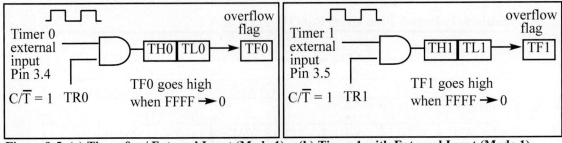

Figure 9-5. (a) Timer 0 w/ External Input (Mode 1)   (b) Timer 1 with External Input (Mode 1)

Example 9-19

Assume that a 1-Hz frequency pulse is connected to input pin 3.4. Write a program to display counter 0 on an LCD. Set the initial value of TH0 to –60.

Solution:

To display the TL count on an LCD, we must convert 8-bit binary data to ASCII. See Chapter 6 for data conversion.

```
            ACALL LCD_SET_UP          ;initialize the LCD
            MOV   TMOD,#000110B       ;Counter 0,mode2,C/T=1
            MOV   TH0,#-60            ;counting 60 pulses
            SETB  P3.4                ;make T0 as input
AGAIN:      SETB  TR0                 ;starts the counter
BACK:       MOV   A,TL0               ;get copy of count TL0
            ACALL CONV                ;convert in R2,R3,R4
            ACALL DISPLAY             ;display on LCD
            JNB   TF0,BACK            ;loop if TF0=0
            CLR   TR0                 ;stop the counter 0
            CLR   TF0                 ;make TF0=0
            SJMP  AGAIN               ;keep doing it

;converting 8-bit binary to ASCII
;upon return, R4,R3,R2 have ASCII data (R2 has LSD)

CONV:       MOV   B,#10      ;divide by 10
            DIV   AB
            MOV   R2,B       ;save low digit
            MOV   B,#10      ;divide by 10 once more
            DIV   AB
            ORL   A,#30H     ;make it ASCII
            MOV   R4,A       ;save MSD
            MOV   A,B
            ORL   A,#30H     ;make 2nd digit an ASCII
            MOV   R3,A       ;save it
            MOV   A,R2
            ORL   A,#30H     ;make 3rd digit an ASCII
            MOV   R2,A       ;save the ASCII
            RET
```

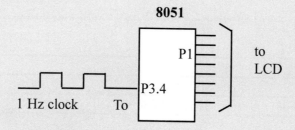

By using 60 Hz we can generate seconds, minutes, hours.

Note that on the first round, it starts from 0, since on RESET, TL0 = 0.
To solve this problem, load TL0 with –60 at the beginning of the program.

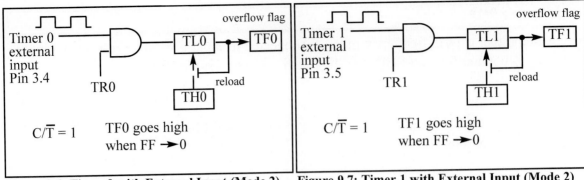

Figure 9.6: Timer 0 with External Input (Mode 2)   **Figure 9.7: Timer 1 with External Input (Mode 2)**

As another example of the application of the timer with C/T = 1, we can feed an external square wave of 60 Hz frequency into the timer. The program will generate the second, the minute, and the hour out of this input frequency and display the result on an LCD. This will be a nice digital clock, but not a very accurate one. This lab experiment can be found in Appendix E.

Before we finish this chapter, we need to state two important points.
1. You might think that the use of the instruction "JNB TFx, target" to monitor the raising of the TFx flag is a waste of the microcontroller's time. You are right. There is a solution to this. The solution is the use of interrupts. In using interrupts we can go about doing other things with the microcontroller. When the TF flag is raised it will inform us. This important and powerful feature of the 8051 is discussed in Chapter 11.
2. You might wonder to what register TR0 and TR1 belong.
   They belong to a register called TCON which is discussed next.

Table 9-2: Equivalent Instructions for the Timer Control Register

For timer 0

SETB TR0	=	SETB TCON.4	
CLR TR0	=	CLR TCON.4	
SETB TF0	=	SETB TCON.5	
CLR TF0	=	CLR TCON.5	

For timer 1

SETB TR1	=	SETB TCON.6	
CLR TR1	=	CLR TCON.6	
SETB TF1	=	SETB TCON.7	
CLR TF1	=	CLR TCON.7	

TCON: Timer/Counter Control Register

TF1	TR1	TF0	TR0	IE1	IT1	IE0	IT0

TCON register

In the examples so far we have seen the use of the TR0 and TR1 flags to turn on or off the timers. These bits are part of a register called TCON (timer control). This register is an 8-bit register. As shown in Table 9-2, the upper four bits are used to store the TF and TR bits of both timer 0 and timer 1. The lower 4 bits are set aside for controlling the interrupt bits, which will be discussed in Chapter 11. We must notice that the TCON register is a bit-addressable register. Instead of using instructions such as "SETB TR1" and "CLR TR1", we could use "SETB TCON.6" and "CLR TCON.6", respectively. Table 9-2 shows replacements of some of the instructions we have seen so far.

The case of GATE = 1 in TMOD

Before we finish this section we need to discuss another case of the GATE bit in the TMOD register. All discussion so far has assumed that GATE = 0. When GATE = 0, the timer is started with instructions "SETB TR0" and "SETB TR1", for timers 0 and 1, respectively. What happens if the GATE bit in TMOD is set to 1? As can be seen in Figures 9-8 and 9-9, if GATE = 1, the start and stop of the timer are done externally through pins P3.2 and P3.3 for timers 0 and 1, respectively. This is in spite of the fact that TRx is turned on by the "SETB TRx" instruction. This allows us to start or stop the timer externally at any time via a simple switch. This hardware way of controlling the stop and start of the timer can have many applications. For example, assume that an 8051 system is used in a product to sound an alarm every second using timer 0, perhaps in addition to many other things. Timer 0 is turned on by the software method of using the "SETB TR0" instruction and is beyond the control of the user of that product. However, a switch connected to pin P3.2 can be used to turn on and off the timer, thereby shutting down the alarm.

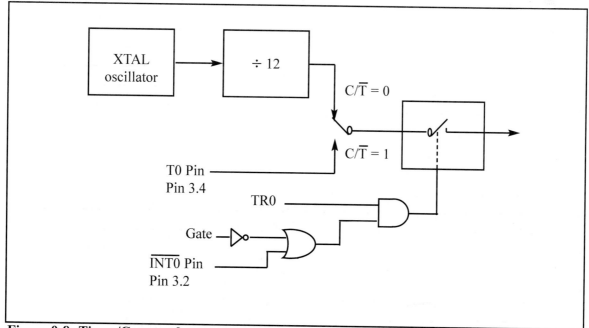

Figure 9-8: Timer/Counter 0

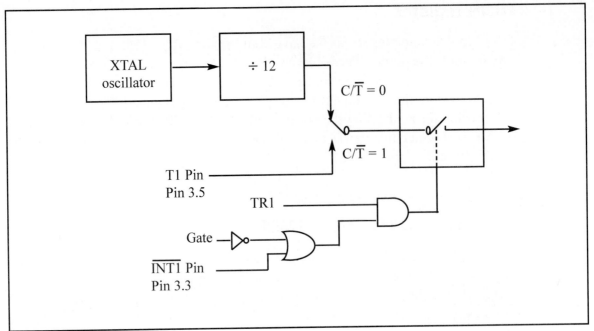

Figure 9-9: Timer/Counter 1

Review Questions

1. Who provides the clock pulses to 8051 timers if $C/\overline{T} = 0$?
2. Who provides the clock pulses to 8051 timers if $C/\overline{T} = 1$?
3. Does the discussion in Section 9.1 apply to timers if $C/\overline{T} = 1$?
4. To allow P3.4 to be used as an input for T1, what must be done, and why?
5. What is the equivalent of the following instruction? "SETB TCON.6"

SUMMARY

The 8051 has two timers/counters. When used as timers they can generate time delays. When used as counters they can serve as event counters. This chapter showed how to program the timers/counters for various modes.

The two timers are accessed as two 8-bit registers, TL0 and TH0 for timer 0, and TL1 and TH1 for timer 1. Both timers use the TMOD register to set timer operation modes. The lower 4 bits of TMOD are used for timer 0 and the upper 4 bits are used for timer 1.

There are different modes that can be used for each timer. Mode 0 sets the timer as a 13-bit timer, mode 1 sets it as a 16-bit timer, and mode 2 sets it as an 8-bit timer.

When the timer/counter is used as a timer, the 8051's crystal is used as the source of the frequency; however, when it is used as a counter, it is a pulse outside of the 8051 that increments the TH, TL registers.

PROBLEMS

SECTION 9.1: PROGRAMMING 8051 TIMERS

1. How many timers do we have in the 8051?
2. The timers of the 8051 are ____-bit and are designated as _____ and _____.
3. The registers of timer 0 are accessed as _____ and _____.
4. The registers of timer 1 are accessed as _____ and _____.
5. In Questions 3 and 4, are the registers bit-addressable?
6. The TMOD register is a(n) ___-bit register.
7. What is the job of the TMOD register?
8. True or false. TMOD is a bit-addressable register.
9. Find the TMOD value for both timer 0 and timer 1, mode 2, software start / stop (gate = 0), with the clock coming from the 8051's crystal.
10. Find the frequency and period used by the timer if the crystal attached to the 8051 has the following values.
    (a) XTAL = 11.0592 MHz   (b) XTAL = 20 MHz
    (c) XTAL = 24 MHz   (d) XTAL = 30 MHz
11. Indicate the size of the timer for each of the following modes.
    (a) mode 0   (b) mode 1   (c) mode 2
12. Indicate the roll-over value (in hex and decimal) of the timer for each of the following modes.
    (a) mode 0   (b) mode 1   (c) mode 2
13. Indicate when the TF1 flag is raised for each of the following modes.
    (a) mode 0   (b) mode 1   (c) mode 2
14. True or false. Both timer 0 and timer 1 have their own TF.
15. True or false. Both timer 0 and timer 1 have their own timer start (TR).
16. Assuming XTAL = 11.0592 MHz, indicate when the TF0 flag is raised for the following program.
```
MOV   TMOD,#01
MOV   TL0,#12H
MOV   TH0,#1CH
SETB  TR0
```
17. Assuming that XTAL = 16 MHz, indicate when the TF0 flag is raised for the following program.
```
MOV   TMOD,#01
MOV   TL0,#12H
MOV   TH0,#1CH
SETB  TR0
```
18. Assuming that XTAL = 11.0592 MHz, indicate when the TF0 flag is raised for the following program.
```
MOV   TMOD,#01
MOV   TL0,#10H
MOV   TH0,#0F2H
SETB  TR0
```

19. Assuming that XTAL = 20 MHz, indicate when the TF0 flag is raised for the following program.

```
MOV   TMOD,#01
MOV   TL0,#12H
MOV   TH0,#1CH
SETB  TR0
```

20. Assume that XTAL = 11.0592 MHz. Find the TH1,TL1 value to generate a time delay of 2 ms. Timer 1 is programmed in mode 1.

21. Assume that XTAL = 16 MHz. find the TH1,TL1 value to generate a time delay of 5 ms. Timer 1 is programmed in mode 1.

22. Assuming that XTAL = 11.0592 MHz, program timer 0 to generate a time delay of 2.5 ms.

23. Assuming that XTAL = 11.0592 MHz, program timer 1 to generate a time delay of 0.2 ms.

24. Assuming that XTAL = 20 MHz, program timer 1 to generate a time delay of 100 μs.

25. Assuming that XTAL = 11.0592 MHz, and we are generating a square wave on pin P1.2, find the lowest square wave frequency that we can generate using mode 1.

26. Assuming that XTAL = 11.0592 MHz, and we are generating a square wave on pin P1.2, find the highest square wave frequency that we can generate using mode 1.

27. Assuming that XTAL = 16 MHz, and we are generating a square wave on pin P1.2, find the lowest square wave frequency that we can generate using mode 1.

28. Assuming that XTAL = 16 MHz, and we are generating a square wave on pin P1.2, find the highest square wave frequency that we can generate using mode 1.

29. In mode 2 assuming that TH1 = F1H, indicate which states timer 2 goes through until TF1 is raised. How many states is that?

30. Program timer 1 to generate a square wave of 1 kHz. Assume that XTAL = 11.0592 MHz.

31. Program timer 0 to generate a square wave of 3 kHz. Assume that XTAL = 11.0592 MHz.

32. Program timer 0 to generate a square wave of 0.5 kHz. Assume that XTAL = 20 MHz.

33. Program timer 1 to generate a square wave of 10 kHz. Assume that XTAL = 20 MHz.

34. Assuming that XTAL = 11.0592 MHz, show a program to generate a 1-second time delay. Use any timer you want.

35. Assuming that XTAL = 16 MHz, show a program to generate a 0.25-second time delay. Use any timer you want.

36. Assuming that XTAL = 11.0592 MHz and that we are generating a square wave on pin P1.3, find the lowest square wave frequency that we can generate using mode 2.

37. Assuming that XTAL = 11.0592 MHz and that we are generating a square wave on pin P1.3, find the highest square wave frequency that we can generate using mode 2.

38. Assuming that XTAL = 16 MHz and that we are generating a square wave on pin P1.3, find the lowest square wave frequency that we can generate using mode 2.

39. Assuming that XTAL = 16 MHz and that we are generating a square wave on pin P1.3, find the highest square wave frequency that we can generate using mode 2.

40. Find the value (in hex) loaded into TH in each of the following.

    (a)   `MOV   TH0,#-12`   (b)   `MOV   TH0,#-22`
    (c)   `MOV   TH0,#-34`   (d)   `MOV   TH0,#-92`
    (e)   `MOV   TH1,#-120`   (f)   `MOV   TH1,#-104`
    (g)   `MOV   TH1,#-222`   (h)   `MOV   TH1,#-67`

41. In Problem 40, indicate by what number the machine cycle frequency of 921.6 kHZ (XTAL = 11.0592 MHz) is divided.

42. In Problem 41, find the time delay for each case from the time the timer starts to the time the TF flag is raised.

SECTION 9.2: COUNTER PROGRAMMING

43. To use the timer as an event counter we must set the C/T bit in the TMOD register to _____ (low, high).

44. Can we use both of the timers as event counters?

45. For counter 0, which pin is used to input clocks?

46. For counter 1, which pin is used to input clocks?

47. Program timer 1 to be an event counter. Use mode 1 and display the binary count on P1 and P2 continuously. Set the initial count to 20,000.

48. Program timer 0 to be an event counter. Use mode 2 and display the binary count on P2 continuously. Set the initial count to 20.

49. Program timer 1 to be an event counter. Use mode 2 and display the decimal count on P2, P1, and P0 continuously. Set the initial count to 99.

50. The TCON register is a(n) _____-bit register.

51. True or false. The TCON register is not a bit-addressable register.

52. Give another instruction to perform the action of "SETB TR0".

ANSWERS TO REVIEW QUESTIONS

SECTION 9.1: PROGRAMMING 8051 TIMERS
1. Two
2. 2, 8
3. 8
4. False
5. 0010 0000 indicates timer 1, mode 2, software start and stop, and using XTAL for frequency.
6. FFFFH to 0000
7. FFH to 00
8. −200 is 38H; therefore, TH1 = 38H
9. 2 ms/1.085 μs = 1843 = 0733H where TH = 07H and TL = 33H.
10. 100 μs/1.085 μs = 92 or 5CH; therefore, TH = 5CH.

SECTION 9.2: COUNTER PROGRAMMING
1. The crystal attached to the 8051
2. The clock source for the timers comes from pins T0 and T1.
3. Yes
4. We must use the instruction "SETB P3.4" to configure the T1 pin as input to allow the clocks to come from an external source. This is due to the fact that all ports are configured as output upon reset.
5. SETB TR1

CHAPTER 10

8051 SERIAL COMMUNICATION

OBJECTIVES

Upon completion of this chapter, you will be able to:

>> Contrast and compare serial versus parallel communication
>> List the advantages of serial communication over parallel
>> Explain serial communication protocol
>> Contrast synchronous versus asynchronous communication
>> Contrast half- versus full-duplex transmission
>> Explain the process of data framing
>> Desribe data transfer rate and bps rate
>> Define the RS232 standard
>> Explain the use of the MAX232 and MAX233 chips
>> Interface the 8051 with an RS232 connector
>> Discuss the baud rate of the 8051
>> Describe serial communication features of the 8051
>> Program the 8051 for serial data communication

Computers transfer data in two ways: parallel and serial. In parallel data transfers, often 8 or more lines (wire conductors) are used to transfer data to a device that is only a few feet away. Examples of parallel transfers are printers and hard disks; each uses cables with many wire strips. Although in such cases a lot of data can be transferred in a short amount of time by using many wires in parallel, the distance cannot be great. To transfer to a device located many meters away, the serial method is used. In serial communication, the data is sent one bit at a time, in contrast to parallel communication, in which the data is sent a byte or more at a time. Serial communication of the 8051 is the topic of this chapter. The 8051 has serial communication capability built into it, thereby making possible fast data transfer using only a few wires.

In this chapter we first discuss the basics of serial communication. In Section 10.2, 8051 interfacing to RS232 connectors via MAX232 line drivers is discussed. Serial communication programming of the 8051 is discussed in Section 10.3

SECTION 10.1: BASICS OF SERIAL COMMUNICATION

When a microprocessor communicates with the outside world, it provides the data in byte-sized chunks. In some cases, such as printers, the information is simply grabbed from the 8-bit data bus and presented to the 8-bit data bus of the printer. This can work only if the cable is not too long, since long cables diminish and even distort signals. Furthermore, an 8-bit data path is expensive. For these reasons, serial communication is used for transferring data between two systems located at distances of hundreds of feet to millions of miles apart. Figure 10-1 diagrams serial versus parallel data transfers.

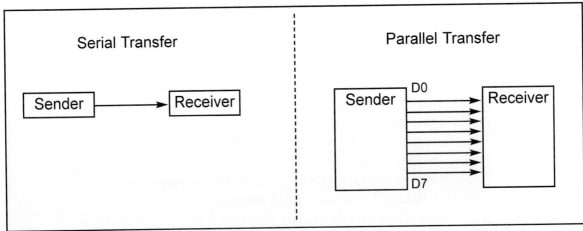

Figure 10-1. Serial versus Parallel Data Transfer

The fact that in serial communication a single data line is used instead of the 8-bit data line of parallel communication makes it not only much cheaper but also makes it possible for two computers located in two different cities to communicate over the telephone.

For serial data communication to work, the byte of data must be converted to serial bits using a parallel-in-serial-out shift register; then it can be transmitted over a single data line. This also means that at the receiving end there must be a serial-in-parallel-out shift register to receive the serial data and pack them into a byte. Of course, if data is to be transferred on the telephone line, it must be converted from 0s and 1s to audio tones, which are sinusoidal-shaped signals. This conversion is performed by a peripheral device called a *modem*, which stands for "modulator/demodulator."

When the distance is short, the digital signal can be transferred as it is on a simple wire and requires no modulation. This is how IBM PC keyboards transfer data to the motherboard. However, for long-distance data transfers using communication lines such as a telephone, serial data communication requires a modem to *modulate* (convert from 0s and 1s to audio tones) and *demodulate* (converting from audio tones to 0s and 1s).

Serial data communication uses two methods, asynchronous and synchronous. The *synchronous* method transfers a block of data (characters) at a time while the *asynchronous* transfers a single byte at a time. It is possible to write software to use either of these methods, but the programs can be tedious and long. For this reason, there are special IC chips made by many manufacturers for serial data communications. These chips are commonly referred to as UART (universal asynchronous receiver-transmitter) and USART (universal synchronous-asynchronous receiver-transmitter). The 8051 chip has a built-in UART, which is discussed in detail in Section 10.3.

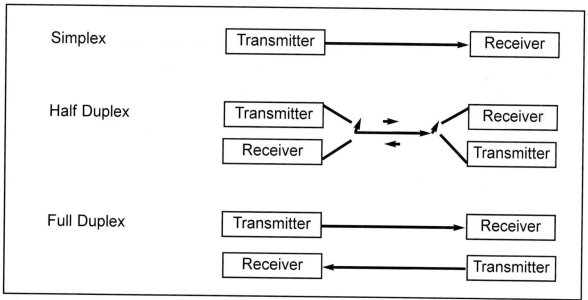

Figure 10-2. Simplex, Half-, and Full-Duplex Transfers

Half- and full-duplex transmission

In data transmission if the data can be transmitted and received, it is a *duplex* transmission. This is in contrast to *simplex* transmissions such as with printers, in which the computer only sends data. Duplex transmissions can be half or full duplex, depending on whether or not the data transfer can be simultaneous. If data is transmitted one way at a time, it is referred to as *half duplex*. If the data can go both ways at the same time, it is *full duplex*. Of course, full duplex requires two wire conductors for the data lines (in addition to the signal ground), one for transmission and one for reception, in order to transfer and receive data simultaneously. See Figure 10-2.

Asynchronous serial communication and data framing

The data coming in at the receiving end of the data line in a serial data transfer is all 0s and 1s; it is difficult to make sense of the data unless the sender and receiver agree on a set of rules, a *protocol*, on how the data is packed, how many bits constitute a character, and when the data begins and ends.

Start and stop bits

Asynchronous serial data communication is widely used for character-oriented transmissions, while block-oriented data transfers use the synchronous method. In the asynchronous method, each character is placed in between start and stop bits. This is called *framing*. In data framing for asynchronous communications, the data, such as ASCII characters, are packed in between a start bit and a stop bit. The start bit is always one bit but the stop bit can be one or two bits. The start bit is always a 0 (low) and the stop bit(s) is 1 (high). For example, look at Figure 10-3 in which the ASCII character "A" (8-bit binary 0100 0001) is framed in between the start bit and a single stop bit. Notice that the LSB is sent out first.

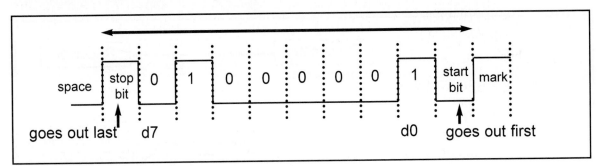

Figure 10-3. Framing ASCII "A" (41H)

Notice in Figure 10-3 that when there is no transfer, the signal is 1 (high), which is referred to as *mark*. The 0 (low) is referred to as *space*. Notice that the transmission begins with a start bit followed by D0, the LSB, then the rest of the bits until the MSB (D7), and finally, the one stop bit indicating the end of the character "A".

In asynchronous serial communications, peripheral chips and modems can be programmed for data that is 7 or 8 bits wide. This is in addition to the number of stop bits, 1 or 2. While in older systems ASCII characters were 7-bit, in recent years due to the extended ASCII characters, 8-bit data has become common. In some older systems, due to the slowness of the receiving mechanical device, two stop bits were used to give the device sufficient time to organize itself before transmission of the next byte. However, in modern PCs the use of one stop bit is standard. Assuming that we are transferring a text file of ASCII characters using 1 stop bit, we have a total of 10 bits for each character: 8 bits for the ASCII code, and 1 bit each for the start and stop bits. Therefore, for each 8-bit character there are an extra 2 bits, which gives 25% overhead.

In some systems in order to maintain data integrity, the parity bit of the character byte is included in the data frame. This means that for each character (7- or 8-bit, depending on the system) we have a single parity bit in addition to start and stop bits. The parity bit is odd or even. In the case of an odd-parity bit the number of data bits, including the parity bit, has an odd number of 1s. Similarly, in an even-parity bit system the total number of bits, including the parity bit, is even. For example, the ASCII character "A", binary 0100 0001, has 0 for the even-parity bit. UART chips allow programming of the parity bit for odd-, even-, and no-parity options.

Data transfer rate

The rate of data transfer in serial data communication is stated in *bps* (bits per second). Another widely used terminology for bps is *baud rate*. However, the baud and bps rates are not necessarily equal. This is due to the fact that baud rate is the modem terminology and is defined as the number of signal changes per second. In modems, there are occasions when a single change of signal transfers several bits of data. As far as the conductor wire is concerned, the baud rate and bps are the same, and for this reason in this book we use the terms bps and baud interchangeably.

The data transfer rate of a given computer system depends on communication ports incorporated into that system. For example, the early IBM PC/XT could transfer data at the rate of 100 to 9600 bps. However in recent years, Pentium-based PCs transfer data at rates as high as 56K bps. It must be noted that in asynchronous serial data communication, the baud rate is generally limited to 100,000 bps.

RS232 standards

To allow compatibility among data communication equipment made by various manufacturers, an interfacing standard called RS232 was set by the Electronics Industries Association (EIA) in 1960. In 1963 it was modified and called RS232A. RS232B and RS232C were issued in 1965 and 1969, respectively. In this book we refer to it simply as RS232. Today, RS232 is the most widely used serial I/O interfacing standard. This standard is used in PCs and numerous types of equipment. However, since the standard was set long before the advent of the TTL logic family, its input and output voltage levels are not TTL compatible. In RS232, a 1 is represented by −3 to −25 V, while a 0 bit is +3 to +25 V, making −3 to +3 undefined. For this reason, to connect any RS232 to a microcontroller system we must use voltage converters such as MAX232 to convert the TTL logic levels to the RS232 voltage level, and vice versa. MAX232 IC chips are commonly referred to as line drivers. RS232 connection to MAX232 is discussed in Section 10.2.

RS232 pins

Table 10-1 provides the pins and their labels for the RS232 cable, commonly referred to as the DB-25 connector. In labeling, DB-25P refers to the plug connector (male) and DB-25S is for the socket connector (female). See Figure 10-4.

CHAPTER 10: 8051 SERIAL COMMUNICATION   187

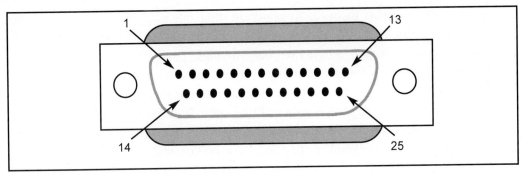

Figure 10-4. RS232 Connector DB-25

Since not all the pins are used in PC cables, IBM introduced the DB-9 version of the serial I/O standard, which uses 9 pins only, as shown in Table 10-2. The DB-9 pins are shown in Figure 10-5.

Data communication classification

Current terminology classifies data communication equipment as DTE (data terminal equipment) or DCE (data communication equipment). DTE refers to terminals and computers that send and receive data, while DCE refers to communication equipment, such as modems, that are responsible for transferring the data. Notice that all the RS232 pin function definitions of Tables 10-1 and 10-2 are from the DTE point of view.

The simplest connection between a PC and microcontroller requires a minimum of three pins, TxD, RxD, and ground, as shown in Figure 10-6. Notice in that figure that the RxD and TxD pins are interchanged.

Examining RS232 handshaking signals

To ensure fast and reliable data transmission between two devices, the data transfer must be coordinated. Just as in the case of the printer, due to the fact that in serial data communication the receiving device may have no room for the data, there must be a way to inform the sender to stop sending data. Many of the pins of the RS-232 connector are used for handshaking signals. Their description is provided below only as a reference and they can be by-passed since they are not supported by the 8051 UART chip.

Table 10-1: RS232 Pins (DB-25)

Pin	Description
1	Protective ground
2	Transmitted data (TxD)
3	Received data (RxD)
4	Request to send ($\overline{\text{RTS}}$)
5	Clear to send ($\overline{\text{CTS}}$)
6	Data set ready ($\overline{\text{DSR}}$)
7	Signal ground (GND)
8	Data carrier detect ($\overline{\text{DCD}}$)
9/10	Reserved for data testing
11	Unassigned
12	Secondary data carrier detect
13	Secondary clear to send
14	Secondary transmitted data
15	Transmit signal element timing
16	Secondary received data
17	Receive signal element timing
18	Unassigned
19	Secondary request to send
20	Data terminal ready ($\overline{\text{DTR}}$)
21	Signal quality detector
22	Ring indicator
23	Data signal rate select
24	Transmit signal element timing
25	Unassigned

1. DTR (data terminal ready). When the terminal (or a PC COM port) is turned on, after going through a self-test, it sends out signal DTR to indicate that it is ready for communication. If there is something wrong with the COM port, this signal will not be activated. This is an active-low signal and can be used to inform the modem that the computer is alive and kicking. This is an output pin from DTE (PC COM port) and an input to the modem.

2. DSR (data set ready). When DCE (modem) is turned on and has gone through the self-test, it asserts DSR to indicate that it is ready to communicate. Thus, it is an output from the modem (DCE) and input to the PC (DTE). This is an active-low signal. If for any reason the modem cannot make a connection to the telephone, this signal remains inactive, indicating to the PC (or terminal) that it cannot accept or send data.

3. RTS (request to send). When the DTE device (such as a PC) has a byte to transmit, it asserts RTS to signal the modem that it has a byte of data to transmit. RTS is an active-low output from the DTE and an input to the modem.

4. CTS (clear to send). In response to RTS, when the modem has room for storing the data it is to receive, it sends out signal CTS to the DTE (PC) to indicate that it can receive the data now. This input signal to the DTE is used by the DTE to start transmission.

5. DCD (carrier detect, or DCD, data carrier detect). The modem asserts signal DCD to inform the DTE (PC) that a valid carrier has been detected and that contact between it and the other modem is established. Therefore, DCD is an output from the modem and an input to the PC (DTE).

6. RI (ring indicator). An output from the modem (DCE) and an input to a PC (DTE) indicates that the telephone is ringing. It goes on and off in synchronization with the ringing sound. Of the 6 handshake signals, this is the least often used, due to the fact that modems take care of answering the phone. However, if in a given system the PC is in charge of answering the phone, this signal can be used.

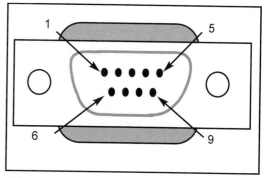

Figure 10-5. DB-9 9-Pin Connector

Table 10-2: IBM PC DB-9 Signals

Pin	Description
1	Data carrier detect ($\overline{DCD}$)
2	Received data (RxD)
3	Transmitted data (TxD)
4	Data terminal ready (DTR)
5	Signal ground (GND)
6	Data set ready ($\overline{DSR}$)
7	Request to send ($\overline{RTS}$)
8	Clear to send ($\overline{CTS}$)
9	Ring indicator (RI)

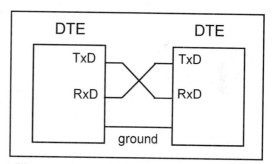

Figure 10-6. Null Modem Connection

From the above description, PC and modem communication can be summarized as follows: While signals DTR and DSR are used by the PC and modem, respectively, to indicate that they are alive and well, it is RTS and CTS that actually control the flow of data. When the PC wants to send data it asserts RTS, and in response, if the modem is ready (has room) to accept the data, it sends back CTS. If, for lack of room, the modem does not activate CTS, the PC will deassert DTR and try again. RTS and CTS are also referred to as hardware control flow signals.

This concludes the description of the 9 most important pins of the RS232 handshake signals plus TxD, RxD, and ground. Ground is also referred to as SG (signal ground).

IBM PC/compatible COM ports

IBM PC/compatible computers based on x86 (8086, 286, 386, 486, and Pentium) microprocessors normally have two COM ports. Both COM ports have RS232-type connectors. Many PCs use one each of the DB-25 and DB-9 RS232 connectors. The COM ports are designated as COM 1 and COM 2. In recent years, COM 1 is used for the mouse and COM 2 is available for devices such as a modem. We can connect the 8051 serial port to the COM 2 port of a PC for serial communication experiments.

With this background in serial communication, we are ready to look at the 8051. In the next section we discuss the physical connection of the 8051 and RS232 connector, and in Section 10.3 we see how to program the 8051 serial communication port.

Review Questions

1. The transfer of data using parallel lines is _____ (faster, slower) but _____ (more expensive, less expensive).
2. True or false. Sending data to a printer is duplex.
3. True or false. In full duplex we must have two data lines, one for transfer and one for receive.
4. The start and stop bits are used in the _____ (synch, asynch) method.
5. Assuming that we are transmitting the ASCII letter "E" (0100 0101 in binary) with no parity bit and one stop bit, show the sequence of bits transferred serially.
6. In Question 5, find the overhead due to framing.
7. Calculate the time it takes to transfer 10,000 characters as in Question 5 if we use 9600 bps. What percentage of time is wasted due to overhead?
8. True or false. RS232 is not TTL-compatible.
9. What voltage levels are used for binary 0 in RS232?
10. True or false. The 8051 has a built-in UART.
11. On the back of x86 PCs, we normally have ____ COM port connectors.
12. The PC COM ports are designated by DOS and Windows as _____ and _____.

SECTION 10.2: 8051 CONNECTION TO RS232

In this section, the details of the physical connections of the 8051 to RS232 connectors are given. As stated in Section 10.2, the RS232 standard is not TTL compatible; therefore, it requires a line driver such as the MAX232 chip to convert RS232 voltage levels to TTL levels, and vice versa. The interfacing of 8051 with RS232 connectors via the MAX232 chip is the main topic of this section.

RxD and TxD pins in the 8051

The 8051 has two pins that are used specifically for transferring and receiving data serially. These two pins are called TxD and RxD and are part of the port 3 group (P3.0 and P3.1). Pin 11 of the 8051 (P3.1) is assigned to TxD and pin 10 (P3.0) is designated as RxD. These pins are TTL compatible; therefore, they require a line driver to make them RS232 compatible. One such line driver is the MAX232 chip. This is discussed next.

MAX232

Since the RS232 is not compatible with today's microprocessors and microcontrollers, we need a line driver (voltage converter) to convert the RS232's signals to TTL voltage levels that will be acceptable to the 8051's TxD and RxD pins. One example of such a converter is MAX232 from Maxim Corp. (www.maxim-ic.com). The MAX232 converts from RS232 voltage levels to TTL voltage levels, and vice versa. One advantage of the MAX232 chip is that it uses a +5 V power source which is the same as the source voltage for the 8051. In other words, with a single +5 V power supply we can power both the 8051 and MAX232, with no need for the dual power supplies that are common in many older systems.

The MAX232 has two sets of line drivers for transferring and receiving data, as shown in Figure 10-7. The line drivers used for TxD are called T1 and T2,

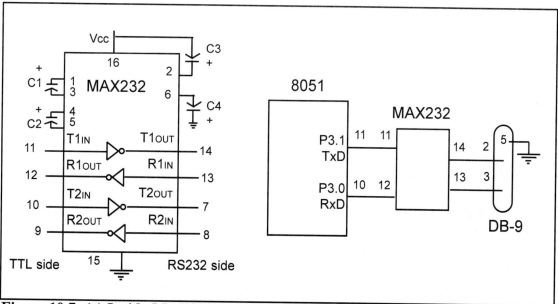

Figure 10.7: (a) Inside MAX232 and (b) its Connection to the 8051 (Null Modem)

while the line drivers for RxD are designated as R1 and R2. In many applications only one of each is used. For example, T1 and R1 are used together for TxD and RxD of the 8051, and the second set is left unused. Notice in MAX232 that the T1 line driver has a designation of T1in and T1out on pin numbers 11 and 14, respectively. The T1in pin is the TTL side and is connected to TxD of the microcontroller, while T1out is the RS232 side that is connected to the RxD pin of the RS232 DB connector. The R1 line driver has a designation of R1in and R1out on pin numbers 13 and 12, respectively. The R1in (pin 13) is the RS232 side that is connected to the TxD pin of the RS232 DB connector, and R1out (pin 12) is the TTL side that is connected to the RxD pin of the microcontroller. See Figure 10-7. Notice the null modem connection where RxD for one is TxD for the other.

MAX232 requires four capacitors ranging from 1 to 22 µF. The most widely used value for these capacitors is 22 µF.

MAX233

To save board space, some designers use the MAX233 chip from Maxim. The MAX233 performs the same job as the MAX232 but eliminates the need for capacitors. However, the MAX233 chip is much more expensive than the MAX232. Notice that MAX233 and MAX232 are not pin compatible. You cannot take a MAX232 out of a board and replace it with a MAX233. See Figure 10-8 for MAX233 with no capacitor used.

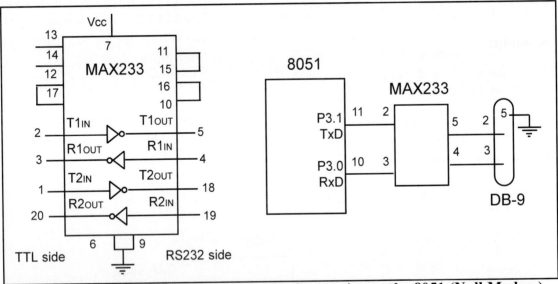

Figure 10.8: (a) Inside MAX233 and (b) Its Connection to the 8051 (Null Modem)

Review Questions

1. True or false. The PC COM port connector is the RS232 type.
2. Which pins of the 8051 are set aside for serial communication, and what are their functions?
3. What are line drivers such as MAX 232 used for?
4. MAX232 can support _____ lines for TxD and _____ lines for RxD.
5. What is the advantage of the MAX233 over the MAX232 chip?

SECTION 10.3: 8051 SERIAL COMMUNICATION PROGRAMMING

Table 10-3: PC Baud Rates

110
150
300
600
1200
2400
4800
9600
19200

Note: Baud rates supported by 486/Pentium IBM PC BIOS.

In this section we discuss the serial communication registers of the 8051 and show how to program them to transfer and receive data serially. Since IBM PC/compatible computers are so widely used to communicate with 8051-based systems, we will emphasize serial communications of the 8051 with the COM port of the PC. To allow data transfer between the PC and an 8051 system without any error, we must make sure that the baud rate of the 8051 system matches the baud rate of the PC's COM port. The baud rates supported by PC BIOS are listed in Table 10-3. You can examine these baud rates by going to the Windows Terminal program and clicking on the Communication Settings option. The terminal.exe program comes with Windows 3.1 and also works in Windows 95 and 98. In Windows 95 and higher you can use the Hyperterminal function. Hyperterminal supports baud rates much higher than the ones listed in Table 10-3.

Example 10-1

With XTAL = 11.0592 MHz, find the TH1 value needed to have the following baud rates.   (a) 9600   (b) 2400   (c) 1200

Solution:

With XTAL = 11.0592 MHz, we have:

The machine cycle frequency of the 8051 = 11.0592 MHz / 12 = 921.6 kHz, and 921.6 kHz/ 32 = 28,800 Hz is the frequency provided by UART to timer 1 to set baud rate.

(a) 28,800 / 3 = 9600   where −3 = FD (hex) is loaded into TH1
(b) 28,800 / 12 = 2400   where −12 = F4 (hex) is loaded into TH1
(c) 28,800 / 24 = 1200   where −24 = E8 (hex) is loaded into TH1

Notice that dividing 1/12th of the crystal frequency by 32 is the default value upon activation of the 8051 RESET pin. We can change this default setting. This is explained at the end of this chapter.

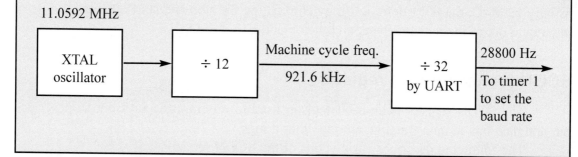

Table 10-4: Timer 1 TH1 Register Values for Various Baud Rates

Baud Rate	TH1 (Decimal)	TH1 (Hex)
9600	-3	FD
4800	-6	FA
2400	-12	F4
1200	-24	E8

Note: XTAL = 11.0592 MHz.

Baud rate in the 8051

The 8051 transfers and receives data serially at many different baud rates. The baud rate in the 8051 is programmable. This is done with the help of timer 1. Before we discuss how to do that, we will look at the relationship between the crystal frequency and the baud rate in the 8051.

As discussed in previous chapters, the 8051 divides the crystal frequency by 12 to get the machine cycle frequency. In the case of XTAL = 11.0592 MHz, the machine cycle frequency is 921.6 kHz (11.0592 MHz / 12 = 921.6 kHz). The 8051's serial communication UART circuitry divides the machine cycle frequency of 921.6 kHz by 32 once more before it is used by timer 1 to set the baud rate. Therefore, 921.6 kHz divided by 32 gives 28,800 Hz. This is the number we will use throughout this section to find the timer 1 value to set the baud rate. When timer 1 is used to set the baud rate it must be programmed in mode 2, that is 8-bit, auto-reload. To get baud rates compatible with the PC, we must load TH1 with the values shown in Table 10-3. Example 10-1 shows how to verify the data in Table 10-3.

SBUF register

SBUF is an 8-bit register used solely for serial communication in the 8051. For a byte of data to be transferred via the TxD line, it must be placed in the SBUF register. Similarly, SBUF holds the byte of data when it is received by the 8051's RxD line. SBUF can be accessed like any other register in the 8051. Look at the following examples of how this register is accessed:

```
MOV SBUF,#'D'    ;load SBUF=44H, ASCII for 'D'
MOV SBUF,A       ;copy accumulator into SBUF
MOV A,SBUF       ;copy SBUF into accumulator
```

The moment a byte is written into SBUF, it is framed with the start and stop bits and transferred serially via the TxD pin. Similarly, when the bits are received serially via RxD, the 8051 deframes it by eliminating the stop and start bits, making a byte out of the data received, and then placing it in the SBUF.

SCON (serial control) register

The SCON register is an 8-bit register used to program the start bit, stop bit, and data bits of data framing, among other things.

The following describes various bits of the SCON register.

SM0	SM1	SM2	REN	TB8	RB8	TI	RI

SM0	SCON.7	Serial port mode specifier
SM1	SCON.6	Serial port mode specifier
SM2	SCON.5	Used for multiprocessor communication. (Make it 0)
REN	SCON.4	Set/cleared by software to enable/disable reception.
TB8	SCON.3	Not widely used.
RB8	SCON.2	Not widely used.
TI	SCON.1	Transmit interrupt flag. Set by hardware at the beginning of the stop bit in mode 1. Must be cleared by software.
RI	SCON.0	Receive interrupt flag. Set by hardware halfway through the stop bit time in mode 1. Must be cleared by software.

Note: Make SM2, TB8, and RB8 = 0.

Figure 10-9. SCON Serial Port Control Register (Bit Addressable)

SM0, SM1

SM0 and SM1 are D7 and D6 of the SCON register, respectively. These two bits determine the framing of data by specifying the number of bits per character, and the start and stop bits. They take the following combinations.

SM0	*SM1*	
0	0	Serial Mode 0
0	1	Serial Mode 1, 8-bit data, 1 stop bit, 1 start bit
1	0	Serial Mode 2
1	1	Serial Mode 3

Of the 4 serial modes, only mode 1 is of interest to us. Further explanation for the other three modes is in Appendix A.3. They are rarely used today. In the SCON register when serial mode 1 is chosen, the data framing is 8 bits, 1 stop bit, and 1 start bit, which makes it compatible with the COM port of IBM/compatible PCs. More importantly, serial mode 1 allows the baud rate to be variable and is set by timer 1 of the 8051. In serial mode 1, for each character a total of 10 bits are transferred, where the first bit is the start bit, followed by 8 bits of data, and finally 1 stop bit.

SM2

SM2 is the D5 bit of the SCON register. This bit enables the multiprocessing capability of the 8051 and is beyond the discussion of this chapter. For our applications, we will make SM2 = 0 since we are not using the 8051 in a multiprocessor environment.

REN

The REN (receive enable) bit is D4 of the SCON register. The REN bit is also referred to as SCON.4 since SCON is a bit-addressable register. When the REN bit is high, it allows the 8051 to receive data on the RxD pin of the 8051. As a result if we want the 8051 to both transfer and receive data, REN must be set to 1. By making REN = 0, the receiver is disabled. Making REN = 1 or REN = 0 can be achieved by the instructions "SETB SCON.4" and "CLR SCON.4", respectively. Notice that these instructions use the bit-addressable features of register SCON. This bit can be used to block any serial data reception, and is an extremely important bit in the SCON register.

TB8

TB8 (transfer bit 8) is bit D3 of SCON. It is used for serial modes 2 and 3. We make TB8 = 0 since it is not used in our applications.

RB8

RB8 (receive bit 8) is bit D2 of the SCON register. In serial mode 1, this bit gets a copy of the stop bit when an 8-bit data is received. This bit (as is the case for TB8) is rarely used anymore. In all our applications we will make RB8 = 0. Like TB8, the RB8 bit is also used in serial modes 2 and 3.

TI

TI (transmit interrupt) is bit D1 of the SCON register. This is an extremely important flag bit in the SCON register. When the 8051 finishes the transfer of the 8-bit character, it raises the TI flag to indicate that it is ready to transfer another byte. The TI bit is raised at the beginning of the stop bit. We will discuss further its role when programming examples of data transmission are given.

RI

RI (receive interrupt) is the D0 bit of the SCON register. This is another extremely important flag bit in the SCON register. When the 8051 receives data serially via RxD, it gets rid of the start and stop bits and places the byte in the SBUF register. Then it raises the RI flag bit to indicate that a byte has been received and should be picked up before it is lost. RI is raised halfway through the stop bit, and we will soon see how this bit is used in programs for receiving data serially.

Programming the 8051 to transfer data serially

In programming the 8051 to transfer character bytes serially, the following steps must be taken.
1. The TMOD register is loaded with the value 20H, indicating the use of timer 1 in mode 2 (8-bit auto-reload) to set the baud rate.
2. The TH1 is loaded with one of the values in Table 10-4 to set the baud rate for serial data transfer (assuming XTAL = 11.0592 MHz).
3. The SCON register is loaded with the value 50H, indicating serial mode 1, where an 8-bit data is framed with start and stop bits.
4. TR1 is set to 1 to start timer 1.
5. TI is cleared by the "CLR TI" instruction.
6. The character byte to be transferred serially is written into the SBUF register.
7. The TI flag bit is monitored with the use of the instruction "JNB TI,xx" to see if the character has been transferred completely.
8. To transfer the next character, go to Step 5.

Example 10-2 shows the program to transfer data serially at 4800 baud. Example 10-3 shows how to transfer "YES" continuously.

Example 10-2

Write a program for the 8051 to transfer letter "A" serially at 4800 baud, continuously.

Solution:

```
            MOV   TMOD,#20H ;timer 1, mode 2 (auto reload)
            MOV   TH1,#-6   ;4800 baud rate
            MOV   SCON,#50H ;8-bit, 1 stop, REN enabled
            SETB  TR1       ;start timer 1
AGAIN:      MOV   SBUF,#"A" ;letter "A" to be transferred
HERE:       JNB   TI,HERE   ;wait for the last bit
            CLR   TI        ;clear TI for next char
            SJMP  AGAIN     ;keep sending A
```

Example 10-3

Write a program to transfer the message "YES" serially at 9600 baud, 8-bit data, 1 stop bit. Do this continuously.

Solution:

```
            MOV   TMOD,#20H ;timer 1, mode 2
            MOV   TH1,#-3   ;9600 baud
            MOV   SCON,#50H ;8-bit, 1 stop bit, REN enabled
            SETB  TR1       ;start timer 1
AGAIN:      MOV   A,#"Y"    ;transfer "Y"
            ACALL TRANS
            MOV   A,#"E"    ;transfer "E"
            ACALL TRANS
            MOV   A,#"S"    ;transfer "S"
            ACALL TRANS
            SJMP  AGAIN     ;keep doing it
;serial data transfer subroutine
TRANS:      MOV   SBUF,A    ;load SBUF
HERE:       JNB   TI,HERE   ;wait for last bit to transfer
            CLR   TI        ;get ready for next byte
            RET
```

Importance of the TI flag

To understand the importance of the role of TI, look at the following sequence of steps that the 8051 goes through in transmitting a character via TxD.
1. The byte character to be transmitted is written into the SBUF register.
2. It transfers the start bit.
3. The 8-bit character is transferred one bit at a time.
4. The stop bit is transferred. It is during the transfer of the stop bit that the 8051 raises the TI flag (TI = 1), indicating that the last character was transmitted and it is ready to transfer the next character.
5. By monitoring the TI flag, we make sure that we are not overloading the SBUF register. If we write another byte into the SBUF register before TI is raised, the untransmitted portion of the previous byte will be lost. In other words when the

8051 finishes transferring a byte, it raises the TI flag to indicate it is ready for the next character.

6. After SBUF is loaded with a new byte, the TI flag bit must be forced to 0 by the "CLR TI" instruction in order for this new byte to be transferred.

From the above discussion we conclude that by checking the TI flag bit, we know whether or not the 8051 is ready to transfer another byte. More importantly, it must be noted that the TI flag bit is raised by the 8051 itself when it finishes the transfer of data, whereas it must be cleared by the programmer with an instruction such as "CLR TI". It also must be noted that if we write a byte into SBUF before the TI flag bit is raised, we risk the loss of a portion of the byte being transferred. The TI flag bit can be checked by the instruction "JNB TI, ..." or we can use an interrupt, as we will see in Chapter 11. In Chapter 11 we will show how to use interrupts to transfer data serially, and avoid tying down the microcontroller with instructions such as "JNB TI, xx".

Programming the 8051 to receive data serially

In the programming of the 8051 to received character bytes serially, the following steps must be taken.

1. The TMOD register is loaded with the value 20H, indicating the use of timer 1 in mode 2 (8-bit auto-reload) to set the baud rate.
2. TH1 is loaded with one of the values in Table 10-4 to set the baud rate (assuming XTAL = 11.0592 MHz).
3. The SCON register is loaded with the value 50H, indicating serial mode 1, where 8-bit data is framed with start and stop bits.
4. TR1 is set to 1 to start timer 1.
5. RI is cleared with the "CLR RI" instruction.
6. The RI flag bit is monitored with the use of the instruction "JNB RI, xx" to see if an entire character has been received yet.
7. When RI is raised, SBUF has the byte. Its contents are moved into a safe place.
8. To receive the next character, go to Step 5.

Example 10-4 shows the coding of the above steps.

Example 10-4

Program the 8051 to receive bytes of data serially, and put them in P1. Set the baud rate at 4800, 8-bit data, and 1 stop bit.

Solution:

```
        MOV   TMOD,#20H  ;timer1, mode 2 (auto reload)
        MOV   TH1,#-6     ;4800 baud
        MOV   SCON,#50H  ;8-bit, 1 stop, REN enabled
        SETB  TR1        ;start timer 1
HERE:   JNB   RI,HERE    ;wait for char to come in
        MOV   A,SBUF     ;save incoming byte in A
        MOV   P1,A       ;send to port 1
        CLR   RI         ;get ready to receive next byte
        SJMP  HERE       ;keep getting data
```

Example 10-5

Assume that the 8051 serial port is connected to the COM port of the IBM PC, and on the PC we are using the terminal.exe program to send and receive data serially. P1 and P2 of the 8051 are connected to LEDs and switches, respectively. Write an 8051 program to (a) send to the PC the message "We Are Ready", (b) receive any data sent by the PC and put it on LEDs connected to P1, and (c) get data on switches connected to P2 and send it to the PC serially. The program should perform part (a) once, but parts (b) and (c) continously. Use the 4800 baud rate.

Solution:

```
                ORG  0
                MOV  P2,#0FFH      ;make P2 an input port
                MOV  TMOD,#20H     ;timer 1,mode 2(auto-reload)
                MOV  TH1,#0FAH     ;4800 baud rate
                MOV  SCON,#50H     ;8-bit,1 stop, REN enabled
                SETB TR1           ;start timer 1
                MOV  DPTR,#MYDATA   ;load pointer for message
H_1:            CLR  A
                MOVC  A,@A+DPTR      ;get the character
                JZ   B_1            ;if last character get out
                ACALL SEND          ;otherwise call transfer
                INC  DPTR           ;next one
                SJMP H_1        ;stay in loop
B_1:            MOV A,P2        ;read data on P2
                ACALL SEND      ;transfer it serially
                ACALL RECV      ;get the serial data
                MOV P1,A        ;display it on LEDs
                SJMP B_1        ;stay in loop indefinitly
;---------------serial data transfer. ACC has the data
SEND:           MOV SBUF,A      ;load the data
H_2:            JNB TI,H_2      ;stay here until last bit gone
                CLR TI          ;get ready for next char
                RET             ;return to caller
;--------------- Receive data serially in Acc
RECV:           JNB RI,RECV     ;wait here for char
                MOV A,SBUF      ;save it in ACC
                CLR RI          ;get ready for next char
                RET             ;return to caller
;-----------------The message
MYDATA:         DB    "We Are Ready",0
                END
```

8051

Importance of the RI flag bit

In receiving bits via its RxD pin, the 8051 goes through the following steps.

1. It receives the start bit indicating that the next bit is the first bit of the character byte it is about to receive.
2. The 8-bit character is received one bit at time. When the last bit is received, a byte is formed and placed in SBUF.
3. The stop bit is received. It is during receiving the stop bit that the 8051 makes RI = 1, indicating that an entire character byte has been received and must be picked up before it gets overwritten by an incoming character.
4. By checking the RI flag bit when it is raised, we know that a character has been received and is sitting in the SBUF register. We copy the SBUF contents to a safe place in some other register or memory before it is lost.
5. After the SBUF contents are copied into a safe place, the RI flag bit must be forced to 0 by the "CLR RI" instruction in order to allow the next received character byte to be placed in SBUF. Failure to do this causes loss of the received character.

From the above discussion we conclude that by checking the RI flag bit we know whether or not the 8051 has received a character byte. If we fail to copy SBUF into a safe place, we risk the loss of the received byte. More importantly, it must be noted that the RI flag bit is raised by the 8051, but it must be cleared by the programmer with an instruction such as "CLR TI". It also must be noted that if we copy SBUF into a safe place before the RI flag bit is raised, we risk copying garbage. The RI flag bit can be checked by the instruction "JNB RI,xx" or by using an interrupt, as we will see in Chapter 11.

Doubling the baud rate in the 8051

There are two ways to increase the baud rate of data transfer in the 8051.

1. To use a higher frequency crystal.
2. To change a bit in the PCON register, shown below.

D7   D0

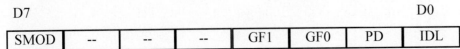

| SMOD | -- | -- | -- | GF1 | GF0 | PD | IDL |

Option 1 is not feasible in many situations since the system crystal is fixed. More importantly, it is not feasible beause the new crystal may not be compatible with the IBM PC serial COM ports baud rate. Therefore, we will explore option 2. There is a software way to double the baud rate of the 8051 while the crystal frequency is fixed. This is done with the register called PCON (power control). The PCON register is an 8-bit register. Of the 8 bits, some are unused, and some are used for the power control capability of the 8051. The bit which is used for the serial communication is D7, the SMOD (serial mode) bit. When the 8051 is powered up, D7 (SMOD bit) of the PCON register is zero. We can set it to high by software and thereby double the baud rate. The following sequence of instructions must be used to set high D7 of PCON, since it is not a bit-addressable register:

```
MOV   A,PCON          ;place a copy of PCON in ACC
SETB  ACC.7           ;make D7=1
MOV   PCON,A          ;now SMOD=1 without
                      ;  changing any other bits
```

To see how the baud rate is doubled with this method, we show the role of the SMOD bit (D7 bit of the PCON register), which can be 0 or 1. We discuss each case.

Baud rates for SMOD = 0

When SMOD = 0, the 8051 divides 1/12 of the crystal frequency by 32 and uses that frequency for timer 1 to set the baud rate. In the case of XTAL = 11.0592 MHz we have:

```
Machine cycle freq. = 11.0592 MHz / 12 = 921.6 kHz
and
921.6 kHz / 32 = 28,800 Hz since SMOD = 0
```

This is the frequency used by timer 1 to set the baud rate. This has been the basis of all the examples so far since it is the default when the 8051 is powered up. The baud rate for SMOD = 0 was listed in Table 10-4.

Baud rates for SMOD = 1

With the fixed crystal frequency, we can double the baud rate by making SMOD = 1. When the SMOD bit (D7 of the PCON register) is set to 1, 1/12 of XTAL is divided by 16 (instead of 32) and that is the frequency used by timer 1 to set the baud rate. In the case of XTAL = 11.0592 MHz, we have:

```
Machine cycle freq. = 11.0592 MHz / 12 = 921.6 kHz
and
921.6 kHz / 16 = 57,600 Hz since SMOD = 1
```

This is the frequency used by timer 1 to set the baud rate.

Table 10-5 shows that the values loaded into TH1 are the same for both cases; however, the baud rates are doubled when SMOD = 1. Look at the following examples to clarify the data given in Table 10-5.

Table 10-5: Baud Rate Comparison for SMOD = 0 and SMOD = 1

TH1 (Decimal)	(Hex)	SMOD = 0	SMOD = 1
–3	FD	9,600	19,200
–6	FA	4,800	9,600
–12	F4	2,400	4,800
–24	E8	1,200	2,400

Note: XTAL = 11.0592 MHz.

Example 10-6

Assuming that XTAL = 11.0592 MHz for the following program, state (a) what this program does, (b) compute the frequency used by timer 1 to set the baud rate, and (c) find the baud rate of the data transfer.

```
        MOV   A,PCON      ;A=PCON
        SETB  ACC.7       ;make D7=1
        MOV   PCON,A      ;SMOD=1,double baud rate
                          ;with same XTAL freq.
        MOV   TMOD,#20H   ;Timer 1, mode 2,auto reload
        MOV   TH1,-3      ;19200(57,600/3 =19200 baud rate
                          ;since SMOD=1)
        MOV   SCON,#50H   ;8-bit data,1 stop bit, RI enabled
        SETB  TR1         ;start Timer 1
        MOV   A,#"B"      ;transfer letter B
A_1:    CLR   TI          ;make sure TI=0
        MOV   SBUF,A      ;transfer it
H_1:    JNB   TI H_1      ;stay here until the last bit is gone
        SJMP  A_1         ;keep sending "B" again and again
```

Solution:

(a) This program transfers ASCII letter B (01000010 binary) continuously.
(b) With XTAL=11.0592 MHz and SMOD = 1 in the above program, we have:

11.0592 / 12 = 921.6 kHz machine cycle frequency.
921.6 / 16 = 57,600 Hz frequency used by timer 1 to set the baud rate.
57,600 / 3 = 19,200, the baud rate.

Example 10-7

Find the TH1 value (in both decimal and hex) to set the baud rate to each of the following.
(a) 9600   (b) 4800 if SMOD = 1   Assume that XTAL = 11.0592 MHz.

Solution:

With XTAL = 11.0592 and SMOD = 1, we have timer 1 frequency = 57,600 Hz.
(a) 57,600 / 9600 = 6; therefore, TH1 = −6 or TH1 = FAH.
(b) 57,600 / 4800 = 12; therefore, TH1 = −12 or TH1 = F4H.

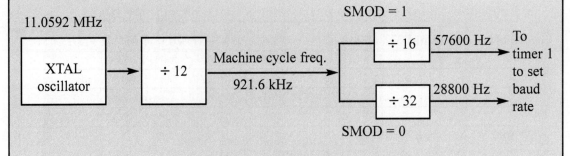

Find the baud rate if TH1 = –2, SMOD = 1, and XTAL = 11.0592 MHz. Is this baud rate supported by IBM/compatible PCs?

Solution:

With XTAL = 11.0592 and SMOD = 1, we have timer 1 frequency = 57,600 Hz. The baud rate is 57,600 / 2 = 28,800. This baud rate is not supported by the BIOS of the PC; however, the PC can be programmed to do data transfer at such a speed. The software of many modems can do this. Also, Hyperterminal in Windows 95 (and higher) supports this and other baud rates.

Interrupt-based data transfer

By now you might have noticed that it is a waste of the microcontroller's time to poll the TI and RI flags. In Chapter 11, we will show how to use interrupts to program the 8051's serial communication port.

Review Questions

1. Which timer of the 8051 is used to set the baud rate?
2. If XTAL = 11.0592 MHz, what frequency is used by the timer to set the baud rate?
3. Which mode of the timer is used to set the baud rate?
4. With XTAL = 11.0592 MHz, what value should be loaded into TH1 to have a 9600 baud rate? Give the answer in both decimal and hex.
5. To transfer a byte of data serially, it must be placed in register _____.
6. SCON stands for _____ and it is a(n) ____-bit register.
7. Which register is used to set the data size and other framing information such as the stop bit?
8. True or false. SCON is a bit-addressable register.
9. When is TI raised?
10. Which register has the SMOD bit, and what is its status when the 8051 is powered up?

SUMMARY

This chapter began with an introduction to the fundamentals of serial communication. Serial communication, in which data is sent one bit a time, is used in situations where data is sent over significant distances since in parallel communication, where data is sent a byte or more a time, great distances can cause distortion of the data. Serial communication has the additional advantage of allowing transmission over phone lines. Serial communication uses two methods: synchronous and asynchronous. In synchronous communication, data is sent in blocks of bytes; in asynchronous, data is sent in bytes. Data communication can be simplex (can send but cannot receive), half-duplex (can send and receive, but not at the same time), or full-duplex (can send and receive at the same time). RS232 is a standard for serial communication connectors.

The 8051's UART was discussed. We showed how to interface the 8051 with an RS232 connector and change the baud rate of the 8051. In addition, we described the serial communication features of the 8051, and programmed the 8051 for serial data communication.

PROBLEMS

SECTION 10.1: BASICS OF SERIAL COMMUNICATION

1. Which is more expensive, parallel or serial data transfer?
2. True or false. 0- and 5-V digital pulses can be transferred on the telephone without being converted (modulated).
3. Show the framing of the letter ASCII "Z" (0101 1010), no parity, 1 stop bit.
4. If there is no data transfer and the line is high, it is called _____ (mark, space).
5. True or false. The stop bit can be 1, 2, or none at all.
6. Calculate the overhead percentage if the data size is 7, 1 stop bit, no parity.
7. True or false. RS232 voltage specification is TTL compatible.
8. What is the function of the MAX 232 chip?
9. True or false. DB-25 and DB-9 are pin compatible for the first 9 pins.
10. How many pins of the RS232 are used by the IBM serial cable, and why?
11. True or false. The longer the cable, the higher the data transfer baud rate.
12. State the absolute minimum number of signals needed to transfer data between two PCs connected serially. What are those?
13. If two PCs are connected through the RS232 without the modem, they are both configured as a _____ (DTE, DCE) -to- _____ (DTE, DCE) connection.
14. State the 9 most important signals of the RS232.
15. Calculate the total number of bits transferred if 200 pages of ASCII data are sent using asynchronous serial data transfer. Assume a data size of 8 bits, 1 stop bit, no parity. Assume each page has 80x25 of text characters.
16. In Problem 15, how long will the data transfer take if the baud rate is 9,600?

SECTION 10.2: 8051 CONNECTION TO RS232

17. The MAX232 DIP package has _____ pins.
18. For the MAX232, indicate the V_{CC} and GND pins.
19. The MAX233 DIP package has _____ pins.
20. For the MAX233, indicate the V_{CC} and GND pins.
21. Is the MAX232 pin compatible with the MAX233?
22. State the advantages and disadvantages of the MAX232 and MAX233.
23. MAX232/233 has _____ line driver(s) for the RxD wire.
24. MAX232/233 has _____ line driver(s) for the TxD wire.

25. Show the connection of pins TxD and RxD of the 8051 to a DB-9 RS232 connector via the second set of line drivers of MAX232.
26. Show the connection of the TxD and RxD pins of the 8051 to a DB-9 RS232 connector via the second set of line drivers of MAX233.
27. Show the connection of the TxD and RxD pins of the 8051 to a DB-25 RS232 connector via MAX232.
28. Show the connection of the TxD and RxD pins of the 8051 to a DB-25 RS232 connector via MAX233.

SECTION 10.3: SERIAL COMMUNICATION PROGRAMMING

29. Which of the following baud rates are supported by the BIOS of 486/Pentium PCs?
    (a) 4,800   (b) 3,600   (c) 9,600
    (d) 1,800   (e) 1,200   (f) 19,200
30. Which timer of the 8051 is used for baud rate programming?
31. Which mode of the timer is used for baud rate programming?
32. What is the role of the SBUF register in serial data transfer?
33. SBUF is a(n) _____-bit register.
34. What is the role of the SCON register in serial data transfer?
35. SCON is a(n) _____-bit register.
36. For XTAL = 11.0592 MHz, find the TH1 value (in both decimal and hex) for each of the following baud rates.
    (a) 9,600   (b) 4,800   (c) 1,200   (d) 300   (e) 150
37. What is the baud rate if we use "MOV TH1, #-1" to program the baud rate?
38. Write an 8051 program to transfer serially the letter "Z" continuously at 1,200 baud rate.
39. Write a 8051 program to transfer serially the message "The earth is but one country and mankind its citizens" continuously at 57,600 baud rate.
40. When is the TI flag bit raised?
41. When is the RI flag bit raised?
42. To which register do RI and TI belong? Is that register bit-addressable?
43. What is the role of the REN bit in SCON register?
44. In a given situation we can not accept reception of any serial data. How do you block such a reception with a single instruction?
45. To which register does the SMOD bit belong? State its role in rate of data transfer.
46. Is SMOD bit high or low when the 8051 is powered up?

In the following questions the baud rates are not compatible with COM ports of PC (x86 IBM/compatible) .
47. Find the baud rate for the following if XTAL=16 MHz. and SMOD=0
    (a) MOV TH1,#-10   (b) MOV TH1,#-25
    (c) MOV TH1,#-200   (d) MOV TH1,#-180
48. Find the baud rate for the following if XTAL=24 MHz. and SMOD=0
    (a) MOV TH1,#-15   (b) MOV TH1,#-24
    (c) MOV TH1,#-100   (d) MOV TH1,#-150

49. Find the baud rate for the following if XTAL = 16 MHz and SMOD = 1.
    (a) MOV TH1,#-10   (b) MOV TH1,#-25
    (c) MOV TH1,#-200   (d) MOV TH1,#-180

50. Find the baud rate for the following if XTAL = 24 MHz and SMOD = 1.
    (a) MOV TH1,#-15   (b) MOV TH1,#-24
    (c) MOV TH1,#-100   (d) MOV TH1,#-150

ANSWERS TO REVIEW QUESTIONS

SECTION 10.1: BASICS OF SERIAL COMMUNICATION
1. Faster, more expensive
2. False; it is simplex.
3. True
4. Asynch
5. With 0100 0101 binary the bits are transmitted in the sequence:
   (a) 0 (start bit) (b) 1 (c) 0 (d) 1 (e) 0 (f) 0 (g) 0 (h) 1 (i) 0 (j) 1 (stop bit)
6. 2 bits (one for the start bit and one for the stop bit). Therefore, for each 8-bit character, a total of 10 bits is transferred.
7. $10000 \times 10 = 100000$ bits total bits transmitted. $100000 / 9600 = 10.4$ seconds; $2 / 10 = 20\%$.
8. True
9. +3 to +25 V
10. True
11. 2
12. COM 1 and COM 2

SECTION 10.2: 8051 CONNECTION TO RS232
1. True
2. Pin 10 and 11. Pin 10 is for TxD and pin 11 for RxD.
3. They are used for converting from RS232 voltage level to TTL voltage levels and vice versa.
4. 2,2
5. It does not need the four capacitors that MAX232 must have.

SECTION 10.3: SERIAL COMMUNICATION PROGRAMMING
1. Timer 1
2. 28,800 Hz
3. Mode 2
4. –3 or FDH since $28,800 / 3 = 9,600$
5. SBUF
6. Serial control, 8
7. SCON
8. False
9. During transfer of stop bit
10. PCON; it is low upon RESET.

CHAPTER 11

INTERRUPTS PROGRAMMING

OBJECTIVES

Upon completion of this chapter, you will be able to:

≫≫   Contrast and compare interrupts versus polling

≫≫   Explain the purpose of the ISR (interrupt service routine)

≫≫   List the 6 interrupts of the 8051

≫≫   Explain the purpose of the interrupt vector table

≫≫   Enable or disable 8051 interrupts

≫≫   Program the 8051 timers using interrupts

≫≫   Describe the two external hardware interrupts of the 8051

≫≫   Contrast edge-triggered with level-triggered interrupts

≫≫   Program the 8051 for interrupt-based serial communication

≫≫   Define the interrupt priority of the 8051

An *interrupt* is an external or internal event that interrupts the microcontroller to inform it that a device needs its service. In this chapter we explore the concept of the interrupt and interrupt programming. In Section 11.1 the basics of 8051 interrupts are discussed. In Section 11.2 interrupts belonging to timers 0 and 1 are discussed. External hardware interrupts are discussed in Section 11.3, while the interrupt related to serial communication is presented in Section 11.4. Finally, in Section 11.5 we cover interrupt priority in the 8051.

SECTION 11.1: 8051 INTERRUPTS

In this section, first we examine the difference between polling and interrupts and then describe the various interrupts of the 8051.

Interrupts vs. polling

A single microcontroller can serve several devices. There are two ways to do that: interrupts or polling. In the *interrupt* method, whenever any device needs its service, the device notifies the microcontroller by sending it an interrupt signal. Upon receiving an interrupt signal, the microcontroller interrupts whatever it is doing and serves the device. The program which is associated with the interrupt is called the *interrupt service routine* (ISR) or *interrupt handler*. In *polling*, the microcontroller continuously monitors the status of a given device; when the condition is met, it performs the service. After that, it moves on to monitor the next device until every one is serviced. Although polling can monitor the status of several devices and serve each of them as certain conditions are met, it is not efficient as far as the use of the microcontroller is concerned. The advantage of interrupts is that the microcontroller can serve many devices (of course, not all at the same time). Each device can get the attention of the microcontroller based on the priority assigned to it. For the polling method, it is not possible to assign priority since it checks all devices in a round-robin fashion. More importantly, in the interrupt method the microcontroller can also ignore (mask) a device request for service. This is again not possible for the polling method. The most important reason that the interrupt method is preferable is that the polling method wastes much of the microcontroller's time by polling devices that do not need service. In order to avoid tying down the microcontroller, interrupts are used. For example, in discussing timers in Chapter 9 we used the instruction "JNB TF,target", and waited until the timer rolled over. In that example, while we were waiting we could not do any thing else. That is a waste of the microcontroller's time that could have been used to perform some useful tasks. In the case of the timer, if we use the interrupt method, the microcontroller can go about doing other tasks, and when the TF flag is raised it will interrupt the microcontroller in whatever it is doing.

Interrupt service routine

For every interrupt, there must be an interrupt service routine (ISR), or interrupt handler. When an interrupt is invoked, the microcontroller runs the interrupt service routine. For every interrupt, there is a fixed location in memory that holds the address of its ISR. The group of memory locations set aside to hold the addresses of ISRs is called the interrupt vector table, shown in Table 11-1.

Steps in executing an interrupt

Upon activation of an interrupt, the microcontroller goes through the following steps.

1. It finishes the instruction it is executing and saves the address of the next instruction (PC) on the stack.
2. It also saves the current status of all the interrupts internally (i.e., not on the stack).
3. It jumps to a fixed location in memory called the interrupt vector table that holds the address of the interrupt service routine.
4. The microcontroller gets the address of the ISR from the interrupt vector table and jumps to it. It starts to execute the interrupt service subroutine until it reaches the last instruction of the subroutine which is RETI (return from interrupt).
5. Upon executing the RETI instruction, the microcontroller returns to the place where it was interrupted. First, it gets the program counter (PC) address from the stack by popping the top two bytes of the stack into the PC. Then it starts to execute from that address.

Notice from Step 5 the critical role of the stack. For this reason, we must be careful in manipulating the stack contents in the ISR. Specifically, in the ISR, just as in any CALL subroutine, the number of pushes and pops must be equal.

Six interrupts in the 8051

There are really five interrupts available to the user in the 8051 but many manufacturer's data sheets state that there are six interrupts since they include reset. The six interrupts in the 8051 are allocated as follows.

1. Reset. When the reset pin is activated, the 8051 jumps to address location 0000. This is the power-up reset discussed in Chapter 4.
2. Two interrupts are set aside for the timers: one for timer 0 and one for timer 1. Memory locations 000BH and 001BH in the interrupt vector table belong to timer 0 and timer 1, respectively.
3. Two interrupts are set aside for hardware external hardware interrupts. Pin numbers 12 (P3.2) and 13 (P3.3) in port 3 are for the external hardware interrupts INT0 and INT1, respectively. These external interrupts are also referred to as EX1 and EX2. Memory locations 0003H and 0013H in the interrupt vector table are assigned to INT0 and INT1, respectively.
4. Serial communication has a single interrupt that belongs to both receive and transfer. The interrupt vector table location 0023H belongs to this interrupt.

Notice in Table 11-1 that there is a limited number of bytes set aside for each interrupt. For example for INT0, external hardware interrupt 0, a total of 8 bytes from location 0003 to 0000A is set aside for it. Similarly, a total of 8 bytes from location 000BH to 0012H is reserved for TF0, timer 0 interrupt. If the service routine for a given interrupt is short enough to fit in the memory space allocated to it, it is placed in the vector table; otherwise, an LJMP instruction is placed in the vector table to point to the address of the ISR. In that case, the rest of the bytes allocated to that interrupt are unused. In next three sections we will see many examples of interrupt programming that clarify these concepts.

From Table 11-1, also notice the fact that there are only three bytes of ROM space assigned to the reset pin. They are ROM address locations 0, 1, and 2. Address location 3 belongs to external hardware interrupt 0. For this reason, in our program we put the LJMP as the first instruction and redirect the processor away from the interrupt vector table as shown in Figure 11-1. In the next section we will see how this works in the context of some examples.

Table 11.1: Interrupt Vector Table for the 8051

Interrupt	ROM Location (Hex)	Pin
Reset	0000	9
External hardware interrupt 0 (INT0)	0003	P3.2 (12)
Timer 0 interrupt (TF0)	000B	
External hardware interrupt 1 (INT1)	0013	P3.3 (13)
Timer 1 interrupt (TF1)	001B	
Serial COM interrupt (RI and TI)	0023	

```
        ORG  0     ;wake-up ROM reset location
        LJMP MAIN ;by-pass interrupt vector table

;---- the wake-up program
        ORG  30H
MAIN:

        ....
        END
```

Figure 11-1: Redirecting the 8051 From the Interrupt Vector Table At Power-Up

Enabling and disabling an interrupt

Upon reset, all interrupts are disabled (masked), meaning that none will be responded to by the microcontroller if they are activated. The interrupts must be enabled by software in order for the microcontroller to respond to them. There is a register called IE (interrupt enable) that is responsible for enabling (unmasking) and disabling (masking) the interrupts. Figure 11-2 shows the IE register. Note that IE is a bit-addressable register.

From Figure 11-2 notice that bit D7 in the IE register is called EA (enable all). This must be set to 1 in order for rest of the register to take effect. D6 is unused. D5 is used by the 8052. The D4 bit is for the serial interrupt and so on.

Steps in enabling an interrupt

To enable an interrupt, we take the following steps:
1. Bit D7 of the IE register (EA) must be set to high to allow the rest of register to take effect.
2. If EA = 1, interrupts are enabled and will be responded to if their corresponding bits in IE are high. If EA = 0, no interrupt will be responded to, even if the associated bit in the IE register is high.

To understand this important point look at Example 11-1.

D7							D0
EA	--	ET2	ES	ET1	EX1	ET0	EX0

EA   IE.7   Disables all interrupts. If EA = 0, no interrupt is acknowledged. If EA = 1, each interrupt source is individually enabled or disabled by setting or clearing its enable bit.

--   IE.6   Not implemented, reserved for future use. *

ET2   IE.5   Enables or disables timer 2 overflow or capture interrupt (8952).

ES   IE.4   Enables or disables the serial port interrupt.

ET1   IE.3   Enables or disables timer 1 overflow interrupt.

EX1   IE.2   Enables or disables external interrupt 1.

ET0   IE.1   Enables or disables timer 0 overflow interrupt.

EX0   IE.0   Enables or disables external interrupt 0.

*   User software should not write 1s to reserved bits. These bits may be used in future Flash microcontrollers to invoke new features.

Figure 11-2. IE (Interrupt Enable) Register

Example 11-1

Show the instructions to (a) enable the serial interrupt, timer 0 interrupt, and external hardware interrupt 1 (EX1), and (b) disable (mask) the timer 0 interrupt, then (c) show how to disable all the interrupts with a single instruction.

Solution:

(a)   MOV IE,#10010110B   ;enable serial, timer 0, EX1
Since IE is a bit-addressable register, we can use the following instructions to access individual bits of the register.
(b)   CLR IE.1   ;mask(disable) timer 0 interrupt only
(c)   CLR IE.7   ;disable all interrupts
Another way to perform the "MOV IE,#10010110B" instruction is by using single-bit instructions as shown below.

```
SETB IE.7          ;EA=1, Global enable
SETB IE.4          ;enable serial interrupt
SETB IE.1          ;enable Timer 0 interrupt
SETB IE.2          ;enable EX1
```

Review Questions

1. Of the interrupt and polling methods, which one avoids tying down the micro-controller?
2. Besides reset, how many interrupts do we have in the 8051?
3. In the 8051, what memory area is assigned to the interrupt vector table? Can the programmer change the memory space assigned to the table?
4. What are the contents of register IE upon reset, and what do these contents mean?
5. Show the instruction to enable the EX0 and timer 0 interrupts.
6. Which pin of the 8051 is assigned to the external hardware interrupt INT1?
7. What address in the interrupt vector table is assigned to the INT1 and timer 1 interrupts?

SECTION 11.2: PROGRAMMING TIMER INTERRUPTS

In Chapter 9 we discussed how to use timer 0 and timer 1 with the polling method. In this section we use interrupts to program the 8051 timers. Please review Chapter 9 before you study this section.

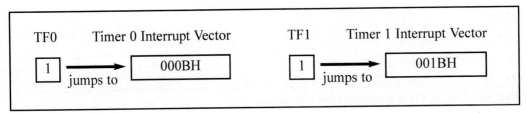

Figure 11-3. TF Interrupt

Roll-over timer flag and interrupt

In Chapter 9 we stated that the timer flag (TF) is raised when the timer rolls over. In that chapter, we also showed how to monitor TF with the instruction "JNB TF, target". In polling TF, we have to wait until the TF is raised. The problem with this method is that the microcontroller is tied down while waiting for TF to be raised, and can not do any thing else. Using interrupts solves this problem and avoids tying down the controller. If the timer interrupt in the IE register is enabled, whenever the timer rolls over, TF is raised, and the microcontroller is interrupted in whatever it is doing, and jumps to the interrupt vector table to service the ISR. In this way, the microcontroller can do other things until it is notified that the timer has rolled over. See Figure 11-3 and Example 11-2.

Notice the following points about the program in Example 11-2.
1. We must avoid using the memory space allocated to the interrupt vector table. Therefore, we place all the initialization codes in memory starting at 30H. The LJMP instruction is the first instruction that the 8051 executes when it is powered up. LJMP redirects the controller away from the interrupt vector table.

2. The ISR for timer 0 is located starting at memory location 000BH since it is small enough to fit the address space allocated to this interrupt.
3. We enabled the timer 0 interrupt with "MOV IE,#10000010B" in MAIN.
4. While the P0 data is brought in and issued to P1 continuously, whenever timer 0 is rolled over, the TF0 flag is raised, and the microcontroller gets out of the "BACK" loop and goes to 0000BH to execute the ISR associated with timer 0.
5. In the ISR for timer 0, notice that there is no need for a "CLR TF0" instruction before the RETI instruction. The reason for this is that that the 8051 clears the TF flag internally upon jumping to the interrupt vector table.

Example 11-2

Write a program that continuously gets 8-bit data from P0 and sends it to P1 while simultaneously creating a square wave of 200 μs period on pin P2.1. Use timer 0 to create the square wave. Assume that XTAL = 11.0592 MHz.

Solution:

We will use timer 0 in mode 2 (auto reload). TH0 = 100/1.085 μs = 92.

```
;--Upon wake-up go to main, avoid using memory space
;allocated to Interrupt Vector Table
        ORG   0000H
        LJMP  MAIN        ;by-pass interrupt vector table
;
;--ISR for Timer 0 to generate square wave
        ORG   000BH       ;Timer 0 interrupt vector table
        CPL   P2.1        ;toggle P2.1 pin
        RETI              ;return from ISR
;
;--The main program for initialization
        ORG   0030H       ;after vector table space
MAIN:   MOV   TMOD,#02H   ;Timer 0,mode 2(auto reload)
        MOV   P0,#0FFH    ;make P0 an input port
        MOV   TH0,#-92    ;TH0=A4H for -92
        MOV   IE,#82H     ;IE=10000010 (bin) enable Timer 0
        SETB  TR0         ;Start Timer 0
BACK:   MOV   A,P0        ;get data from P0
        MOV   P1,A        ;issue it to P1
        SJMP  BACK        ;keep doing it
                          ;loop unless interrupted by TF0
        END
```

In Example 11-2, the interrupt service routine was short enough that it could be placed in memory locations allocated to the timer 0 interrupt. However, that is not always the case. See Example 11-3.

Example 11-3

Rewrite Example 11-2 to create a square wave that has a high portion of 1085 μs and a low portion of 15 μs. Assume XTAL = 11.0592 MHz. Use timer 1.

Solution:

Since 1085 μs is 1000 × 1.085 we need to use mode 1 of timer 1.

```
;--Upon wake up go to main, avoid using memory space
;allocated to Interrupt Vector Table
          ORG   0000H
          LJMP MAIN        ;by-pass interrupt vector table
;
;--ISR for Timer 1 to generate square wave
          ORG   001BH      ;timer 1 interrupt vector table
          LJMP ISR_T1      ;jump to ISR
;
;--The main program for initialization
          ORG   0030H      ;after vector table
MAIN:     MOV   TMOD,#10H ;timer 1, mode 1
          MOV   P0,#0FFH  ;make P0 an input port
          MOV   TL1,#018H ;TL1=18 the low byte of -1000
          MOV   TH1,#0FCH ;TH1=FC the high byte of -1000
          MOV   IE,#88H   ;IE=10001000 enable timer 1 int.
          SETB TR1         ;start timer 1
BACK:     MOV   A,P0       ;get data from P0
          MOV   P1,A       ;issue it to P1
          SJMP BACK        ;keep doing it
;
;--Timer 1 ISR. Must be reloaded since not auto-reload
ISR_T1:   CLR   TR1        ;stop Timer 1
          CLR   P2.1       ;P2.1=0, start of low portion
          MOV   R2,#4      ;                        2 MC
HERE:     DJNZ R2,HERE     ;   4x2 machine cycle(MC)8 MC
          MOV   TL1,#18H   ;load T1 low byte value  2 MC
          MOV   TH1,#0FCH ;load T1 high byte value 2 MC
          SETB TR1         ;starts timer 1          1 MC
          SETB P2.1        ;P2.1=1, back to high    1 MC
          RETI             ;return to main
          END
```

Notice that the low portion of the pulse is created by the 14 MC (machine cycle) where each MC = 1.085 μs and 14 × 1.085 μs = 15.19 μs.

Example 11-4

Write a program to generate a square wave of 50 Hz frequency on pin P1.2. This is similar to Example 9-12 except that it uses an interrupt for timer 0. Assume that XTAL = 11.0592 MHz.

Solution:

```
        ORG 0
        LJMP MAIN
        ORG 000BH           ;ISR for Timer 0
        CPL P1.2
        MOV TL0,#00
        MOV TH0,#0DCH
        RETI
        ORG 30H
;------main program for initialization
MAIN:   MOV TMOD,#00000001B        ;Timer 0, Mode 1
        MOV TL0,#00
        MOV TH0,#0DCH
        MOV IE,#82H         ;enable Timer 0 interrupt
        SETB TR0
HERE:    SJMP HERE
        END
```

8051

50 Hz square wave

Review Questions

1. True or false. There is only a single interrupt in the interrupt vector table assigned to both timer 0 and timer 1.
2. What address in the interrupt vector table is assigned to timer 0?
3. Which bit of IE belongs to the timer interrupt? Show how both are enabled.
4. Assume that timer 1 is programmed in mode 2, TH1 = F5H, and the IE bit for timer 1 is enabled. Explain how the interrupt for the timer works.
5. True or false. The last two instructions of the ISR for timer 0 are:
```
        CLR  TF0
        RETI
```

SECTION 11.3: PROGRAMMING EXTERNAL HARDWARE INTERRUPTS

The 8051 has two external hardware interrupts. Pin 12 (P3.2) and pin 13 (P3.3) of the 8051, designated as INT0 and INT1, are used as external hardware interrupts. Upon activation of these pins, the 8051 gets interrupted in whatever it is doing and jumps to the vector table to perform the interrupt service routine. In this section we study these two external hardware interrupts of the 8051 with some examples.

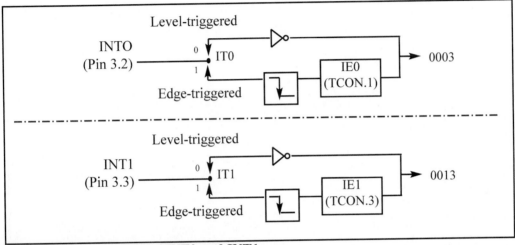

Figure 11-4. Activation of INT0 and INT1

External interrupts INT0 and INT1

There are only two external hardware interrupts in the 8051: INT0 and INT1. They are located on pins P3.2 and P3.3 of port 3, respectively. The interrupt vector table locations 0003H and 0013H are set aside for INT0 and INT1, respectively. As mentioned in Section 11.1, they are enabled and disabled using the IE register. How are they activated? There are two activation levels for the external hardware interrupts: (1) level triggered, and (2) edge triggered. Let's look at each one. First, we see how the level-triggered interrupt works.

Level-triggered interrupt

In the level-triggered mode, INT0 and INT1 pins are normally high (just like all I/O port pins) and if a low-level signal is applied to them, it triggers the interrupt. Then the microcontroller stops whatever it is doing and jumps to the interrupt vector table to service that interrupt. This is called a *level-triggered* or *level-activated* interrupt and is the default mode upon reset of the 8051. The low-level signal at the INT pin must be removed before the execution of the last instruction of the interrupt service routine, RETI; otherwise, another interrupt will be generated. In other words, if the low-level interrupt signal is not removed before the ISR is finished it is interpreted as another interrupt and the 8051 jumps to the vector table to execute the ISR again. Look at Example 11-5.

Example 11-5

Assume that the INT1 pin is connected to a switch that is normally high. Whenever it goes low, it should turn on an LED. The LED is connected to P1.3 and is normally off. When it is turned on it should stay on for a fraction of a second. As long as the switch is pressed low, the LED should stay on.

Solution:

```
        ORG  0000H
        LJMP MAIN         ;by-pass interrupt vector table
;--ISR for hardware interrupt INT1 to turn on the LED
        ORG  0013H        ;INT1 ISR
        SETB P1.3         ;turn on LED
        MOV  R3,#255
BACK:   DJNZ R3,BACK      ;keep LED on for a while
        CLR  P1.3         ;turn off the LED
        RETI              ;return from ISR
;-- MAIN program for initialization
        ORG  30H
MAIN:   MOV  IE,#10000100B  ;enable external INT 1
HERE:   SJMP HERE         ;stay here until get interrupted
        END
```

Pressing the switch will cause the LED to be turned on. If it is kept activated, the LED stays on.

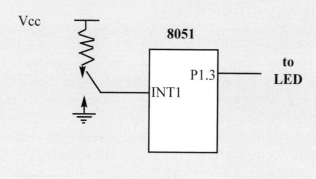

In this program, the microcontroller is looping continuously in the HERE loop. Whenever the switch on INT1 (pin P3.3) is activated, the microcontroller gets out of the loop and jumps to vector location 0013H. The ISR for INT1 turns on the LED, keeps it on for a while, and turns it off before it returns. If by the time it executes the RETI instruction, the INT1 pin is still low, the microcontroller initiates the interrupt again. Therefore, to end this problem, the INT1 pin must be brought back to high by the time RETI is executed.

Sampling the low level-triggered interrupt

Pins P3.2 and P3.3 are used for normal I/O unless the INT0 and INT1 bits in the IE registers are enabled. After the hardware interrupts in the IE register are enabled, the controller keeps sampling the INT*n* pin for a low-level signal once each machine cycle. According to one manufacturer's data sheet "the pin must be held in a low state until the start of the execution of ISR. If the INT*n* pin is brought back to a logic high before the start of the execution of ISR there will be no interrupt." However, upon activation of the interrupt due to the low level, it must be brought back to high before the execution of RETI. Again, according to one manufacturer's data sheet, "If the INT*n* pin is left at a logic low after the RETI instruction of the ISR, another interrupt will be activated after one instruction is executed." Therefore, to ensure the activation of the hardware interrupt at the INT*n* pin, make sure that the duration of the low-level signal is around 4 machine cycles, but no more. This is due to the fact that the level-triggered interrupt is not latched. Thus the pin must be held in a low state until the start of the ISR execution.

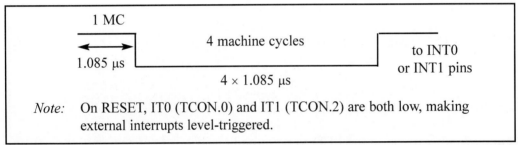

Figure 11-5. Minimum Duration of the Low Level-Triggered Interrupt (XTAL=11.0592 MHz)

Edge-triggered interrupts

As stated before, upon reset the 8051 makes INT0 and INT1 low-level triggered interrupts. To make them edge-triggered interrupts, we must program the bits of the TCON register. The TCON register holds, among other bits, the IT0 and IT1 flag bits that determine level- or edge-triggered mode of the hardware interrupts. IT0 and IT1 are bits D0 and D2 of the TCON register, respectively. They are also referred to as TCON.0 and TCON.2 since the TCON register is bit-addressable. Upon reset, TCON.0 (IT0) and TCON.2 (IT1) are both 0s, meaning that the external hardware interrupts of INT0 and INT1 pins are low-level triggered. By making the TCON.0 and TCON.2 bits high with instructions such as "SETB TCON.0" and "SETB TCON.2", the external hardware interrupts of INT0 and INT1 become edge-triggered. For example, the instruction "SETB TCON.2" makes INT1 what is called an *edge-triggered interrupt*, in which, when a high-to-low signal is applied to pin P3.3, in this case, the controller will be interrupted and forced to jump to location 0013H in the vector table to service the ISR. That is assuming that the interrupt bit is enabled in the IE register.

```
        D7                                                                           D0
        ┌────────┬────────┬────────┬────────┬────────┬────────┬────────┬────────┐
        │  TF1   │  TR1   │  TF0   │  TR0   │  IE1   │  IT1   │  IE0   │  IT0   │
        └────────┴────────┴────────┴────────┴────────┴────────┴────────┴────────┘
```

TF1   TCON.7   Timer 1 overflow flag. Set by hardware when timer/counter 1 overflows. Cleared by hardware as the processor vectors to the interrupt service routine.

TR1   TCON.6   Timer 1 run control bit. Set/cleared by software to turn timer/counter 1 on/off.

TF0   TCON.5   Timer 0 overflow flag. Set by hardware when timer/counter 0 overflows. Cleared by hardware as the processor vectors to the service routine.

TR0   TCON.4   Timer 0 run control bit. Set/cleared by software to turn timer/counter 0 on/off.

IE1   TCON.3   External interrupt 1 edge flag. Set by CPU when the external interrupt edge (H-to-L transition) is detected. Cleared by CPU when the interrupt is processed. *Note:* This flag does not latch low-level triggered interrupts.

IT1   TCON.2   Interrupt 1 type control bit. Set/cleared by software to specify falling edge/low-level triggered external interrupt.

IE0   TCON.1   External interrupt 0 edge flag. Set by CPU when external interrupt (H-to-L transition) edge detected. Cleared by CPU when interrupt is processed. *Note:* This flag does not latch low-level triggered interrupts.

IT0   TCON.0   Interrupt 0 type control bit. Set/cleared by software to specify falling edge/low-level triggered external interrupt.

Figure 11-6. TCON (Timer/Counter) Register (Bit-addressable)

Look at Example 11-6. Notice that the only difference between this program and the program in Example 11-5 is in the first line of MAIN where the instruction "SETB TCON.2" makes INT1 an edge-triggered interrupt. When the falling edge of the signal is applied to pin INT1, the LED will be turned on momentarily. The LED's on-state duration depends on the time delay inside the ISR for INT1. To turn on the LED again, another high-to-low pulse must be applied to pin 3.3. This is the opposite of Example 11-5. In Example 11-5, due to the level-triggered nature of the interrupt, as long as INT1 is kept at a low level, the LED is kept in the on state. But in this example, to turn on the LED again, the INT1 pulse must be brought back high and then forced low to create a falling edge to activate the interrupt.

Example 11-6

Assuming that pin 3.3 (INT1) is connected to a pulse generator, write a program in which the falling edge of the pulse will send a high to P1.3, which is connected to an LED (or buzzer). In other words, the LED is turned on and off at the same rate as the pulses are applied to the INT1 pin. This is an edge-triggered version of Example 11-5.

Solution:

```
          ORG   0000H
          LJMP MAIN
;--ISR for hardware interrupt INT1 to turn on the LED
          ORG   0013H      ;INT1 ISR
          SETB P1.3        ;turn on the LED
          MOV  R3,#255
BACK:     DJNZ R3,HERE     ;keep the buzzer on for a while
          CLR  P1.3        ;turn off the buzzer
          RETI             ;return from Buzzer
;--- MAIN program for initialization
          ORG   30H
MAIN:     SETB TCON.2      ;make INT1 edge-trigger interrupt
          MOV  IE,#10000100B  ;enable External INT 1
HERE:     SJMP HERE        ;stay here until get interrupted
          END
```

Sampling the edge-triggered interrupt

Before ending this section, we need to answer the question of how often the edge-triggered interrupt is sampled. In edge-triggered interrupts, the external source must be held high for at least one machine cycle, and then held low for at least one machine cycle to ensure that the transition is seen by the microcontroller.

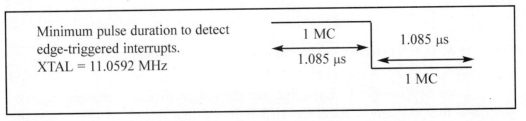

Minimum pulse duration to detect edge-triggered interrupts. XTAL = 11.0592 MHz

The falling edge is latched by the 8051 and is held by the TCON register. The TCON.1 and TCON.3 bits hold the latched falling edge of pins INT0 and INT1, respectively. TCON.1 and TCON.3 are also called IE0 and IE1, respectively as shown in Figure 11-6. They function as interrupt-in-service flags. When an interrupt-in-service flag is raised, it indicates to the external world that the interrupt is being serviced now and on this INT*n* pin no new interrupt will be responded to until this service is finished. This is just like the busy signal you get if calling a telephone number which is in use. Regarding the IT0 and IT1 bits in the TCON register, the following two points must be emphasized.

1. The first point is that when the ISRs are finished (that is, upon execution of instruction RETI), these bits (TCON.1 and TCON.3) are cleared, indicating that the interrupt is finished and the 8051 is ready to respond to another interrupt on that pin. For another interrupt to be recognized, the pin must go back to a logic high state and be brought back low to be considered an edge-triggered interrupt.

2. The second point is that during the time that the interrupt service routine is being executed, the INT*n* pin is ignored, no matter how many times it makes a high-to-low transition. In reality it is one of the functions of the RETI instruction to clear the corresponding bit in the TCON register (TCON.1 or TCON.3). This informs us that the service routine is no longer in progress and has finished being serviced. For this reason, TCON.1 and TCON.3 in the TCON register are called interrupt-in-service flags. The interrupt-in-service flag goes high whenever a falling edge is detected at the INT pin and stays high during the entire execution of the ISR. It is only cleared by RETI, the last instruction of the ISR. Because of this, there is no need for an instruction such as "CLR TCON.1" (or "CLR TCON.3" for INT1) before the RETI in the ISR associated with the hardware interrupt INT0. As we will see in the next section, this is not the case for the serial interrupt.

Example 11-7

What is the difference between the RET and RETI instructions? Explain why we cannot use RET instead of RETI as the last instruction of an ISR.

Solution:

Both perform the same actions of popping off the top two bytes of the stack into the program counter, and making the 8051 return to where it left off. However, RETI also performs an additional task of clearing the interrupt-in-service flag, indicating that the servicing of the interrupt is over and the 8051 now can accept a new interrupt on that pin. If you use RET instead of RETI as the last instruction of the interrupt service routine, you simply block any new interrupt on that pin after the first interrupt, since the pin status would indicate that the interrupt is still being serviced. In the cases of TF0, TF1, TCON.1, and TCON.3, they are cleared due to the execution of RETI.

More about the TCON register

Next we look at the TCON register more closely to understand its role in handling interrupts. Figure 11-6 shows the bits of the TCON register.

IT0 and IT1

TCON.0 and TCON.2 are referred to as IT0 and IT1, respectively. These two bits set the low-level or edge-triggered modes of the external hardware interrupts of the INT0 and INT1 pins. They are both 0 upon reset, which makes them low-level triggered. The programmer can make any one of them high to make the external hardware interrupt edge-triggered. In a given system based on the 8051, once they are set to 0 or 1 they will not be altered again since the designer has fixed the interrupt either as edge- or level-triggered.

IE0 and IE1

TCON.1 and TCON.3 are referred to as IE0 and IE1, respectively. These bits are used by the 8051 to keep track of the edge-triggered interrupt only. In other words if the IT0 and IT1 are 0, meaning that the hardware interrupts are low-level triggered, IE0 and IE1 are not used at all. The IE0 and IE1 bits are used by the 8051 only to latch the high-to-low edge transition on the INT0 and INT1 pins. Upon the edge transition pulse on the INT0 (or INT1) pin, the 8051 marks (sets high) the IE*x* bit in the TCON register, jumps to the vector in the interrupt vector table, and starts to execute the ISR. While it is executing the ISR, no H-to-L pulse transition on the INT0 (or INT1) is recognized, thereby preventing any interrupt inside the interrupt. Only the execution of the RETI instruction at the end of the ISR will clear the IE*x* bit, indicating that a new H-to-L pulse will activate the interrupt again. From this discussion we can see that the IE0 and IE1 bits are used internally by the 8051 to indicate whether or not an interrupt is in use. In other words, the programmer is not concerned with these bits since it is solely for internal use.

TR0 and TR1

These are the D4 (TCON.4) and D6 (TCON.6) bits of the TCON register. We were introduced to these bits in Chapter 9. They are used to start or stop timers 0 and 1, respectively. Although we have used syntax such as "SETB TRx" and "CLR Trx", we could have used instructions such as "SETB TCON.4" and "CLR TCON.4" since TCON is a bit-addressable register.

TF0 and TF1

These are the D5 (TCON.5) and D7 (TCON.7) bits of the TCON register. We were introduced to these bits in Chapter 9. They are used by timers 0 and 1, respectively, to indicate if the timer has rolled over. Although we have used the syntax "JNB TFx,target" and "CLR Trx", we could have used instructions such as "JNB TCON.5,target" and "CLR TCON.5" since TCON is bit-addressable.

Review Questions

1. True or false. There is a single interrupt in the interrupt vector table assigned to both external hardware interrupts IT0 and IT1.
2. What address in the interrupt vector table is assigned to INT0 and INT1? How about the pin numbers on port 3?
3. Which bit of IE belongs to the external hardware interrupts? Show how both are enabled.
4. Assume that the IE bit for the external hardware interrupt EX1 is enabled and is active low. Explain how this interrupt works when it is activated.
5. True or false. Upon reset, the external hardware interrupt is low-level triggered.
6. In Question 5, how do we make sure that a single interrupt is not recognized as multiple interrupts?
7. True or false. The last two instructions of the ISR for INT0 are:
   ```
   CLR   TCON.1
   RETI
   ```
8. Explain the role that each of the two bits TCON.0 and TCON.2 play in the execution of external interrupt 0.

SECTION 11.4: PROGRAMMING THE SERIAL COMMUNICATION INTERRUPT

In Chapter 10 we studied the serial communication of the 8051. All examples in that chapter used the polling method. In this section we explore interrupt-based serial communication which allows the 8051 to do many things, in addition to sending and receiving data from the serial communication port.

RI and TI flags and interrupts

As you recall from Chapter 10, TI (transfer interrupt) is raised when the last bit of the framed data, the stop bit, is transferred, indicating that the SBUF register is ready to transfer the next byte. In the case of RI (received interrupt), it is raised when the entire frame of data, including the stop bit, is received. In other words, when the SBUF register has a byte, RI is raised to indicate that the received byte needs to be picked up before it is lost (overrun) by new incoming serial data. As far as serial communication is concerned, all the above concepts apply equally when using either polling or an interrupt. The only difference is in how the serial communication needs are served. In the polling method, we wait for the flag (TI or RI) to be raised; while we wait we cannot do anything else. In the interrupt method, we are notified when the 8051 has received a byte, or is ready to send the next byte; we can do other things while the serial communication needs are served.

In the 8051 there is only one interrupt set aside for serial communication. This interrupt is used to both send and receive data. If the interrupt bit in the IE register (IE.4) is enabled, when RI or TI is raised the 8051 gets interrupted and jumps to memory address location 0023H to execute the ISR. In that ISR we must examine the TI and RI flags to see which one caused the interrupt and respond accordingly. See Example 11-8.

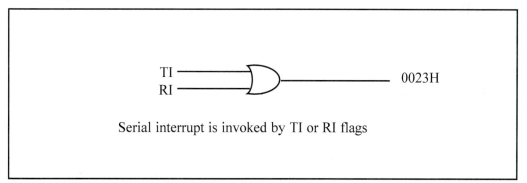

Serial interrupt is invoked by TI or RI flags

Figure 11-7. Single Interrupt for both TI and RI

Use of serial COM in the 8051

In the vast majority of applications, the serial interrupt is used mainly for receiving data and is never used for sending data serially. This is like getting a telephone call in which we need a ring to be notified. If we need to make a phone call there are other ways to remind ourselves and there is no need for ringing. However in receiving the phone call, we must respond immediately no matter what we are doing or we will miss the call. Similarly, we use the serial interrupt in receiving incoming data so that it is not lost. Look at Example 11-9.

Example 11-8

Write a program in which the 8051 reads data from P1 and writes it to P2 continuously while giving a copy of it to the serial COM port to be transferred serially. Assume that XTAL = 11.0592. Set the baud rate at 9600.

Solution:

```
            ORG  0
            LJMP MAIN
            ORG  23H
            LJMP SERIAL       ;jump to serial interrupt ISR
            ORG  30H
MAIN:       MOV  P1,#0FFH      ;make P1 an input port
            MOV  TMOD,#20H     ;timer 1, mode 2 (auto reload)
            MOV  TH1,#0FDH     ;9600 baud rate
            MOV  SCON,#50H     ;8-bit, 1 stop, ren enabled
            MOV  IE,#10010000B   ;enable serial interrupt
            SETB TR1                ;start timer 1
BACK:       MOV  A,P1          ;read data from port 1
            MOV  SBUF,A        ;give a copy to SBUF
            MOV  P2,A          ;send it to P2
            SJMP BACK          ;stay in loop indefinitely
;
;-------------------SERIAL PORT ISR
            ORG  100H
SERIAL:     JB   TI,TRANS      ;jump if TI is high
            MOV  A,SBUF        ;otherwise due to receive
            CLR  RI            ;clear RI since CPU does not
            RETI               ;return from ISR
TRANS:      CLR  TI            ;clear TI since CPU does not
            RETI               ;return from ISR
            END
```

In the above problem notice the role of TI and RI. The moment a byte is written into SBUF it is framed and transferred serially. As a result, when the last bit (stop bit) is transferred the TI is raised, and that causes the serial interrupt to be invoked since the corresponding bit in the IE register is high. In the serial ISR, we check for both TI and RI since both could have invoked interrupt. In other words, there is only one interrupt for both transmit and receive.

Example 11-9

Write a program in which the 8051 gets data from P1 and sends it to P2 continuously while incoming data from the serial port is sent to P0. Assume that XTAL = 11.0592. Set the baud rate at 9600.

Solution:

```
            ORG  0
            LJMP MAIN
            ORG  23H
            LJMP SERIAL      ;jump to serial ISR
            ORG  30H
MAIN:       MOV  P1,#0FFH     ;make P1 an input port
            MOV  TMOD,#20H    ;timer 1, mode 2 (auto reload)
            MOV  TH1,#0FDH    ;9600 baud rate
            MOV  SCON,#50H    ;8-bit,1 stop, REN enabled
            MOV  IE,#10010000B  ;enable serial interrupt
            SETB TR1            ;start timer 1
BACK:       MOV  A,P1           ;read data from port 1
            MOV  P2,A           ;send it to P2
            SJMP BACK           ;stay in loop indefinitely
;-------------------SERIAL PORT ISR
            ORG  100H
SERIAL:     JB   TI,TRANS       ;jump if TI is high
            MOV  A,SBUF         ;otherwise due to receive
            MOV  P0,A           ;send incoming data to P0
            CLR  RI             ;clear RI since CPU doesn't
            RETI                ;return from ISR
TRANS:      CLR  TI             ;clear TI since CPU doesn't
            RETI                ;return from ISR
            END
```

Clearing RI and TI before the RETI instruction

Notice in Example 11-9 that the last instruction before the RETI is the clearing of the RI or TI flags. This is necessary since there is only one interrupt for both receive and transmit, and the 8051 does not know who generated it; therefore, it is the job of the ISR to clear the flag. Contrast this with the external and timer interrupts where it is the job of the 8051 to clear the interrupt flags. By contrast, in serial communication the RI (or TI) must be cleared by the programmer using software instructions such as "CLR TI" and "CLR RI" in the ISR. See Example 11-10. Notice that the last two instructions of the ISR are clearing the flag, followed by RETI.

Example 11-10

Write a program using interrupts to do the following:

(a) Receive data serially and sent it to P0,

(b) Have P1 port read and transmitted serially, and a copy given to P2,

(c) Make timer 0 generate a square wave of 5 kHz frequency on P0.1.

Assume that XTAL = 11.0592. Set the baud rate at 4800.

Solution:

```
            ORG  0
            LJMP MAIN
            ORG  000BH                  ;ISR FOR TIMER 0
            CPL  P0.1                   ;TOGGLE P0.1
            RETI                        ;RETURN FROM ISR
            ORG  23H
            LJMP SERIAL            ;JUMP TO SERIAL INT. ISR
            ORG  30H
MAIN:       MOV  P1,#0FFH               ;MAKE P1 AN INPUT PORT
            MOV  TMOD,#22H  ;TIMER 0&1,MODE 2, AUTO RELOAD
            MOV  TH1,#0F6H        ;4800 BAUD RATE
            MOV  SCON,#50H        ;8-BIT,1 STOP,REN ENABLED
            MOV  TH0,#-92         ;FOR 5 KHZ WAVE.
            MOV  IE,#10010010B   ;ENABLE SERIAL,TIMER 0 INT.
            SETB TR1                  ;START TIMER 1
            SETB TR0                  ;START TIMER 0
BACK:       MOV  A,P1              ;READ DATA FROM PORT 1
            MOV  SBUF,A           ;GIVE A COPY TO SBUF
            MOV  P2,A             ;WRITE IT TO P2
            SJMP BACK             ;STAY IN LOOP INDEFINITELY
;------------------SERIAL PORT ISR
            ORG  100H
SERIAL:     JB   TI,TRANS         ;JUMP IF TI IS HIGH
            MOV  A,SBUF           ;OTHERWISE DUE TO RECEIVE
            MOV  P0,A             ;SEND SERIAL DATA TO P0
            CLR  RI           ;CLEAR RI SINCE CPU DOES NOT
            RETI                 ;RETURN FROM ISR
TRANS:      CLR  TI           ;CLEAR TI SINCE CPU DOES NOT
            RETI                 ;RETURN FROM ISR
            END
```

Before finishing this section notice the list of all interrupt flags given in Table 11-2. While the TCON register holds four of the interrupt flags, in the 8051 the SCON register has the RI and TI flags.

Table 11-2: Interrupt Flag Bits

Interrupt	Flag	SFR Register Bit
External 0	IE0	TCON.1
External 1	IE1	TCON.3
Timer 0	TF0	TCON.5
Timer 1	TF1	TCON.7
Serial port	TI	SCON.1
Timer 2	TF2	T2CON.7 (AT89C52)
Timer 2	EXF2	T2CON.6 (AT89C52)

Review Questions

1. True or false. There is a single interrupt in the interrupt vector table assigned to both the TI and RI interrupts.
2. What address in the interrupt vector table is assigned to the serial interrupt?
3. Which bit of the IE register belongs to the serial interrupt? Show how it is enabled.
4. Assume that the IE bit for the serial interrupt is enabled. Explain how this interrupt gets activated and also explain its actions upon activation.
5. True or false. Upon reset, the serial interrupt is active and ready to go.
6. True or false. The last two instructions of the ISR for the receive interrupt are:
   ```
   CLR  RI
   RETI
   ```
7. Answer Question 6 for the send interrupt.

SECTION 11.5: INTERRUPT PRIORITY IN THE 8051

The next topic that we must deal with is what happens if two interrupts are activated at the same time? Which of these two interrupts is responded to first? Interrupt priority is the main topic of discussion in this section.

Interrupt priority upon reset

When the 8051 is powered up, the priorities are assigned according to Table 11-3. From Table 11-3 we see, for example, that if external hardware interrupts 0 and 1 are activated at the same time, external interrupt 0

Table 11-3: 8051 Interrupt Priority Upon Reset

Highest to Lowest Priority	
External Interrupt 0	(INT0)
Timer Interrupt 0	(TF0)
External Interrupt 1	(INT1)
Timer Interrupt 1	(TF1)
Serial Communication	(RI + TI)

(INT0) is responded to first. Only after INT0 has been serviced is INT1 serviced, since INT1 has the lower priority. In reality, the priority scheme in the table is nothing but an internal polling sequence in which the 8051 polls the interrupts in the sequence listed in Table 11-3, and responds accordingly.

Example 11-11

Discuss what happens if interrupts INT0, TF0, and INT1 are activated at the same time. Assume priority levels were set by the power-up reset and that the external hardware interrupts are edge-triggered.

Solution:

If these three interrupts are activated at the same time, they are latched and kept internally. Then the 8051 checks all five interrupts according to the sequence listed in Table 11-3. If any is activated, it services it in sequence. Therefore, when the above three interrupts are activated, IE0 (external interrupt 0) is serviced first, then timer 0 (TF0), and finally IE1 (external interrupt 1).

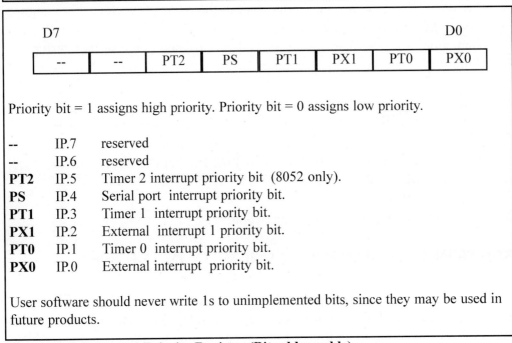

Priority bit = 1 assigns high priority. Priority bit = 0 assigns low priority.

--	IP.7	reserved
--	IP.6	reserved
PT2	IP.5	Timer 2 interrupt priority bit (8052 only).
PS	IP.4	Serial port interrupt priority bit.
PT1	IP.3	Timer 1 interrupt priority bit.
PX1	IP.2	External interrupt 1 priority bit.
PT0	IP.1	Timer 0 interrupt priority bit.
PX0	IP.0	External interrupt priority bit.

User software should never write 1s to unimplemented bits, since they may be used in future products.

Figure 11-8. Interrupt Priority Register (Bit-addressable)

Setting interrupt priority with the IP register

We can alter the sequence of Table 11-3 by assigning a higher priority to any one of the interrupts. This is done by programming a register called IP (interrupt priority). Figure 11-8 shows the bits of the IP register. Upon power-up reset, the IP register contains all 0s, making the priority sequence based on Table 11-3. To give a higher priority to any of the interrupts, we make the corresponding bit in the IP register high. Look at Example 11-12.

Another point that needs to be clarified is the interrupt priority when two or more interrupt bits in the IP register are set to high. In this case, while these interrupts have a higher priority than others, they are serviced according to the sequence of Table 11-3. See Example 11-13.

Example 11-12

(a) Program the IP register to assign the highest priority to INT1 (external interrupt 1), then (b) discuss what happens if INT0, INT1, and TF0 are activated at the same time. Assume that the interrupts are both edge-triggered.

Solution:

(a) `MOV IP,#00000100B ;IP.2=1 to assign INT1 higher priority`
The instruction "`SETB IP.2`" also will do the same thing as the above line since IP is bit-addressable.
(b) The instruction in Step (a) assigned a higher priority to INT1 than the others; therefore, when INT0, INT1, and TF0 interrupts are activated at the same time, the 8051 services INT1 first, then it services INT0, then TF0. This is due to the fact that INT1 has a higher priority than the other two because of the instruction in Step (a). The instruction in Step (a) makes both the INT0 and TF0 bits in the IP register 0. As a result, the sequence in Table 11-3 is followed which gives a higher priority to INT0 over TF0.

Example 11-13

Assume that after reset, the interrupt priority is set by the instruction "`MOV IP,#00001100B`". Discuss the sequence in which the interrupts are serviced.

Solution:

The instruction "`MOV IP,#00001100B`" (B is for binary) sets the external interrupt 1 (INT1) and timer 1 (TF1) to a higher priority level compared with the rest of the interrupts. However, since they are polled according to Table 11-3, they will have the following priority.

Highest Priority	External Interrupt 1	(INT1)
	Timer Interrupt 1	(TF1)
	External Interrupt 0	(INT0)
	Timer Interrupt 0	(TF0)
Lowest Priority	Serial Communication	(RI + TI)

Interrupt inside an interrupt

What happens if the 8051 is executing an ISR belonging to an interrupt and another interrupt is activated? In such cases, a high-priority interrupt can interrupt a low-priority interrupt. This is an interrupt inside an interrupt. In the 8051 a low-priority interrupt can be interrupted by a higher-priority interrupt but not by another low-priority interrupt. Although all the interrupts are latched and kept internally, no low-priority interrupt can get the immediate attention of the CPU until the 8051 has finished servicing the high-priority interrupts.

Triggering the interrupt by software

There are times when we need to test an ISR by way of simulation. This can be done with simple instructions to set the interrupts high and thereby cause the 8051 to jump to the interrupt vector table. For example, if the IE bit for timer 1 is set, an instruction such as "`SETB TF1`" will interrupt the 8051 in whatever it is doing and will force it to jump to the interrupt vector table. In other words, we do not need to wait for timer 1 to roll over to have an interrupt. We can cause an interrupt with an instruction which raises the interrupt flag.

Review Questions

1. True or false. Upon reset, all interrupts have the same priority.
2. What register keeps track of interrupt priority in the 8051? Is it a bit-addressable register?
3. Which bit of IP belongs to the serial interrupt priority? Show how to assign it the highest priority.
4. Assume that the IP register contains all 0s. Explain what happens if both INT0 and INT1 are activated at the same time.
5. Explain what happens if a higher-priority interrupt is activated while the 8051 is serving a lower-priority interrupt (that is, executing a lower-priority ISR).

SUMMARY

An interrupt is an external or internal event that interrupts the microcontroller to inform it that a device needs its service. Every interrupt has a program associated with it called the ISR, interrupt service routine. The 8051 has 6 interrupts, 5 of which are user-accessible. The interrupts are for reset: two for the timers, two for external hardware interrupts, and a serial communication interrupt.

The 8051 can be programmed to enable or disable an interrupt, and the interrupt priority can be altered. This chapter showed how to program interrupts in the 8051.

PROBLEMS

SECTION 11.1: 8051 INTERRUPTS

1. Which technique, interrupt or polling, avoids tying down the microcontroller?
2. Including reset, how many interrupts do we have in the 8051?
3. In the 8051 what memory area is assigned to the interrupt vector table?
4. True or false. The 8051 programmer cannot change the memory space assigned to the interrupt vector table.
5. What memory address in the interrupt vector table is assigned to INT0?
6. What memory address in the interrupt vector table is assigned to INT1?
7. What memory address in the interrupt vector table is assigned to timer 0?
8. What memory address in the interrupt vector table is assigned to timer 1?

9. What memory address in the interrupt vector table is assigned to the serial COM interrupt?
10. Why do we put an LJMP instruction at address 0?
11. What are the contents of the IE register upon reset, and what do these values mean?
12. Show the instruction to enable the EX1 and timer 1 interrupts.
13. Show the instruction to enable every interrupt of the 8051.
14. Which pin of the 8051 is assigned to the external hardware interrupts INT0 and INT1?
15. How many bytes of address space in the interrupt vector table are assigned to the INT0 and INT1 interrupts?
16. How many bytes of address space in the interrupt vector table are assigned to the timer 0 and timer 1 interrupts?
17. To put the entire interrupt service routine in the interrupt vector table, it must be no more than _____ bytes in size.
18. True or false. The IE register is not a bit-addressable register.
19. With a single instruction, show how to disable all the interrupts.
20. With a single instruction, show how to disable the EX1 interrupt.
21. True or false. Upon reset, all interrupts are enabled by the 8051.
22. In the 8051, how many bytes of ROM space are assigned to the reset interrupt, and why?

SECTION 11.2: PROGRAMMING TIMER INTERRUPTS

23. True or false. For both timer 0 and timer 1, there is an interrupt assigned to it in the interrupt vector table.
24. What address in the interrupt vector table is assigned to timer 1?
25. Which bit of IE belongs to the timer 0 interrupt? Show how it is enabled.
26. Which bit of IE belongs to the timer 1 interrupt? Show how it is enabled.
27. Assume that timer 0 is programmed in mode 2, TH1 = F0H, and the IE bit for timer 0 is enabled. Explain how the interrupt for the timer works.
28. True or false. The last two instructions of the ISR for timer 1 are:
```
        CLR   TF1
        RETI
```
29. Assume that timer 1 is programmed for mode 1, TH0 = FFH, TL1 = F8H, and the IE bit for timer 1 is enabled. Explain how the interrupt is activated.
30. If timer 1 is programmed for interrupts in mode 2, explain when the interrupt is activated.
31. Write a program to create a square wave of T = 160 ms on pin P2.2 while at the same time the 8051 is sending out 55H and AAH to P1 continuously.
32. Write a program in which every 2 seconds, the LED connected to P2.7 is turned on and off four times, while at the same time the 8051 is getting data from P1 and sending it to P0 continuously. Make sure the on and off states are 50 ms in duration.

33. True or false. A single interrupt is assigned to each of the external hardware interrupts EX0 and EX1.
34. What address in the interrupt vector table is assigned to INT0 and INT1? How about the pin numbers on port 3?
35. Which bit of IE belongs to the EX0 interrupt? Show how it is enabled.
36. Which bit of IE belongs to the EX1 interrupt? Show how it is enabled.
37. Show how to enable both external hardware interrupts.
38. Assume that the IE bit for external hardware interrupt EX0 is enabled and is low-level triggered. Explain how this interrupt works when it is activated. How can we make sure that a single interrupt is not interpreted as multiple interrupts?
39. True or false. Upon reset, the external hardware interrupt is edge-triggered.
40. In Question 39, how do we make sure that a single interrupt is not recognized as multiple interrupts?
41. Which bits of TCON belong to EX0?
42. Which bits of TCON belong to EX1?
43. True or false. The last two instructions of the ISR for INT1 are:
```
CLR   TCON.3
RETI
```
44. Explain the role of TCON.0 and TCON.2 in the execution of external interrupt 0.
45. Explain the role of TCON.1 and TCON.3 in the execution of external interrupt 1.
46. Assume that the IE bit for external hardware interrupt EX1 is enabled and is edge-triggered. Explain how this interrupt works when it is activated. How can we make sure that a single interrupt is not interpreted as multiple interrupts?
47. Write a program using interrupts to get data from P1 and send it to P2 while timer 0 is generating a square wave of 3 kHz.
48. Write a program using interrupts to get data from P1 and send it to P2 while timer 1 is turning on and off the LED connected to P0.4 every second.
49. Explain the difference between the low-level and edge-triggered interrupts.
50. How do we make the hardware interrupt edge-triggered?
51. Which interrupts are latched, low-level, or edge-triggered?
52. Which register keeps the latched interrupt for INT0 and INT1?

SECTION 11.4: PROGRAMMING THE SERIAL COMMUNICATION INTERRUPT

53. True or false. There are two interrupts assigned to interrupts TI and RI.
54. What address in the interrupt vector table is assigned to the serial interrupt? How many bytes are assigned to it?
55. Which bit of the IE register belongs to the serial interrupt? Show how it is enabled.
56. Assume that the IE bit for the serial interrupt is enabled. Explain how this interrupt gets activated and also explain its working upon activation.
57. True or false. Upon reset, the serial interrupt is blocked.
58. True or false. The last two instructions of the ISR for the receive interrupt are:
```
CLR   TI
RETI
```

59. Answer Question 58 for the receive interrupt.
60. Assuming that the interrupt bit in the IE register is enabled, when is TI raised, what happens subsequently?
61. Assuming that the interrupt bit in the IE register is enabled, when RI is raised, what happens subsequently?
62. Write a program using interrupts to get data serially and send it to P2 while at the same time, timer 0 is generating a square wave of 5 kHz.
63. Write a program using interrupts to get data serially and send it to P2 while timer 0 is turning the LED connected to P1.6 on and off every second.

SECTION 11.5: INTERRUPT PRIORITY IN THE 8051

64. True or false. Upon reset, EX1 has the highest priority.
65. What register keeps track of interrupt priority in the 8051? Explain its role.
66. Which bit of IP belongs to the EX2 interrupt priority? Show how to assign it the highest priority.
67. Which bit of IP belongs to the timer 1 interrupt priority? Show how to assign it the highest priority.
68. Which bit of IP belongs to the EX1 interrupt priority? Show how to assign it the highest priority.
69. Assume that the IP register has all 0s. Explain what happens if both INT0 and INT1 are activated at the same time.
70. Assume that the IP register has all 0s. Explain what happens if both TF0 and TF1 are activated at the same time.
71. If both TF0 and TF1 in the IP are set to high, what happens if both are activated at the same time?
72. If both INT0 and INT1 in the IP are set to high, what happens if both are activated at the same time?
73. Explain what happens if a low-priority interrupt is activated while the 8051 is serving a higher-priority interrupt.

ANSWERS TO REVIEW QUESTIONS

SECTION 11.1: 8051 INTERRUPTS

1. Interrupts
2. 5
3. Address locations 0000 to 25H. No. They are set when the processor is designed.
4. All 0s means that all interrupts are masked, and as a result no interrupts will be responded to by the 8051.
5. `MOV IE,#10000011B`
6. P3.3, which is pin 13 on the 40-pin DIP package
7. 0013H for INT1 and 001BH for timer 1

SECTION 11.2: PROGRAMMING TIMER INTERRUPTS

1. False. There is an interrupt for each of timer 0 and timer 1.   2. 000BH
3. Bits D1 and D3 and "MOV IE, #10001010B" will enable both of the timer interrupts.
4. After timer 1 is started with instruction "SETB TR1", the timer will count up from F5H to FFH on its own while the 8051 is executing other tasks. Upon rolling over from FFH to 00, the TF1 flag is raised which will interrupt the 8051 in whatever it is doing and force it to jump to memory location 001BH to execute the ISR belonging to this interrupt.
5. False. There is no need for "CLR TF0" since the RETI instruction does that for us.

SECTION 11.3: PROGRAMMING EXTERNAL HARDWARE INTERRUPTS

1. False. There is an interrupt for each of the external hardware interrupts of INT0 and INT1.
2. 0003H and 0013H. The pins numbered 12 (P3.2) and 13 (P3.3) on the DIP package.
3. Bits D0 and D2 and "MOV IE, #10000101B" will enable both of the external hardware interrupts.
4. Upon application of a low pulse (4-machine cycles wide) to pin P3.3, the 8051 is interrupted in whatever it is doing and jumps to ROM location 0013H to execute the ISR.
5. True
6. Make sure that the low pulse applied to pin INT1 is no wider than 4 machine cycles . Or, make sure that the INT1 pin is brought back to high by the time the 8051 executes the RETI instruction in the ISR.
7. False. There is no need for the "CLR TCON.0" since the RETI instruction does that for us.
8. TCON.0 is set to high to make INT0 an edge-triggered interrupt. If INT0 is edge-triggerred (that is TCON.0 is set), whenever a high-to-low pulse is applied to the INT0 pin it is captured (latched) and kept by the TCON.2 bit by making TCON.2 high. While the the ISR for INT0 is being serviced, TCON.2 stays high no matter how many times a H-to-L pulse is applied to pin INT0. Upon the execution of the last instruction of the ISR, which is RETI, the TCON.2 bit is cleared indicating that the INT0 pin can respond to another interrupt.

SECTION 11.4: PROGRAMMING THE SERIAL COMMUNICATION INTERRUPT

1. True. There is only one interrupt for both the transfer and receive.   2. 23H
3. Bit D4 (IE.4) and "MOV IE, #10010000B" will enable the serial interrupt.
4. The RI (received interrupt) flag is raised when the entire frame of data including the stop bit is received. As a result the received byte is delivered to the SBUF register and the 8051 jumps to memory location 0023H to execute the ISR belonging to this interrupt. In the serial COM interrupt service routine, we must save the SBUF contents before it is lost by the incoming data.
5. False
6. True. We must do it since the RETI instruction will not do it for the serial interrupt.
7.
```
CLR    TI
RETI
```

SECTION 11.5: INTERRUPT PRIORITY IN THE 8051

1. False. They are assigned priority according to Table 11-3.
2. IP (interrupt priority) register. Yes, it is bit-addressable.
3. Bit D4 (IP.4) and the instruction "MOV IP, #00010000B" will do it.
4. If both are activated at the same time, INT0 is serviced first since it has a higher priority. After INT0 is serviced, INT1 is serviced. That is assuming that the external interrupts are edge-triggered and H-to-L transitions are latched. In the case of low-level triggered interrupts, if both are activated at the same time, the INT0 is serviced first; and then after the 8051 has finished servicing the INT0, when it scans the INT0 and INT1 pins again, if the INT1 pin is still high, it will be serviced.
5. Then we have an interrupt inside an interrupt meaning that the lower-priority interrupt is put on hold and the higher one is serviced. After finishing servicing this higher-priority interrupt, the 8051 resumes servicing the lower-priority ISR.

CHAPTER 12

REAL-WORLD INTERFACING I: LCD, ADC, AND SENSORS

OBJECTIVES

Upon completion of this chapter, you will be able to:

>> List reasons that LCDs are gaining wide-spread use, replacing LEDs
>> Describe the functions of the pins of a typical LCD
>> List instruction command codes for programming an LCD
>> Interface an LCD to the 8051
>> Program an LCD by sending data or commands to it from the 8051
>> Interface ADC (analog-to-digital converter) chips to the 8051
>> Interface temperature sensors to the 8051
>> Explain the process of data acquisition using ADC chips
>> Describe factors to consider in selecting an ADC chip
>> Describe the function of the pins of a typical ADC chip
>> Explain the function of precision IC temperature sensors
>> Describe signal conditioning and its role in data acquisition

This chapter explores some real-world applications of the 8051. We explain how to interface the 8051 to devices such as an LCD, ADC, and sensors. In Section 12.1, we show LCD interfacing with the 8051. In Section 12.2, we describe analog-to-digital converter (ADC) connection with sensors and the 8051.

SECTION 12.1: INTERFACING AN LCD TO THE 8051

This section describes the operation modes of LCDs, then describes how to program and interface an LCD to an 8051.

LCD operation

In recent years the LCD is finding widespread use replacing LEDs (seven-segment LEDs or other multisegment LEDs). This is due to the following reasons:
1. The declining prices of LCDs.
2. The ability to display numbers, characters, and graphics. This is in contrast to LEDs, which are limited to numbers and a few characters.
3. Incorporation of a refreshing controller into the LCD, thereby relieving the CPU of the task of refreshing the LCD. In contrast, the LED must be refreshed by the CPU (or in some other way) to keep displaying the data.
4. Ease of programming for characters and graphics.

LCD pin descriptions

The LCD discussed in this section has 14 pins. The function of each pin is given in Table 12-1. Figure 12-1 shows the pin positions for various LCDs.

V_{CC}, V_{SS}, and V_{EE}

While V_{CC} and V_{SS} provide +5V and ground, respectively, V_{EE} is used for controlling LCD contrast.

RS, register select

There are two very important registers inside the LCD. The RS pin is used for their selection as follows. If RS = 0, the instruction command code register is selected, allowing the user to send a command such as clear display, cursor at home, etc. If RS = 1 the data register is selected, allowing the user to send data to be displayed on the LCD.

R/W, read/write

R/W input allows the user to write information to the LCD or read information from it. R/W = 1 when reading; R/W = 0 when writing.

E, enable

The enable pin is used by the LCD to latch information presented to its data pins.

Table 12-1. Pin Descriptions for LCD

Pin	Symbol	I/O	Description
1	V_{SS}	--	Ground
2	V_{CC}	--	+5V power supply
3	V_{EE}	--	Power supply to control contrast
4	RS	I	RS=0 to select command register, RS=1 to select data register
5	R/W	I	R/W=0 for write, R/W=1 for read
6	E	I/O	Enable
7	DB0	I/O	The 8-bit data bus
8	DB1	I/O	The 8-bit data bus
9	DB2	I/O	The 8-bit data bus
10	DB3	I/O	The 8-bit data bus
11	DB4	I/O	The 8-bit data bus
12	DB5	I/O	The 8-bit data bus
13	DB6	I/O	The 8-bit data bus
14	DB7	I/O	The 8-bit data bus

When data is supplied to data pins, a high-to-low pulse must be applied to this pin in order for the LCD to latch in the data present at the data pins. This pulse must be a minimum of 450 ns wide.

D0 - D7

The 8-bit data pins, D0 - D7, are used to send information to the LCD or read the contents of the LCD's internal registers.

To display letters and numbers, we send ASCII codes for the letters A - Z, a - z, and numbers 0 - 9 to these pins while making RS = 1.

There are also instruction command codes that can be sent to the LCD to clear the display or force the cursor to the home position or blink the cursor. Table 12-2 lists the instruction command codes.

We also use RS = 0 to check the busy flag bit to see if the LCD is ready to receive information. The busy flag is D7 and can be read when R/W = 1 and RS = 0, as follows: if R/W = 1, RS = 0. When D7 = 1 (busy flag = 1), the LCD is busy taking care of internal operations and will not accept any new information. When D7 = 0, the LCD is ready to receive new information. *Note:* It is recommended to check the busy flag before writing any data to the LCD.

Table 12-2: LCD Command Codes

Code (Hex)	Command to LCD Instruction Register
1	Clear display screen
2	Return home
4	Decrement cursor (shift cursor to left)
6	Increment cursor (shift cursor to right)
5	Shift display right
7	Shift display left
8	Display off, cursor off
A	Display off, cursor on
C	Display on, cursor off
E	Display on, cursor blinking
F	Display on, cursor blinking
10	Shift cursor position to left
14	Shift cursor position to right
18	Shift the entire display to the left
1C	Shift the entire display to the right
80	Force cursor to beginning of 1st line
C0	Force cursor to beginning of 2nd line
38	2 lines and 5x7 matrix

Note: This table is extracted from Table 12-4.

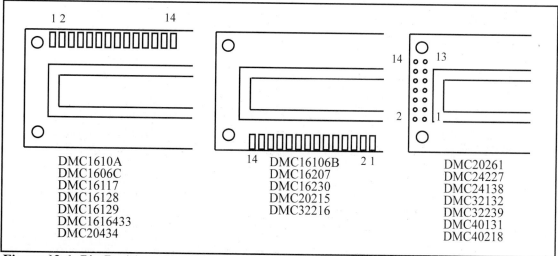

Figure 12-1. Pin Positions for Various LCDs from Optrex

Sending commands and data to LCDs with a time delay

To send any of the commands from Table 12-2 to the LCD, make pin RS = 0. For data, make RS = 1. Then send a high-to-low pulse to the E pin to enable the internal latch of the LCD. This is shown in the code below. See Figure 12-2.

```
;calls a time delay before sending next data/command
; P1.0-P1.7 are connected to LCD data pins D0-D7
; P2.0 is connected to RS pin of LCD
; P2.1 is connected to R/W pin of LCD
; P2.2 is connected to E pin of LCD
            ORG
            MOV     A,#38H   ;init. LCD 2 lines,5x7 matrix
            ACALL   COMNWRT      ;call command subroutine
            ACALL   DELAY        ;give LCD some time
            MOV     A,#0EH       ;display on, cursor on
            ACALL   COMNWRT      ;call command subroutine
            ACALL   DELAY        ;give LCD some time
            MOV     A,#01        ;clear LCD
            ACALL   COMNWRT      ;call command subroutine
            ACALL   DELAY        ;give LCD some time
            MOV     A,#06H       ;shift cursor right
            ACALL   COMNWRT      ;call command subroutine
            ACALL   DELAY        ;give LCD some time
            MOV     A,#84H       ;cursor at line 1,pos. 4
            ACALL   COMNWRT      ;call command subroutine
            ACALL   DELAY        ;give LCD some time
            MOV     A,#'N'       ;display letter N
            ACALL   DATAWRT      ;call display subroutine
            ACALL   DELAY        ;give LCD some time
            MOV     A,#'O'       ;display letter O
            ACALL   DATAWRT      ;call display subroutine
AGAIN:      SJMP    AGAIN        ;stay here
COMNWRT:                         ;send command to LCD
            MOV     P1,A         ;copy reg A to port1
            CLR     P2.0         ;RS=0 for command
            CLR     P2.1         ;R/W=0 for write
            SETB    P2.2         ;E=1 for high pulse
            CLR     P2.2         ;E=0 for H-to-L pulse
            RET
DATAWRT:                         ;write data to LCD
            MOV     P1,A         ;copy reg A to port1
            SETB    P2.0         ;RS=1 for data
            CLR     P2.1         ;R/W=0 for write
            SETB    P2.2         ;E=1 for high pulse
            CLR     P2.2         ;E=0 for H-to-L pulse
            RET
DELAY:      MOV     R3,#50    ;50 or higher for fast CPUs
HERE2:      MOV     R4,#255   ;R4=255
HERE:       DJNZ    R4,HERE   ;stay until R4 becomes 0
            DJNZ    R3,HERE2
            RET
            END
```

Sending code or data to the LCD with checking busy flag

The above code showed how to send commands to the LCD without checking the busy flag. Notice that we must put a long delay in between issuing data or commands to the LCD. However, a much better way is to monitor the busy flag before issuing a command or data to the LCD. This is shown below.

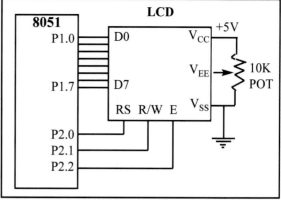

Figure 12-2. LCD Connection

```
;Check busy flag before sending data, command to LCD
;P1=data pin
;P2.0 connected to RS pin
;P2.1 connected to R/W pin
;P2.2 connected to E pin
            ORG
            MOV     A,#38H  ;init. LCD 2 lines,5x7 matrix
            ACALL   COMMAND         ;issue command
            MOV     A,#0EH          ;LCD on, cursor on
            ACALL   COMMAND         ;issue command
            MOV     A,#01H          ;clear LCD command
            ACALL   COMMAND         ;issue command
            MOV     A,#06H          ;shift cursor right
            ACALL   COMMAND         ;issue command
            MOV     A,#86H          ;cursor: line 1, pos. 6
            ACALL   COMMAND         ;command subroutine
            MOV     A,#'N'          ;display letter N
            ACALL   DATA_DISPLAY
            MOV     A,#'O'          ;display letter O
            ACALL   DATA_DISPLAY
HERE:       SJMP    HERE            ;STAY HERE
COMMAND:    ACALL   READY           ;is LCD ready?
            MOV     P1,A            ;issue command code
            CLR     P2.0            ;RS=0 for command
            CLR     P2.1            ;R/W=0 to write to LCD
            SETB    P2.2            ;E=1 for H-to-L pulse
            CLR     P2.2            ;E=0 ,latch in
            RET
DATA_DISPLAY:
            ACALL   READY           ;is LCD ready?
            MOV     P1,A            ;issue data
            SETB    P2.0            ;RS=1 for data
            CLR     P2.1            ;R/W=0 to write to LCD
            SETB    P2.2            ;E=1 for H-to-L pulse
            CLR     P2.2            ;E=0, latch in
            RET
```

```
READY:
      SETB       P1.7                ;make P1.7 input port
      CLR        P2.0                ;RS=0 access command reg
      SETB       P2.1                ;R/W=1 read command reg
;read command reg and check busy flag
BACK:CLR         P2.2                ;E=1 for H-to-L pulse
      SETB       P2.2                ;E=0 H-to-L pulse
      JB         P1.7,BACK           ;stay until busy flag=0
      RET
      END
```

Notice in the above program that the busy flag is D7 of the command register. To read the command register we make R/W = 1, RS = 0, and a H-to-L pulse for the E pin will provide us the command register. After reading the command register, if bit D7 (the busy flag) is high, the LCD is busy and no information (command or data) should be issued to it. Only when D7 = 0 can we send data or commands to the LCD. Notice in this method that there are no time delays used since we are checking the busy flag before issuing commands or data to the LCD.

LCD data sheet

In the LCD, one can put data at any location. The following shows address locations and how they are accessed.

RS	R/W	DB7	DB6	DB5	DB4	DB3	DB2	DB1	DB0
0	0	1	A	A	A	A	A	A	A

where AAAAAAA = 0000000 to 0100111 for line 1 and AAAAAAA = 1000000 to 1100111 for line 2. See Table 12-3.

Table 12-3: LCD Addressing

	DB7	DB6	DB5	DB4	DB3	DB2	DB1	DB0
Line 1 (min)	1	0	0	0	0	0	0	0
Line 1 (max)	1	0	1	0	0	1	1	1
Line 2 (min)	1	1	0	0	0	0	0	0
Line 2 (max)	1	1	1	0	0	1	1	1

The upper address range can go as high as 0100111 for the 40-character-wide LCD while for the 20-character-wide LCD it goes up to 010011 (19 decimal = 10011 binary). Notice that the upper range 0100111 (binary) = 39 decimal which corresponds to locations 0 to 39 for the LCDs of 40x2 size.

From the above discussion we can get the addresses of cursor positions for various sizes of LCDs. See Figure 12-3. Note that all the addresses are in hex. Figure 12-4 gives a diagram of LCD timing. Table 12-4 provides a detailed list of LCD commands and instructions. Table 12-2 is extracted from this table.

16 x 2 LCD	80	81	82	83	84	85	86 through 8F
	C0	C1	C2	C3	C4	C5	C6 through CF
20 x 1 LCD	80	81	82	83	through 93		
20 x 2 LCD	80	81	82	83	through 93		
	C0	C1	C2	C3	through D3		
20 x 4 LCD	80	81	82	83	through 93		
	C0	C1	C2	C3	through D3		
	94	95	96	97	through A7		
	D4	D5	D6	D7	through E7		
40 x 2 LCD	80	81	82	83	through A7		
	C0	C1	C2	C3	through E7		

Note: All data is in hex.

Figure 12-3 Cursor Addresses for Some LCDs

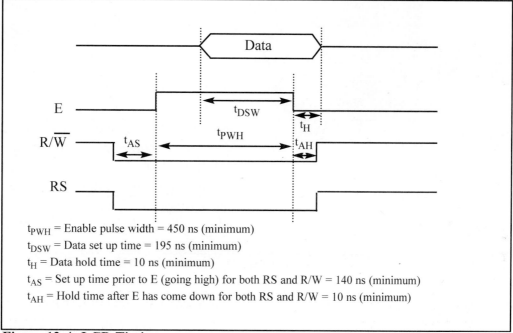

t_{PWH} = Enable pulse width = 450 ns (minimum)

t_{DSW} = Data set up time = 195 ns (minimum)

t_H = Data hold time = 10 ns (minimum)

t_{AS} = Set up time prior to E (going high) for both RS and R/W = 140 ns (minimum)

t_{AH} = Hold time after E has come down for both RS and R/W = 10 ns (minimum)

Figure 12-4: LCD Timing

Review Questions

1. The RS pin is an _____ (input, output) pin for the LCD.
2. The E pin is an _____ (input, output) pin for the LCD.
3. The E pin requires an _____ (H-to-L, L-to-H) pulse to latch in information at the data pins of the LCD.
4. For the LCD to recognize information at the data pins as data, RS must be set to _____ (high, low).
5. Give the command codes for line 1, first character, and line 2, first character.

Table 12-4: List of LCD Instructions

Instruction	RS	R/W	DB7	DB6	DB5	DB4	DB3	DB2	DB1	DB0	Description	Execution Time (Max)
Clear Display	0	0	0	0	0	0	0	0	0	1	Clears entire display and sets DD RAM address 0 in address counter	1.64 ms
Return Home	0	0	0	0	0	0	0	0	1	–	Sets DD RAM address 0 as address counter. Also returns display being shifted to original position. DD RAM contents remain unchanged.	1.64 ms
Entry Mode Set	0	0	0	0	0	0	0	1	1/D	S	Sets cursor move direction and specifies shift of display. These operations are performed during data write and read.	40 μs
Display On/ Off Control	0	0	0	0	0	0	1	D	C	B	Sets On/Off of entire display (D), cursor On/Off (C), and blink of cursor position character (B).	40 μs
Cursor or Display Shift	0	0	0	0	0	1	S/C	R/L	–	–	Moves cursor and shifts display without changing DD RAM contents.	40 μs
Function Set	0	0	0	0	1	DL	N	F	–	–	Sets interface data length (DL), number of display lines (L) and character font (F).	40 μs
Set CG RAM Address	0	0	0	1			AGC				Sets CG RAM address. CG RAM data is sent and received after this setting.	40 μs
Set DD RAM Address	0	0	1				ADD				Sets DD RAM address. DD RAM data is sent and received after this setting.	40 μs
Read Busy Flag & Address	0	1	BF				AC				Reads Busy flag (BF) indicating internal operation is being performed and reads address counter contents.	40 μs
Write Data CG or DD RAM	1	0				Write Data					Writes data into DD or CG RAM.	40 μs
Read Data CG or DD RAM	1	1				Read Data					Reads data from DD or CG RAM.	40 μs

Notes:

1. Execution times are maximum times when fcp or fosc is 250 kHz.
2. Execution time changes when frequency changes. Ex: When fcp or fosc is 270 kHz: 40 μs × 250 / 270 = 37 μs.
3. Abbreviations:

DD RAM	Display data RAM		
CG RAM	Character generator RAM		
ACC	CG RAM address		
ADD	DD RAM address, corresponds to cursor address		
AC	Address counter used for both DD and CG RAM addresses.		
1/D = 1	Increment	1/D = 0	Decrement
S = 1	Accompanies display shift		
S/C = 1	Display shift;	S/C = 0	Cursor move
R/L = 1	Shift to the right;	R/L = 0	Shift to the left
DL = 1	8 bits, DL = 0: 4 bits		
N = 1	1line, N = 0 : 1 line		
F = 1	5 x 10 dots, F = 0 : 5 x 7 dots		
BF = 1	Internal operation;	BF = 0	Can accept instruction

SECTION 12.2: 8051 INTERFACING TO ADC, SENSORS

This section will explore interfacing ADC (analog-to-digital converter) chips and temperature sensors to the 8051. First, we describe ADC chips, then show how to interface an ADC to the 8051. Then we examine the characteristics of the LM35 temperature sensor and show how to interface it to the 8051.

ADC devices

Analog-to-digital converters are among the most widely used devices for data acquisition. Digital computers use binary (discrete) values, but in the physical world everything is analog (continuous). Temperature, pressure (wind or liquid), humidity, and velocity are a few examples of physical quantities that we deal with every day. A physical quantity is converted to electrical (voltage, current) signals using a device called a *transducer*. Transducers are also referred to as *sensors*. Although there are sensors for temperature, velocity, pressure, light, and many other natural quantities, they produce an output that is voltage (or current). Therefore, we need an analog-to-digital converter to translate the analog signals to digital numbers so that the microcontroller can read them. A widely used ADC chip is the ADC804.

ADC804 chip

The ADC804 IC is an analog-to-digital converter in the family of the ADC800 series from National Semiconductor. It is also available from many other manufacturers. It works with +5 volts and has a resolution of 8 bits. In addition to resolution, conversion time is another major factor in judging an ADC. *Conversion time* is defined as the time it takes the ADC to convert the analog input to a digital (binary) number. In the ADC804, the conversion time varies depending on the clocking signals applied to the CLK R and CLK IN pins, but it cannot be faster than 110 μs. The ADC804 pin descriptions follow.

CS

Chip select is an active low input used to activate the ADC804 chip. To access the ADC804, this pin must be low.

RD (read)

This is an input signal and is active low. The ADC converts the analog input to its binary equivalent and holds it in an internal register. RD is used to get the converted data out of the ADC804 chip. When CS = 0, if a high-to-low pulse is applied to the RD pin, the 8-bit digital output shows up at the D0 - D7 data pins. The RD pin is also referred to as output enable.

WR (write; a better name might be "start conversion")

This is an active low input used to inform the ADC804 to start the conversion process. If CS = 0 when WR makes a low-to-high transition, the ADC804 starts converting the analog input value of Vin to an 8-bit digital number. The amount of time it takes to convert varies depending on the CLK IN and CLK R values explained below. When the data conversion is complete, the INTR pin is forced low by the ADC804.

CLK IN and CLK R

CLK IN is an input pin connected to an external clock source when an external clock is used for timing. However, the 804 has an internal clock generator. To use the internal clock generator (also called self-clocking) of the ADC804, the CLK IN and CLK R pins are connected to a capacitor and a resistor, as shown in Figure 12-5. In that case the clock frequency is determined by the equation:

$$f = \frac{1}{1.1\,RC}$$

Typical values are R = 10K ohms and C = 150 pF. Substituting in the above equation, we get f = 606 kHz. In that case, the conversion time is 110 μs.

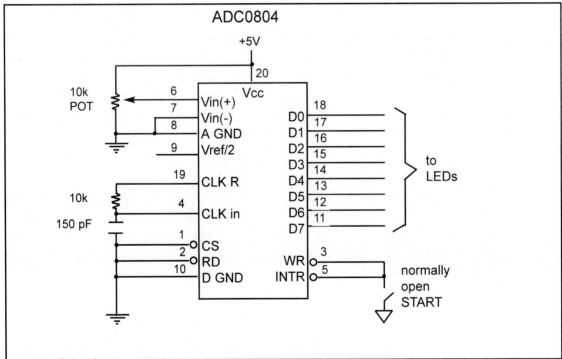

Figure 12-5. Testing ADC804 in Free Running Mode

INTR (interrupt; a better name might be "end of conversion")

This is an output pin and is active low. It is a normally high pin and when the conversion is finished, it goes low to signal the CPU that the converted data is ready to be picked up. After INTR goes low, we make CS = 0 and send a high-to-low pulse to the RD pin to get the data out of the ADC804 chip.

V_{in} (+) and V_{in} (–)

These are the differential analog inputs where $V_{in} = V_{in}\,(+) - V_{in}\,(-)$. Often the $V_{in}\,(-)$ pin is connected to ground and the $V_{in}\,(+)$ pin is used as the analog input to be converted to digital.

V_{CC}

This is the +5 volt power supply. It is also used as a reference voltage when the $V_{ref}/2$ input (pin 9) is open (not connected). This is discussed next.

$V_{ref}/2$

Pin 9 is an input voltage used for the reference voltage. If this pin is open (not connected), the analog input voltage for the ADC804 is in the range of 0 to 5 volts (the same as the V_{CC} pin). However, there are many applications where the analog input applied to Vin needs to be other than the 0 to +5 V range. $V_{ref}/2$ is used to implement analog input voltages other than 0 – 5 V. For example, if the analog input range needs to be 0 to 4 volts, $V_{ref}/2$ is connected to 2 volts. Table 12-5 shows the V_{in} range for various $V_{ref}/2$ inputs.

Table 12-5: $V_{ref}/2$ Relation to V_{in} Range

$V_{ref}/2$ (V)	V_{in} (V)	Step Size (mV)
not connected*	0 to 5	5/256 = 19.53
2.0	0 to 4	4/255 = 15.62
1.5	0 to 3	3/256 = 11.71
1.28	0 to 2.56	2.56/256 = 10
1.0	0 to 2	2/256 = 7.81
0.5	0 to 1	1/256 = 3.90

Notes: V_{CC} = 5 V

* When not connected (open), $V_{ref}/2$ is measured at 2.5 volts for V_{CC} = 5 V.
Step Size (resolution) is the smallest change that can be discerned by an ADC.

D0 - D7

D0 - D7 (where D7 is the MSB, D0 the LSB) are the digital data output pins. These are tri-state buffered and the converted data is accessed only when CS = 0 and RD is forced low. To calculate the output voltage, use the following formula.

$$D_{out} = \frac{V_{in}}{step\ size}$$

where D_{out} = digital data output (in decimal), V_{in} = analog input voltage, and step size (resolution) is the smallest change, which is $(2 \times V_{ref}/2)/256$ for an 8-bit ADC.

Analog ground and digital ground

These are the input pins providing the ground for both the analog signal and the digital signal. Analog ground is connected to the ground of the analog V_{in} while digital ground is connected to the ground of the V_{CC} pin. The reason that we have two ground pins is to isolate the analog V_{in} signal from transient voltages caused by digital switching of the output D0 - D7. Such isolation contributes to the accuracy of the digital data output. In our discussion, both are connected to the same ground; however, in the real world of data acquisition the analog and digital grounds are handled separately.

From this discussion we conclude that the following steps must be followed for data conversion by the ADC804 chip.

1. Make CS = 0 and send a low-to-high pulse to pin WR to start the conversion.
2. Keep monitoring the INTR pin. If INTR is low, the conversion is finished and we can go to the next step. If INTR is high, keep polling until it goes low.
3. After the INTR has become low, we make CS = 0 and send a high-to-low pulse to the RD pin to get the data out of the ADC804 IC chip. The timing for this process is shown in Figure 12-6.

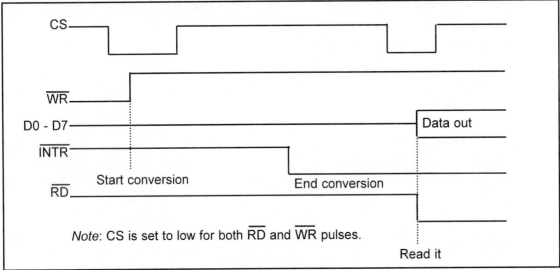

Figure 12-6. Read and Write Timing for ADC804

Testing the ADC804

One can test the ADC804 using the circuit shown in Figure 12-5. This setup is called free running test mode and is recommended by the manufacturer. Figure 12-5 shows a potentiometer used to apply a 0-to-5 V analog voltage to the input V_{in} (+) of the 804 ADC. The binary outputs are monitored on the LEDs of the digital trainer. It must be noted that in free running test mode the CS input is grounded and the WR input is connected to the INTR output. However, according to National Semiconductor's databook "the WR and INTR node should be momentarily forced to low following a power-up cycle to guarantee operation."

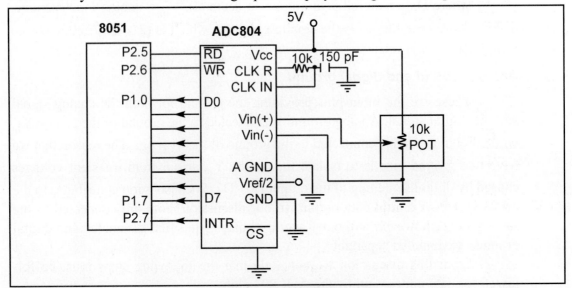

Figure 12-7. 8051 Connection to ADC804 with Self-Clocking

Example 12-1

Examine the ADC804 connection to the 8051 in Figure 12-7. Write a program to monitor the INTR pin and bring an analog input into register A. Then call a hex-to-ASCII conversion and data display subroutines. Do this continuously.

Solution:

```
;P2.6 = WR (start conversion needs to L-to-H pulse)
;P2.7 When low, end-of-conversion)
;P2.5 = RD (a H-to-L will read the data from ADC chip)
;P1,0 - P1.7 = D0 - D7 of the ADC804
;
            MOV  P1,#0FFH  ;make P1 = input
BACK:       CLR  P2.6      ;WR=0
            SETB P2.6      ;WR=1 L-to-H to start conversion
HERE:       JB   P2.7,HERE ;wait for end of conversion
            CLR  P2.5      ;conversion finished,enable RD
            MOV  A,P1      ;read the data
            ACALL CONVERSION   ;hex-to-ASCII conversion
            ACALL DATA_DISPLAY ;display the data
            SETB P2.5      ;make RD=1 for next round
            SJMP BACK
```

Note: For a hex-to-ASCII conversion subroutine, see Chapter 7.

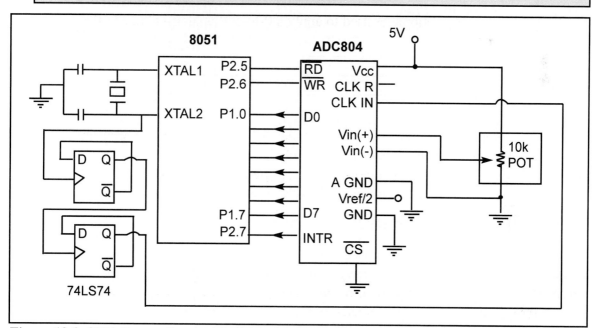

Figure 12-8. 8051 Connection to ADC804 with Clock from XTAL2 of the 8051

In Figure 12-8, notice that the clock in for the ADC804 is coming from the crystal of the microcontroller. Since this frequency is too high, we use two D flip-flops (74LS74) to divide the frequency by 4. A D flip-flop divides the frequency by 2 if we connect its $\overline{Q}$ to the D input. For a higher frequency, use more flip-flops.

Interfacing a temperature sensor to the 8051

Transducers convert physical data such as temperature, light intensity, flow, and speed to electrical signals. Depending on the transducer, the output produced is in the form of voltage, current, resistance, or capacitance. For example, temperature is converted to electrical signals using a transducer called a *thermistor*. A thermistor responds to temperature change by changing resistance, but its response is not linear, as seen in Table 12-6.

Table 12-6. Thermistor Resistance vs. Temperature

Temperature (C)	Tf (K ohms)
0	29.490
25	10.000
50	3.893
75	1.700
100	0.817

From William Kleitz, *Digital Electronics*

Table 12-7. LM34 Temperature Sensor Series Selection Guide

Part Scale	Temperature Range	Accuracy	Output
LM34A	−50 F to +300 F	+2.0 F	10 mV/F
LM34	−50 F to +300 F	+3.0 F	10 mV/F
LM34CA	−40 F to +230 F	+2.0 F	10 mV/F
LM34C	−40 F to +230 F	+3.0 F	10 mV/F
LM34D	−32 F to +212 F	+4.0 F	10 mV/F

Note: Temperature range is in degrees Fahrenheit.

Table 12-8. LM35 Temperature Sensor Series Selection Guide

Part	Temperature Range	Accuracy	Output Scale
LM35A	−55 C to +150 C	+1.0 C	10 mV/F
LM35	−55 C to +150 C	+1.5 C	10 mV/F
LM35CA	−40 C to +110 C	+1.0 C	10 mV/F
LM35C	−40 C to +110 C	+1.5 C	10 mV/F
LM35D	0 C to +100 C	+2.0 C	10 mV/F

Note: Temperature range is in degrees Celsius.

The complexity associated with writing software for such nonlinear devices has led many manufacturers to market the linear temperature sensor. Simple and widely used linear temperature sensors include the LM34 and LM35 series from National Semiconductor Corp. They are discussed next.

LM34 and LM35 temperature sensors

The sensors of the LM34 series are precision integrated-circuit temperature sensors whose output voltage is linearly proportional to the Fahrenheit temperature. See Table 12-7. The LM34 requires no external calibration since it is inherently calibrated. It outputs 10 mV for each degree of Fahrenheit temperature. Table 12-7 is the selection guide for the LM34.

The LM35 series sensors are precision integrated-circuit temperature sensors whose output voltage is linearly proportional to the Celsius (centigrade) temperature. The LM35 requires no external calibration since it is inherently calibrated. It outputs 10 mV for each degree of centigrade temperature. Table 12-8 is the selection guide for the LM35 (For further information see www.national.com.).

Signal conditioning and interfacing the LM35 to the 8051

Signal conditioning is a widely used term in the world of data acquisition. The most common transducers produce an output in the form of voltage, current, charge, capacitance, and resistance. However, we need to convert these signals to voltage in order to send input to an A-to-D converter. This conversion (modification) is commonly called *signal conditioning*. Signal conditioning can be a current-to-voltage conversion or a signal amplification. For example, the thermistor changes resistance with temperature. The change of resistance must be translated into voltages in order to be of any use to an ADC. Look at the case of connecting an LM35 to an ADC804. Since the ADC804 has 8-bit resolution with a maximum of 256 (2^8) steps and the LM35 (or LM34) produces 10 mV for every degree of temperature change, we can condition V_{in} of the ADC804 to produce a V_{out} of 2560 mV (2.56V) for full-scale output. Therefore, in order to produce the full-scale V_{out} of 2.56 V for the ADC804, we need to set $V_{ref}/2 = 1.28$. This makes V_{out} of the ADC804 correspond directly to the temperature as monitored by the LM35. See Table 12-9. $V_{ref}/2$ values are given in Table 12-5.

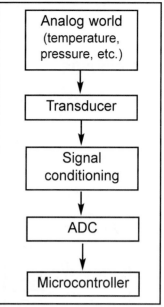

Figure 12-9. Getting Data From the Analog World

Table 12-9. Temperature v. Vout of the ADC804

Temp. (C)	Vin (mV)	Vout (D7 - D0)
0	0	0000 0000
1	10	0000 0001
2	20	0000 0010
3	30	0000 0011
10	100	0000 1010
30	300	0001 1110

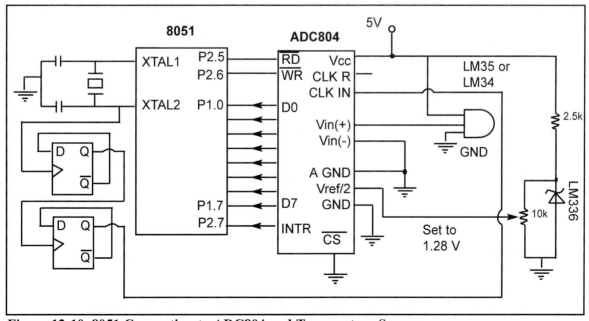

Figure 12-10. 8051 Connection to ADC804 and Temperature Sensor

Figure 12-10 shows connection of a temperature sensor to the ADC804. Notice that we use the LM336-2.5 zener diode to fix the voltage across the 10K pot at 2.5 volts. The use of the LM336-2.5 should overcome any fluctuations in the power supply.

ADC808/809 chip with 8 analog channels

Another useful chip is the ADC808/809 from National Semiconductor. See Figure 12-11. While the ADC804 has only one analog input, this chip has 8 of them. The ADC808/809 chip allows us to monitor up to 8 different transducers using only a single chip. Notice that the ADC808/809 has an 8-bit data output just like the ADC804. The 8 analog input channels are multiplexed and selected according to Table 12-10 using three address pins, A, B, and C.

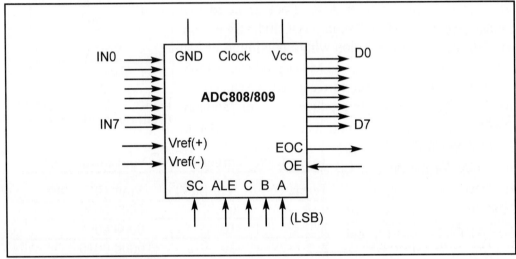

Figure 12-11. ADC808/809

Table 12-10: ADC808 Analog Channel Selection

Selected Analog Channel	C	B	A
IN0	0	0	0
IN1	0	0	1
IN2	0	1	0
IN3	0	1	1
IN4	1	0	0
IN5	1	0	1
IN6	1	1	0
IN7	1	1	1

In the ADC808/809, $V_{ref}(+)$ and $V_{ref}(-)$ set the reference voltage. If $V_{ref}(-)$ = Gnd and $V_{ref}(+)$ = 5 V, the step size is 5 V/256 = 19.53 mV. Therefore, to get a 10 mV step size we need to set $V_{ref}(+)$ = 2.56 V and $V_{ref}(-)$ = Gnd. From Figure 12-11, notice the ALE pin. We use A, B, and C addresses to select IN0 - IN7, and activate ALE to latch in the address. SC is for start conversion. EOC is for end-of-conversion, and OE is for output enable (READ). Next, we give the steps for programming this chip.

Steps to program the ADC808/809

The following are the steps to get data from analog input of ADC808/809 into the microcontroller.

1. Select an analog channel by providing bits to A, B, and C addresses according to Table 12-10.
2. Activate the ALE (address latch enable) pin. It needs an L-to-H pulse to latch in the address.
3. Activate SC (start conversion) by an H-to-L pulse to initiate conversion.
4. Monitor EOC (end of conversion) to see whether conversion is finished. H-to-L output indicates that the data is converted and is ready to be picked up.
5. Activate OE (output enable) to read data out of the ADC chip. An H-to-L pulse to the OE pin will bring digital data out of the chip.

Notice in the ADC808/809 that there is no self-clocking and the clock must be provided from an external source to the CLK pin. Although the speed of conversion depends on the frequency of the clock connected to CLK pin, it cannot be faster than 100 ms.

Review Questions

1. In the ADC804, the INTR signal is an _____ (input, output).
2. To begin conversion, send a(n) _____ pulse to pin _____.
3. Which pin of the ADC804 indicates end-of-conversion?
4. True or false. The transducer must be connected to signal conditioning circuitry before it is sent to the ADC.
5. The LM35 provides _____ mV for each degree of _____ (Fahrenheit, Celsius) temperature.
6. Both the ADC804 and ADC808 are _____-bit converters.
7. Indicate the direction (out, in) for each of the following pins of the ADC808/809.
   (a) A, B, C   (b) SC   (c) EOC

SUMMARY

This chapter showed how to interface real-world devices such as LCDs, ADC chips, and sensors, to the 8051. First, we described the operation modes of LCDs, then described how to program the LCD by sending data or commands to it via its interface to the 8051.

Next we explored ADC chips and temperature sensors. Getting data from the analog world to a digital device is called *signal conditioning*. It is an essential feature of data acquisition systems.

PROBLEMS

SECTION 12.1: INTERFACING AN LCD TO THE 8051

1. The LCD discussed in this section has _____ (4, 8) data pins.
2. Describe the function of pins E, R/W, and RS in the LCD.
3. What is the difference between the V_{CC} and V_{EE} pins on the LCD?
4. "Clear LCD" is a _____ (command code, data item) and its value is ___ hex.
5. What is the hex value of the command code for "display on, cursor on"?
6. Give the state of RS, E, and R/W when sending a command code to the LCD.
7. Give the state of RS, E, and R/W when sending data character 'Z' to the LCD.
8. Which of the following is needed on the E pin in order for a command code (or data) to be latched in by the LCD?
   (a) H-to-L pulse (b) L-to-H pulse
9. True or false. For the above to work, the value of the command code (data) must be already at the D0 - D7 pins.
10. In sending streams of characters to the LCD there are two methods: (1) checking the busy flag, or (2) putting some time delay in between sending each character without checking the busy flag. Explain the difference and the advantages and disadvantages of each method. Also explain how we monitor the busy flag.
11. For a 16x2 LCD, the location of the last character of line 1 is 8FH (its command code). Show how this value was calculated.
12. For a 16x2 LCD, the location of the first character of line 2 is C0H (its command code). Show how this value was calculated.
13. For a 20x2 LCD, the location of the last character of line 2 is 93H (its command code). Show how this value was calculated.
14. For a 20x2 LCD, the location of the third character of line 2 is C2H (its command code). Show how this value was calculated.
15. For a 40x2 LCD, the location of the last character of line 1 is A7H (its command code). Show how this value was calculated.
16. For a 40x2 LCD, the location of the last character of line 2 is E7H (its command code). Show how this value was calculated.
17. Show the value (in hex) for the command code for the 10th location, line 1 on a 20x2 LCD. Show how you got your value.
18. Show the value (in hex) for the command code for the 20th location, line 2 on a 40x2 LCD. Show how you got your value.
19. Rewrite the COMNWRT subroutine. Assume connections P1.4 = RS, P1.5 = R/W, P1.6 = E.
20. Repeat the Problem 19 for the data write subroutine. Send the string "Hello" to the LCD with checking the busy flag. Use instruction MOVC.

---

21. Give the status of CS and WR in order to start conversion for the ADC804.
22. Give the status of CS and WR in order to get data from the ADC804.
23. In the ADC804, what happens to the converted analog data? How do we know if the ADC is ready to provide us the data?
24. In the ADC804, what happens to the old data if we start conversion again before we pick up the last data?
25. In the ADC804, INTR is an _____ (input, output) signal. What is its function in the ADC804?
26. For an ADC804 chip, find the step size for each of the following $V_{ref}/2$ values.
    (a) $V_{ref}/2 = 1.28$ V   (b) $V_{ref}/2 = 1$ V   (c) $V_{ref}/2 = 1.9$ V
27. In the ADC804, what should be the $V_{ref}/2$ value for a step size of 20 mv?
28. In the ADC804, what should be the $V_{ref}/2$ value for a step size of 5 mv?
29. In the ADC804, what is the role of pins $V_{in}(+)$ and $V_{in}(-)$?
30. With a step size of 19.53 mV, what is the analog input voltage if all outputs are 1?
31. With $V_{ref}/2 = 0.64$ V, find the V_{in} for the following outputs.
    (a) D7 - D0 = 11111111   (b) D7 - D0 = 10011001   (c) D7 - D0 = 1101100
32. What does it mean when it is said that a given sensor has a linear output?
33. The LM34 sensor produces _____ mV for each degree of temperature.
34. What is signal conditioning?
35. What is the purpose of the LM336 Zener diode around the pot setting the $V_{ref}/2$ in Figure 12-10?
36. True or false. ADC804 is an 8-bit ADC.
37. Indicate the direction (in, out) for each of the following ADC808 pins.
    (a) SC   (b) EOC   (c) A, B, C
    (d) ALE   (e) OE   (f) IN0 - IN7
    (g) D0 - D7
38. Explain the role of the ALE pin in the ADC808 and show how to select channel 5 analog input.
39. In the ADC808, assume $V_{ref}(-)$ = Gnd. Give the $V_{ref}(+)$ voltage value if we want the following step sizes:
    (a) 20 mv   (b) 5 mv   (c) 10 mv
    (d) 15 mv   (e) 2 mv   (f) 25 mv
40. In the ADC808, assume $V_{ref}(-)$ = Gnd. Find the step size for the following values of $V_{ref}(+)$:
    (a) 1.28 V   (b) 1 V   (c) 0.64 V

ANSWERS TO REVIEW QUESTIONS

SECTION 12.1: INTERFACING AN LCD TO THE 8051
1. Input   2. Input   3. H-to-L
4. High   5. 80H and C0H

SECTION 12.2: 8051 INTERFACING TO ADC, SENSORS

1. Output   2. L-to-H, WR   3. INTR
4. True   5. 10, both.
6. 8   7. (a) all in   (b) in (c) out

CHAPTER 13

REAL-WORLD INTERFACING II: STEPPER MOTOR, KEYBOARD, AND DAC

OBJECTIVES

Upon completion of this chapter, you will be able to:

- ➢➢ Describe the basic operation of a stepper motor
- ➢➢ Interface the 8051 with a stepper motor
- ➢➢ Code 8051 programs to control and operate a stepper motor
- ➢➢ Define stepper motor operation in terms of step angle, steps per revolution, tooth pitch, rotation speed, and RPM
- ➢➢ Explain the basic operation of a keyboard
- ➢➢ Describe the key press and detection mechanisms
- ➢➢ Interface a 4x4 keypad to the 8051
- ➢➢ Describe the basic operation of a DAC (digital-to-analog converter) chip
- ➢➢ Interface a DAC chip to the 8051
- ➢➢ Program a DAC chip to produce a sine wave on an oscilloscope

This chapter discusses 8051 interfacing with stepper motors, keyboards, and DAC (digital-to-analog converter) devices. In Section 13.1, the basics of the stepper motor are described. Then we show the stepper motor interfacing with the 8051. In Section 13.2, keyboard interfacing with the 8051 is shown. The characteristics of the DAC are discussed in Section 13.3, along with interfacing to the 8051.

SECTION 13.1: INTERFACING A STEPPER MOTOR

This section begins with an overview of the basic operation of stepper motors. Then we describe how to interface a stepper motor to the 8051. Finally, we use Assembly language programs to demonstrate control of the angle and direction of stepper motor rotation.

Stepper motors

A *stepper motor* is a widely used device that translates electrical pulses into mechanical movement. In applications such as disk drives, dot matrix printers, and robotics, the stepper motor is used for position control. Every stepper motor has a permanent magnet *rotor* (also called the *shaft*) surrounded by a *stator* (see Figure 13-1). The most common stepper motors have four stator windings that are paired with a center-tapped common as shown in Figure 13-2. This type of stepper motor is commonly referred to as a *four-phase* stepper motor. The center tap allows a change of current direction in each of two coils when a winding is grounded, thereby resulting in a polarity change of the stator. Notice that while a conventional motor shaft runs freely, the stepper motor shaft moves in a fixed repeatable increment which allows one to move it to a precise position. This repeatable fixed movement is possible as a result of basic magnetic theory where poles of the same polarity repel and opposite poles attract. The direction of the rotation is dictated by

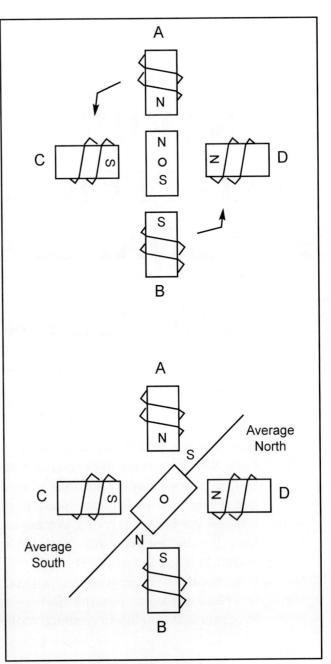

Figure 13-1. Rotor Alignment

the stator poles. The stator poles are determined by the current sent through the wire coils. As the direction of the current is changed, the polarity is also changed causing the reverse motion of the rotor. The stepper motor discussed here has a total of 6 leads: 4 leads representing the four stator windings and 2 commons for the center tapped leads. As the sequence of power is applied to each stator winding, the rotor will rotate. There are several widely used sequences where each has a different degree of precision. Table 13-1 shows the normal 4-step sequence.

Table 13-1. Normal 4-Step Sequence

Clockwise	Step #	Winding A	Winding B	Winding C	Winding D	Counter-clockwise
	1	1	0	0	1	
	2	1	1	0	0	
	3	0	1	1	0	
	4	0	0	1	1	

Table 13-2. Stepper Motor Step Angles

Step Angle	Steps per Revolution
0.72	500
1.8	200
2.0	180
2.5	144
5.0	72
7.5	48
15	24

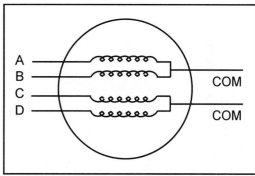

Figure 13-2. Stator Windings Configuration

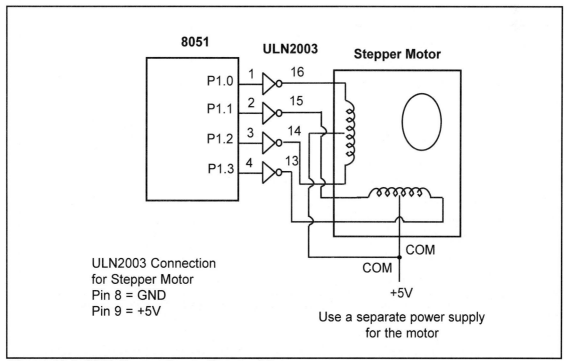

Figure 13-3. 8051 Connection to Stepper Motor

It must be noted that although we can start with any of the sequences in Table 13-1, once we start we must continue in the proper order. For example, if we start with step 3 (0110), we must continue in the sequence of steps 4, 1, 2, etc.

Example 13-1

Describe the 8051 connection to the stepper motor of Figure 13-3 and code a program to rotate it continuously.

Solution:

The following steps show the 8051 connection to the stepper motor and its programming.

1. Use an ohmmeter to measure the resistance of the leads. This should identify which COM leads are connected to which winding leads.
2. The common wire(s) are connected to the positive side of the motor's power supply. In many motors, +5 V is sufficient.
3. The four leads of the stator winding are controlled by four bits of the 8051 port (P1.0 - P1.3). However, since the 8051 lacks sufficient current to drive the stepper motor windings, we must use a driver such as the ULN2003 to energize the stator. Instead of the ULN2003, we could have used transistors as drivers, as shown in Figure 13-4. However, notice that if transistors are used as drivers, we must also use diodes to take care of inductive current generated when the coil is turned off. One reason that the ULN2003 is preferable to the use of transistors as drivers is that the ULN2003 has an internal diode to take care of back EMF.

```
              MOV   A,#66H      ;load step sequence
     BACK:    MOV   P1,A        ;issue sequence to motor
              RR    A           ;rotate right clockwise
              ACALL DELAY       ;wait
              SJMP  BACK        ;keep going

              . . .
     DELAY
              MOV   R2,#100
     H1:      MOV   R3,#255
     H2:      DJNZ  R3,H2
              DJNZ  R2,H1
              RET
```

Change the value of DELAY to set the speed of rotation.
We can use the single-bit instructions SETB and CLR instead of RRA to create the sequences.

Step angle

How much movement is associated with a single step? This depends on the internal construction of the motor, in particular the number of teeth on the stator and the rotor. The *step angle* is the minimum degree of rotation associated with a single step. Various motors have different step angles. Table 13-2 shows some step angles for various motors. In Table 13-2, notice the term *steps per revolution*. This is the total number of steps needed to rotate one complete rotation or 360 degrees (e.g., 180 steps × 2 degrees = 360).

It must be noted that perhaps contrary to one's initial impression, a stepper motor does not need to have more terminal leads for the stator to achieve smaller steps. All the stepper motors discussed in this section have 4 leads for the stator winding and 2 COM wires for the center tap. Although some manufacturers set aside only one lead for the common signal instead of two, they always have 4 leads for the stators. Next we discuss some associated terminology in order to understand the stepper motor further.

Steps per second and RPM relation

The relation between RPM (revolutions per minute), steps per revolution, and steps per second is intuitive and is as follows.

$$Steps\ per\ second = \frac{RPM \times Steps\ per\ revolution}{60}$$

The four-step sequence and number of teeth on rotor

The switching sequence shown earlier in Table 13-1 is called the 4-step switching sequence since after four steps the same two windings will be "ON". How much movement is associated with these four steps? After completing every four steps, the rotor moves only one tooth pitch. Therefore, in a stepper motor with 200 steps per revolution, its rotor has 50 teeth since 4 × 50 = 200 steps are needed to complete one revolution. This leads to the conclusion that the minimum step angle is always a function of the number of teeth on the rotor. In other words, the smaller the step angle the more teeth the rotor passes. See Example 13-2.

Example 13-2

Give the number of times the four-step sequence in Table 13-1 must be applied to a stepper motor to make an 80-degree move if the motor has a 2-degree step angle.

Solution:
A motor with a 2-degree step angle has the following characteristics:

Step angle:	2 degrees	Steps per revolution:	180
Number of rotor teeth:	45	Movement per 4-step sequence:	8 degrees

To move the rotor 80 degrees, we need to send 10 four-step sequences right after each other, since 10 × 4 steps × 2 degrees = 80 degrees.

Looking at Example 13-2, one might wonder what happens if we want to move 45 degrees, since the steps are 2 degrees each. To allow for finer resolutions, all stepper motors allow what is called an *8-step* switching sequence. The 8-step

---

sequence is also called *half-stepping,* since in following the 8-step sequence each step is half of the normal step angle. For example, a motor with a 2-degree step angle can be used as a 1-degree step angle if the sequence of Table 13-3 is applied.

Table 13-3. Half-Step 8-Step Sequence

Clockwise	Step #	Winding A	Winding B	Winding C	Winding D	Counter-clockwise
	1	1	0	0	1	
	2	1	0	0	0	
	3	1	1	0	0	
	4	0	1	0	0	
	5	0	1	1	0	
	6	0	0	1	0	
	7	0	0	1	1	
	8	0	0	0	1	

Motor speed

The motor speed, measured in steps per second (steps/s), is a function of the switching rate. Notice in Example 13-1 that by changing the length of the time delay loop, we can achieve various rotation speeds.

Holding torque

The following is a definition of holding torque: "With the motor shaft at standstill or zero RPM condition, the amount of torque, from an external source, required to break away the shaft from its holding position. This is measured with rated voltage and current applied to the motor." The unit is ounce-inch (or kg-cm).

Wave drive 4-step sequence

In addition to the 8-step and the 4-step sequences discussed earlier, there is another sequence called the wave drive 4-step sequence. It is shown in Table 13-4. Notice that the 8-step sequence of Table 13-3 is simply the combination of the wave drive 4-step and normal 4-step normal sequences shown in Tables 13-4 and 13-1, respectively. Experimenting with the wave drive 4-step is left to the reader.

Table 13-4. Wave Drive 4-Step Sequence

Clockwise	Step #	Winding A	Winding B	Winding C	Winding D	Counter-clockwise
	1	1	0	0	0	
	2	0	1	0	0	
	3	0	0	1	0	
	4	0	0	0	1	

Review Questions

1. Give the 4-step sequence of a stepper motor if we start with 0110.
2. A stepper motor with a step-angle of 5 degrees has _____ steps per revolution.
3. Why do we put a driver in between the microcontroller and the stepper motor?

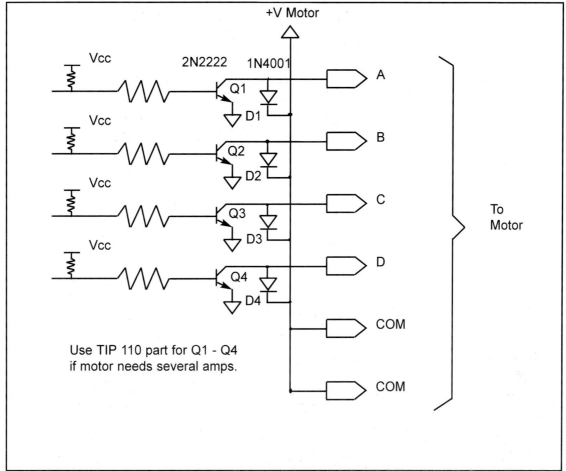

+V Motor

Vcc   2N2222   1N4001

Q1   D1   A

Vcc

Q2   D2   B

Vcc

Q3   D3   C

Vcc

Q4   D4   D

COM

COM

To
Motor

Use TIP 110 part for Q1 - Q4
if motor needs several amps.

Figure 13-4. Using Transistors for Stepper Motor Driver

SECTION 13.2: 8051 INTERFACING TO THE KEYBOARD

Keyboards and LCDs are the most widely used input/output devices of the 8051, and a basic understanding of them is essential. In this section, we first discuss keyboard fundamentals, along with key press and key detection mechanisms. Then we show how a keyboard is interfaced to an 8051.

Interfacing the keyboard to the 8051

At the lowest level, keyboards are organized in a matrix of rows and columns. The CPU accesses both rows and columns through ports; therefore, with two 8-bit ports, an 8 x 8 matrix of keys can be connected to a microprocessor. When a key is pressed, a row and a column make a contact; otherwise, there is no connection between rows and columns. In IBM PC keyboards, a single microcontroller (consisting of a microprocessor, RAM and EPROM, and several ports all on a single chip) takes care of hardware and software interfacing of the keyboard. In such systems, it is the function of programs stored in the EPROM of the microcontroller to scan the keys continuously, identify which one has been activated, and present it to the motherboard. In this section we look at the mechanism by which the 8051 scans and identifies the key.

Scanning and identifying the key

Figure 13-5 shows a 4 x 4 matrix connected to two ports. The rows are connected to an output port and the columns are connected to an input port. If no key has been pressed, reading the input port will yield 1s for all columns since they are all connected to high (V_{CC}). If all the rows are grounded and a key is pressed, one of the columns will have 0 since the key pressed provides the path to ground. It is the function of the microcontroller to scan the keyboard continuously to detect and identify the key pressed. How it is done is explained next.

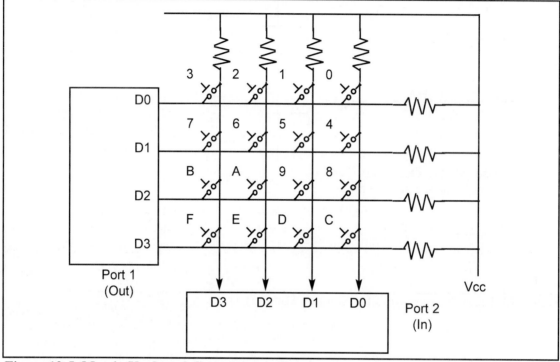

Figure 13-5. Matrix Keyboard Connection to Ports

Grounding rows and reading the columns

To detect a pressed key, the microcontroller grounds all rows by providing 0 to the output latch, then it reads the columns. If the data read from the columns is D3 - D0 = 1111, no key has been pressed and the process continues until a key press is detected. However, if one of the column bits has a zero, this means that a key press has occurred. For example, if D3 - D0 = 1101, this means that a key in the D1 column has been pressed. After a key press is detected, the microcontroller will go through the process of identifying the key. Starting with the top row, the microcontroller grounds it by providing a low to row D0 only; then it reads the columns. If the data read is all 1s, no key in that row is activated and the process is moved to the next row. It grounds the next row, reads the columns, and checks for any zero. This process continues until the row is identified. After identification of the row in which the key has been pressed, the next task is to find out which column the pressed key belongs to. This should be easy since the microcontroller knows at any time which row and column are being accessed. Look at Example 13-3.

Example 13-3

From Figure 13-5, identify the row and column of the pressed key for each of the following.
(a) D3 - D0 = 1110 for the row, D3 - D0 = 1011 for the column
(b) D3 - D0 = 1101 for the row, D3 - D0 = 0111 for the column

Solution:

From Figure 13-5 the row and column can be used to identify the key.
(a) The row belongs to D0 and the column belongs to D2; therefore, key number 2 was pressed.
(b) The row belongs to D1 and the column belongs to D3; therefore, key number 7 was pressed.

Program 13-1 is the 8051 Assembly language program for detection and identification of key activation. In this program, it is assumed that P1 and P2 are initialized as output and input, respectively. Program 13-1 goes through the following four major stages:

1. To make sure that the preceding key has been released, 0s are output to all rows at once, and the columns are read and checked repeatedly until all the columns are high. When all columns are found to be high, the program waits for a short amount of time before it goes to the next stage of waiting for a key to be pressed.

2. To see if any key is pressed, the columns are scanned over and over in an infinite loop until one of them has a 0 on it. Remember that the output latches connected to rows still have their initial zeros (provided in stage 1), making them grounded. After the key press detection, it waits 20 ms for the bounce and then scans the columns again. This serves two functions: (a) it ensures that the first key press detection was not an erroneous one due to a spike noise, and (b) the 20-ms delay prevents the same key press from being interpreted as a multiple key press. If after the 20-ms delay the key is still pressed, it goes to the next stage to detect which row it belongs to; otherwise, it goes back into the loop to detect a real key press.

3. To detect which row the key press belongs to, it grounds one row at a time, reading the columns each time. If it finds that all columns are high, this means that the key press cannot belong to that row; therefore, it grounds the next row and continues until it finds the row the key press belongs to. Upon finding the row that the key press belongs to, it sets up the starting address for the look-up table holding the scan codes (or the ASCII value) for that row and goes to the next stage to identify the key.

4. To identify the key press, it rotates the column bits, one bit at a time, into the carry flag and checks to see if it is low. Upon finding the zero, it pulls out the ASCII code for that key from the look-up table; otherwise, it increments the pointer to point to the next element of the look-up table. Figure 13-6 flowcharts this process.

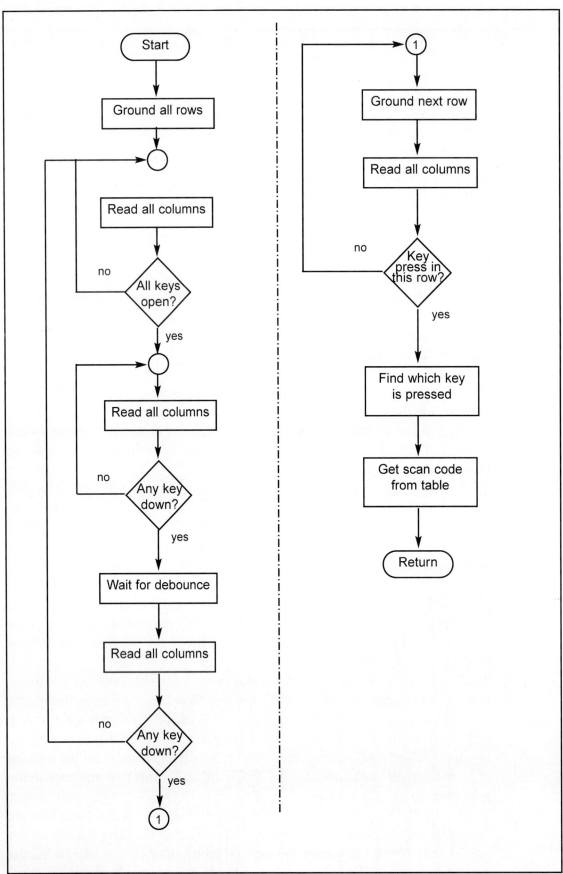

Figure 13-6. Flowchart for Program 13-1

While the key press detection is standard for all keyboards, the process for determining which key is pressed varies. The look-up table method shown in Program 13-1 can be modified to work with any matrix up to 8 x 8. Figure 13-6 provides the flowchart for Program 13-1 for scanning and identifying the pressed key.

There are IC chips such as National Semiconductor's MM74C923 that incorporate keyboard scanning and decoding all in one chip. Such chips use combinations of counters and logic gates (no microcontroller) to implement the underlying concepts presented in Program 13-1.

```
;Keyboard subroutine. This program sends the ASCII code
;for pressed key to P0.1
;P1.0-P1.3 connected to rows P2.0-P2.3 connected to columns
      MOV  P2,#0FFH  ;make P2 an input port
K1:   MOV  P1,#0        ;ground all rows at once
      MOV  A,P2          ;read all col. (ensure all keys open)
      ANL  A,00001111B       ;masked unused bits
      CJNE A,#00001111B,K1      ;check til all keys released
K2:   ACALL DELAY              ;call 20 msec delay
      MOV  A,P2               ;see if any key is pressed
      ANL  A,#00001111B        ;mask unused bits
      CJNE A,#00001111B,OVER  ;key pressed, await closure
      SJMP K2                 ;check il key pressed
OVER: ACALL DELAY             ;wait 20 msec debounce time
      MOV  A,P2               ;check key closure
      ANL  A,#00001111B        ;mask unused bits
      CJNE A,#00001111B,OVER1 ;key pressed, find row
      SJMP K2                 ;if none, keep polling
OVER1: MOV P1,#11111110B      ;ground row 0
      MOV  A,P2               ;read all columns
      ANL  A,#00001111B        ;mask unused bits
      CJNE A,#00001111B,ROW_0 ;key row 0, find the col.
      MOV  P1,#11111101B       ;ground row 1
      MOV  A,P2               ;read all columns
      ANL  A,#00001111B        ;mask unused bits
      CJNE A,#00001111B,ROW_1 ;keyrow 1, find the col.
      MOV  P1,#11111011B       ;ground row 2
      MOV  A,P2               ;read all columns
      ANL  A,#00001111B        ;mask unused bits
      CJNE A,#00001111B,ROW_2 ;key row 2, find the col.
      MOV  P1,#11110111B       ;ground row 3
      MOV  A,P2               ;read all columns
      ANL  A,#00001111B        ;mask unused bits
      CJNE A,#00001111B,ROW_3 ;keyrow 3, find the col.
      LJMP K2                 ;if none, false input, repeat
```

Program 13-1. *(continued on next page)*

```
ROW_0:  MOV DPTR,#KCODE0        ;set DPTR=start of row 0
        SJMP FIND               ;find col. key belongs to
ROW_1:  MOV DPTR,#KCODE1        ;set DPTR=start of row 1
        SJMP FIND               ;find col. key belongs to
ROW_2:  MOV DPTR,#KCODE2        ;set DPTR=start of row 2
        SJMP FIND               ;find col. key belongs to
ROW_3:  MOV DPTR,#KCODE3        ;set DPTR=start of row 3
FIND:   RRC A                   ;see if any CY bit low
        JNC MATCH               ;if zero, get the ASCII code
        INC DPTR                ;point to next col. address
        SJMP FIND               ;keep searching
MATCH:  CLR A                   ;set A=0 (match is found)
        MOVC A,@A+DPTR          ;get ASCII code from table
        MOV  P0,A               ;display pressed key
        LJMP  K1
;ASCII LOOK-UP TABLE FOR EACH ROW
        ORG       300H
KCODE0: DB        '0','1','2','3'         ;ROW 0
KCODE1: DB        '4','5','6','7'         ;ROW 1
KCODE2: DB        '8','9','A','B'         ;ROW 2
KCODE3: DB        'C','D','E','F'         ;ROW 3
        END
```

Program 13-1. *(continued from previous page)*

Review Questions

1. True or false. To see if any key is pressed, all rows are grounded.
2. If D3 - D0 = 0111 is the data read from the columns, which column does the pressed key belong to?
3. True or false. Key press detection and key identification require two different processes.
4. In Figure 13-5, if the row has D3 - D0 = 1110 and the columns are D3 - D0 = 1110, which key is pressed?
5. True or false. To identify the pressed key, one row at a time is grounded.

SECTION 13.3: INTERFACING A DAC TO THE 8051

This section will show how to interface a DAC (digital-to-analog converter) to the 8051. Then we demonstrate how to generate a sine wave on the scope using the DAC.

Digital-to-analog (DAC) converter

The digital-to-analog converter (DAC) is a device widely used to convert digital pulses to analog signals. In this section we discuss the basics of interfacing a DAC to the 8051.

Recall from your digital electronics book the two methods of making the DAC: binary weighted and R/2R ladder. The vast majority of integrated circuit

DACs, including the MC1408 (DAC808) used in this section, use the R/2R method since it can achieve a much higher degree of precision. The first criterion for judging a DAC is its resolution, which is a function of the number of binary inputs. The common ones are 8, 10, and 12 bits. The number of data bit inputs decides the resolution of the DAC since the number of analog output levels is equal to 2^n, where n is the number of data bit inputs. Therefore, an 8-input DAC such as the DAC808 provides 256 discrete voltage (or current) levels of output. Similarly, the 12-bit DAC provides 4096 discrete voltage levels. There are also 16-bit DACs, but they are expensive.

MC1408 DAC (or DAC 808)

In the MC1408 (DAC808), the digital inputs are converted to current (I_{out}) and by connecting a resistor to the I_{out} pin, we convert the result to voltage. The total current provided by the I_{out} pin is a function of the binary numbers at D0 - D7 inputs of the DAC808 and the reference current (I_{ref}) and is as follows.

$$I_{out} = I_{ref} \left(\frac{D7}{2} + \frac{D6}{4} + \frac{D5}{8} + \frac{D4}{16} + \frac{D3}{32} + \frac{D2}{64} + \frac{D1}{128} + \frac{D0}{256} \right)$$

where D0 is the LSB, and D7 is the MSB for the inputs, and I_{ref} is the input current that must be applied to pin 14. The I_{ref} current is generally set to 2.0 mA. Figure 13-7 shows the generation of current reference (setting I_{ref} = 2 mA) by using the standard 5-V power supply and 1K, 1.5K ohm standard resistors. Some also use the zener diode (LM336), which overcomes any fluctuation associated with the power supply voltage. Now assuming that I_{ref} = 2 mA, if all the inputs to the DAC are high, the maximum output current is 1.99 mA (verify this for yourself).

Converting I$_{out}$ to voltage in DAC808

We connect the output pin I_{out} to a resistor and convert this current to voltage and monitor the output on the scope. However, in real life this can cause inaccuracy since the input resistance of the load where it is connected will also affect the output voltage. For this reason, the I_{ref} current output is isolated by connecting it to an op-amp such as the 741 with Rf = 5K ohms for the feedback resistor. Assuming that R = 5K ohms, by changing the binary input, the output voltage changes as shown in Example 13-4.

Example 13-4

Assuming that R = 5K and I_{ref} = 2 mA, calculate V_{out} for the following binary inputs:
(a) 10011001 binary (99H)   (b) 11001000 (C8H)

Solution:

(a) I_{out} = 2 mA (153/255) = 1.195 mA and V_{out} = 1.195 mA × 5K = 5.975 V
(b) I_{out} = 2 mA (200/256) = 1.562 mA and V_{out} = 1.562 mA × 5K = 7.8125 V

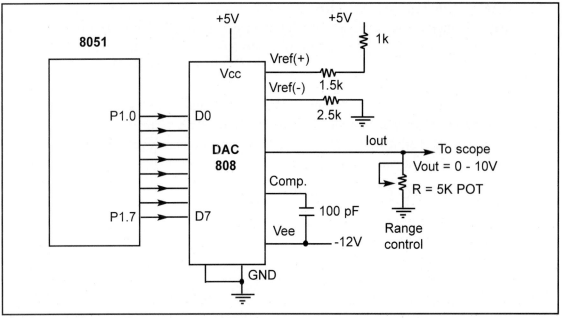

Figure 13-7. 8051 Connection to DAC808

Example 13-5

In order to generate a stair-step ramp, set up the circuit in Figure 13-7 and connect the output to an oscilloscope. Then write a program to send data to the DAC to generate a stair-step ramp.

Solution:

```
        CLR   A
AGAIN:  MOV   P1,A          ;send data to DAC
        INC   A
        ACALL DELAY         ;let DAC recover
        SJMP  AGAIN
```

Example 13-6

Verify the values given for the following angles: (a) 30   (b) 60.

Solution:

(a) $V_{out} = 5\ V + (5\ V \times \sin \theta) = 5\ V + 5 \times \sin 30 = 5\ V + 5 \times 0.5 = 7.5\ V$
    DAC input values = 7.5 V × 25.6 = 192 (decimal)

(b) $V_{out} = 5\ V + (5\ V \times \sin \theta) = 5\ V + 5 \times \sin 60 = 5\ V + 5 \times 0.866 = 9.33\ V$
    DAC input values = 9.33 V × 25.6 = 238 (decimal)

Table 13-5: Angle v. Voltage Magnitude for Sine Wave

Angle θ (degrees)	Sin θ	Vout (Voltage Magnitude) 5 V + (5 V × sin θ)	Values Sent to DAC (decimal) (Voltage Mag × 25.6)
0	0	5	128
30	0.5	7.5	192
60	0.866	9.33	238
90	1.0	10	255
120	0.866	9.33	238
150	0.5	7.5	192
180	0	5	128
210	−0.5	2.5	64
240	−0.866	0.669	17
270	−1.0	0	0
300	−0.866	0.669	17
330	−0.5	2.5	64
360	0	5	128

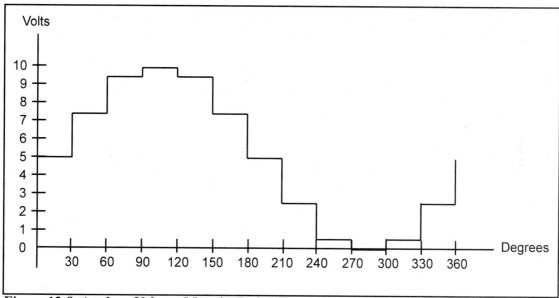

Figure 13-8. Angle v. Voltage Magnitude for Sine Wave

Generating a sine wave

To generate a sine wave, we first need a table whose values represent the magnitude of the sine of angles between 0 and 360 degrees. The values for the sine function vary from −1.0 to +1.0 for 0 to 360 degree angles. Therefore, the table values are integer numbers representing the voltage magnitude for the sine of theta. This method ensures that only integer numbers are output to the DAC by the 8051 microcontroller. Table 13-5 shows the angles, the sine values, the voltage magnitude, and the integer values representing the voltage magnitude for each angle with 30-degree increments. To generate Table 13-5, we assumed the full-scale voltage of 10 V for DAC output (as designed in Figure 13-7). Full-scale output of the DAC is achieved when all the data inputs of the DAC are high. Therefore, to achieve the full-scale 10 V output, we use the following equation. V_{out} of DAC for various sine degrees is calculated as follows.

$$V_{out} = 5\ V + (5 \times \sin \theta)$$

To find the value sent to DAC for various angles, we simply multiply the V_{out} voltage by 25.60 because there are 256 steps and full-scale V_{out} is 10 volts. Therefore, 256 steps / 10 V = 25.6 steps per volt. To further clarify this, look at the following code. This program sends the values to DAC continuously (in an infinite loop) to produce a crude sine wave. See Figure 13-7.

```
AGAIN:     MOV  DPTR,#TABLE
           MOV  R2,#COUNT
BACK:      CLR  A
           MOVC A,@A+DPTR
           MOV  P1,A
           INC  DPTR
           DJNZ R2,BACK
           SJMP AGAIN
           ORG  300
TABLE:     DB   128,192,238,255,238,192 ;see Table 13-5
           DB   128,64,17,0,17,64,128
;To get a better looking sine wave, regenerate
;Table 13-5 for 2-degree angles
```

Review Questions

1. In a DAC, input is _____ (digital, analog) and output is _____ (digital, analog). Answer for ADC input and output as well.
2. DAC808 is an ____-bit D-to-A converter.
3. (a) The output of DAC808 is in _____ (current, voltage).
   (b) True or false. This is ideal to drive a motor.

SUMMARY

This chapter continued showing how to interface the 8051 with real-world devices. Devices covered in this chapter were stepper motors, keyboards, and DAC (digital-to-analog converter) devices.

First, the basic operation of stepper motors was defined, along with key terms used in describing and controlling stepper motor operation. Then the 8051 was interfaced with a stepper motor. The stepper motor was then controlled via 8051 programming.

Keyboards are one of the most widely used input devices for the 8051. This chapter described the operation of keyboards, including key press and detection mechanisms. Then the 8051 was shown interfacing with a keyboard. 8051 programs were written to return the ASCII code for the pressed key.

Finally, the 8051 was interfaced with DAC devices. A typical DAC chip will take digital inputs and convert them to current. By connecting a resistor to the ouput pin, the result is converted to voltage. This voltage level can be monitored on an oscilloscope. This chapter showed how to interface the 8051 with a DAC chip. Then, a simple program was written to produce a crude sine wave on a scope.

PROBLEMS

SECTION 13.1: INTERFACING A STEPPER MOTOR

1. If a motor takes 90 steps to make one complete revolution, what is the step angle for this motor?
2. Calculate the number of steps per revolution for a step angle of 7.5 degrees.
3. Finish the normal four-step sequence clockwise if the first step is 0011 (binary).
4. Finish the normal four-step sequence clockwise if the first step is 1100 (binary).
5. Finish the normal four-step sequence counterclockwise if the first step is 1001 (binary).
6. Finish the normal four-step sequence counterclockwise if the first step is 0110 (binary).
7. What is the purpose of the ULN2003 placed between the 8051 and the stepper motor? Can we use that for 3A motors?
8. Which of the following cannot be a sequence in the normal 4-step sequence for a stepper motor?
   (a) CCH   (b) DDH   (c) 99H   (d) 33H
9. What is the effect of a time delay in between issuing each step?
10. In Question 9, how can we make a stepper motor go faster?

SECTION 13.2: 8051 INTERFACING TO THE KEYBOARD

11. In reading the columns of a keyboard matrix, if no key is pressed we should get all _____ (1s, 0s).
12. In Figure 13-5, to detect the key press, which of the following is grounded?
    (a) all rows   (b) one row at time   (c) both (a) and (b)
13. In Figure 13-5, to identify the key pressed, which of the following is grounded?
    (a) all rows   (b) one row at time   (c) both (a) and (b)
14. For Figure 13-5, indicate the column and row for each of the following.
    (a) D3 - D0 = 0111   (b) D3 - D0 = 1110
15. Indicate the steps to detect the key press.
16. Indicate the steps to identify the key pressed.
17. Modify Program 13-1 and Figure 13-5 for a 4 x 5 keyboard (4 rows, 5 columns).
18. Modify Program 13-1 and Figure 13-5 for a 6 x 6 keyboard.

---

19. Indicate an advantage and a disadvantage of using an IC chip for keyboard scanning and decoding instead of using a microcontroller.
20. What is the best compromise for the answer to Problem 19?

SECTION 13.3: INTERFACING A DAC TO THE 8051

21. True or false. DAC808 is the same as DAC1408. Are they pin compatible?
22. Find the number of discrete voltages provided by the n-bit DAC for the following.
    (a) n = 8 (b) n = 10 (c) n = 12
23. For DAC1408, if I_{ref} = 2 mA show how to get the I_{out} of 1.99 when all inputs are high.
24. Find the I_{out} for the following inputs. Assume I_{ref} = 2 mA for DAC808.
    (a) 10011001 (b) 11001100 (c) 11101110
    (d) 00100010 (e) 00001001 (f) 10001000
25. To get a smaller step, we need a DAC with _____ (more, less) digital inputs.
26. To get full-scale output what should be the inputs for DAC?

ANSWERS TO REVIEW QUESTIONS

SECTION 13.1: INTERFACING A STEPPER MOTOR

1. 0110, 0011, 1001, 1100 for clockwise, and 0110, 1100, 1001, 0011 for counter-clockwise
2. 72
3. Because the microcontroller pins do not provide sufficient current to drive the stepper motor

SECTION 13.2: 8051 INTERFACING TO THE KEYBOARD

1. True   2. Column 3   3. True   4. 0   5. True

SECTION 13.3: INTERFACING A DAC TO THE 8051

1. Digital, analog. In ADC the input is analog, the output is digital.
2. 8   3. (a) current   (b) true

CHAPTER 14

8031/51 INTERFACING TO EXTERNAL MEMORY

OBJECTIVES

Upon completion of this chapter, you will be able to:

≫≫   Contrast and compare various types of semiconductor memories
in terms of their capacity, organization, and access time
≫≫   Describe the relationship between the number of memory locations
on a chip, the number of data pins, and the chip capacity
≫≫   Define ROM memory and describe its use in 8051-based systems
≫≫   Contrast and compare PROM, EPROM, UV-EPROM, EEPROM,
flash memory EPROM, and mask ROM memories
≫≫   Define RAM memory and describe its use in 8051-based systems
≫≫   Contrast and compare SRAM, NV-RAM, check-sum byte,
and DRAM memories
≫≫   List the steps a CPU follows in memory address decoding
≫≫   Contrast and compare address decoding techniques such as the use
of logic gates, 3-to-8 decoders, and programmable logic
≫≫   Explain how to interface ROM with the 8031
≫≫   Explain how to use both on-chip and off-chip memory with the 8051
≫≫   Code 8051 programs accessing the 64K-byte data memory space

In this chapter we discuss how to interface the 8031/51 to external memory. In Section 14.1 we study semiconductor memory concepts with emphasis on different types of ROMs. In Section 14.2 address decoding concepts are discussed. In Section 14.3 we explore 8031 interfacing with external ROM. Interfacing the 8031/51 with external RAM is detailed in Section 14.4.

SECTION 14.1: SEMICONDUCTOR MEMORY

In this section we discuss various types of semiconductor memories and their characteristics such as capacity, organization, and access time. In the design of all microprocessor-based systems, semiconductor memories are used as primary storage for code and data. Semiconductor memories are connected directly to the CPU and they are the memory that the CPU first asks for information (code and data). For this reason, semiconductor memories are sometimes referred to as *primary memory*. The main requirement of primary memory is that it must be fast in responding to the CPU; only semiconductor memories can do that. The most widely used semiconductor memories are ROM and RAM. Before we discuss different types of RAM and ROM, we discuss some important terminology common to all semiconductor memories, such as capacity, organization, and speed.

Memory capacity

The number of bits that a semiconductor memory chip can store is called chip *capacity*. It can be in units of Kbits (kilobits), Mbits (megabits), and so on. This must be distinguished from the storage capacity of computer systems. While the memory capacity of a memory IC chip is always given in bits, the memory capacity of a computer system is given in bytes. For example, an article in a technical journal may state that the 16M chip has become popular. In that case, although it is not mentioned that 16M means 16 megabits, it is understood since the article is referring to an IC memory chip. However, if an advertisement states that a computer comes with 16M memory, since it is referring to a computer system it is understood that 16M means 16 megabytes.

Memory organization

Memory chips are organized into a number of locations within the IC. Each location can hold 1 bit, 4 bits, 8 bits, or even 16 bits, depending on how it is designed internally. The number of bits that each location within the memory chip can hold is always equal to the number of data pins on the chip. How many locations exist inside a memory chip? That depends on the number of address pins. The number of locations within a memory IC always equals 2 to the power of the number of address pins. Therefore, the total number of bits that a memory chip can store is equal to the number of locations times the number of data bits per location. To summarize:

1. A memory chip contains 2^x locations, where x is the number of address pins.
2. Each location contains y bits, where y is the number of data pins on the chip.
3. The entire chip will contain $2^x \times y$ bits, where x is the number of address pins and y is the number of data pins on the chip.

Speed

Table 14-1: Powers of 2

One of the most important characteristics of a memory chip is the speed at which its data can be accessed. To access the data, the address is presented to the address pins, the READ pin is activated, and after a certain amount of time has elapsed, the data shows up at the data pins. The shorter this elapsed time, the better, and consequently, the more expensive the memory chip. The speed of the memory chip is commonly referred to as its *access time*. The access time of memory chips varies from a few nanoseconds to hundreds of nanoseconds, depending on the IC technology used in the design and fabrication process.

The three important memory characteristics of capacity, organization, and access time will be used extensively in this chapter. Table 14-1 serves as a reference for the calculation of memory characteristics. Examples 14-1 and 14-2 demonstrate these concepts.

x	2^x
10	1K
11	2K
12	4K
13	8K
14	16K
15	32K
16	64K
17	128K
18	256K
19	512K
20	1M
21	2M
22	4M
23	8M
24	16M

Example 14-1

A given memory chip has 12 address pins and 4 data pins. Find:
(a) the organization, and (b) the capacity.

Solution:

(a)   This memory chip has 4096 locations ($2^{12} = 4096$), and each location can hold 4 bits of data. This gives an organization of 4096 × 4, often represented as 4Kx4.

(b)   The capacity is equal to 16K bits since there is a total of 4K locations and each location can hold 4 bits of data.

Example 14-2

A 512K memory chip has 8 pins for data. Find:
(a) the organization, and (b) the number of address pins for this memory chip.

Solution:

(a)   A memory chip with 8 data pins means that each location within the chip can hold 8 bits of data. To find the number of locations within this memory chip, divide the capacity by the number of data pins. 512K/8 = 64K; therefore, the organization for this memory chip is 64Kx8.

(b)   The chip has 16 address lines since $2^{16} = 64$K.

ROM (read-only memory)

ROM is a type of memory that does not lose its contents when the power is turned off. For this reason, ROM is also called *nonvolatile* memory. There are different types of read-only memory, such as PROM, EPROM, EEPROM, flash EPROM, and mask ROM. Each is explained below.

PROM (programmable ROM)

PROM refers to the kind of ROM that the user can burn information into. In other words, PROM is a user-programmable memory. For every bit of the PROM, there exists a fuse. PROM is programmed by blowing the fuses. If the information burned into PROM is wrong, that PROM must be discarded since its internal fuses are blown permanently. For this reason, PROM is also referred to as OTP (one-time programmable). Programming ROM, also called *burning* ROM, requires special equipment called a ROM burner or ROM programmer.

EPROM (erasable programmable ROM) and UV-EPROM

EPROM was invented to allow making changes in the contents of PROM after it is burned. In EPROM, one can program the memory chip and erase it thousands of times. This is especially necessary during development of the prototype of a microprocessor-based project. A widely used EPROM is called UV-EPROM where UV stands for ultra-violet. The only problem with UV-EPROM is that erasing its contents can take up to 20 minutes. All UV-EPROM chips have a window that is used to shine ultraviolet (UV) radiation to erase its contents. For this reason, EPROM is also referred to as UV-erasable EPROM or simply UV-EPROM. Figure 14-1 shows the pins for a UV-EPROM chip.

To program a UV-EPROM chip, the following steps must be taken:

1. Its contents must be erased. To erase a chip, it is removed from its socket on the system board and placed in EPROM erasure equipment to expose it to UV radiation for 15 - 20 minutes.
2. Program the chip. To program a UV-EPROM chip, place it in the ROM burner (programmer). To burn code or data into EPROM, the ROM burner uses 12.5 volts or higher, depending on the EPROM type. This voltage is referred to as V_{PP} in the UV-EPROM data sheet.
3. Place the chip back into its socket on the system board.

As can be seen from the above steps, in the same way that there is an EPROM programmer (burner), there is also separate EPROM erasure equipment. The main problem, and indeed the major disadvantage of UV-EPROM, is that it cannot be programmed while in the system board. To find a solution to this problem, EEPROM was invented.

Notice the patterns of the IC numbers in Table 14-2. For example, part number 27128-25 refers to UV-EPROM that has a capacity of 128K bits and access time of 250 nanoseconds. The capacity of the memory chip is indicated in the part number and the access time is given with a zero dropped. In part numbers, C refers to CMOS technology. Notice that 27XX always refers to UV-EPROM chips. For a comprehensive list of available memory chips see JAMECO (jameco.com) or JDR (jdr.com) catalogs.

Table 14-2: Some UV-EPROM Chips

Part #	Capacity	Org.	Access	Pins	V_{PP}
2716	16K	2Kx8	450 ns	24	25 V
2732	32K	4Kx8	450 ns	24	25 V
2732A-20	32K	4Kx8	200 ns	24	21 V
27C32-1	32K	4Kx8	450 ns	24	12.5 V CMOS
2764-20	64K	8Kx8	200 ns	28	21 V
2764A-20	64K	8Kx8	200 ns	28	12.5 V
27C64-12	64K	8Kx8	120 ns	28	12.5 V CMOS
27128-25	128K	16Kx8	250 ns	28	21 V
27C128-12	128K	16Kx8	120 ns	28	12.5 V CMOS
27256-25	256K	32Kx8	250 ns	28	12.5 V
27C256-15	256K	32Kx8	150 ns	28	12.5 V CMOS
27512-25	512K	64Kx8	250 ns	28	12.5 V
27C512-15	512K	64Kx8	150 ns	28	12.5 V CMOS
27C010-15	1024K	128Kx8	150 ns	32	12.5 V CMOS
27C020-15	2048K	256Kx8	150 ns	32	12.5 V CMOS
27C040-15	4096K	512Kx8	150 ns	32	12.5 V CMOS

Example 14-3

For ROM chip 27128, find the number of data and address pins.

Solution:

The 27128 has a capacity of 128K bits. It has 16Kx8 organization (all ROMs have 8 data pins), which indicates that there are 8 pins for data, and 14 pins for address ($2^{14} = 16K$).

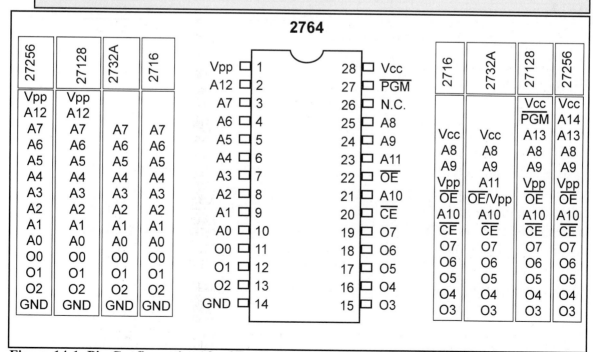

Figure 14-1. Pin Configurations for 27xx ROM Family

EEPROM (electrically erasable programmable ROM)

EEPROM has several advantages over EPROM, such as the fact that its method of erasure is electrical and therefore instant, as opposed to the 20-minute erasure time required for UV-EPROM. In addition, in EEPROM one can select which byte to be erased, in contrast to UV-EPROM, in which the entire contents of ROM are erased. However, the main advantage of EEPROM is the fact that one can program and erase its contents while it is still in the system board. It does not require physical removal of the memory chip from its socket. In other words, unlike UV-EPROM, EEPROM does not require an external erasure and programming device. To utilize EEPROM fully, the designer must incorporate into the system board the circuitry to program the EEPROM, using 12.5 V for V_{PP}. EEPROM with V_{PP} of 5 V is available, but it is more expensive. In general, the cost per bit for EEPROM is much higher than for UV-EPROM.

Table 14-3: Some EEPROM and Flash chips

EEPROMs

Part No.	Capacity	Org.	Speed	Pins	V_{PP}
2816A-25	16K	2Kx8	250 ns	24	5 V
2864A	64K	8Kx8	250 ns	28	5 V
28C64A-25	64K	8Kx8	250 ns	28	5 V CMOS
28C256-15	256K	32Kx8	150 ns	28	5 V
28C256-25	256K	32Kx8	250 ns	28	5 V CMOS

Flash

Part No.	Capacity	Org.	Speed	Pins	V_{PP}
28F256-20	256K	32Kx8	200 ns	32	12 V CMOS
28F010-15	1024K	128Kx8	150 ns	32	12 V CMOS
28F020-15	2048K	256Kx8	150 ns	32	12 V CMOS

Flash memory EPROM

Since the early 1990s, flash EPROM has become a popular user-programmable memory chip, and for good reasons. First, the process of erasure of the entire contents takes less than a second, or one might say in a flash, hence its name, flash memory. In addition, the erasure method is electrical and for this reason it is sometimes referred to as flash EEPROM. To avoid confusion, it is commonly called flash memory. The major difference between EEPROM and flash memory is the fact that when flash memory's contents are erased the entire device is erased, in contrast to EEPROM, where one can erase a desired section or byte. Although there are some flash memories recently made available in which the contents are divided into blocks and the erasure can be done block by block, unlike EEPROM, no byte erasure option is available. Due to the fact that flash memory can be programmed while it is in its socket on the system board, it is becoming widely used as a way to upgrade the BIOS ROM of the PC. Some designers believe that flash memory will replace the hard disk as a mass storage medium. This would increase the performance of the computer tremendously, since flash memory is semiconductor memory with access time in the range of 100 ns compared with disk access time in the range of tens of milliseconds. For this to happen, flash memory's program/erase cycles must become infinite, just like hard disks. Program/erase cycle refers to the number of times that a chip can be erased and programmed before it becomes unusable. At this time, the program/erase cycle is 100,000 for flash and EEPROM, 1000 for UV-EPROM, and infinite for RAM and disks.

Mask ROM

Mask ROM refers to a kind of ROM in which the contents are programmed by the IC manufacturer. In other words, it is not a user-programmable ROM. The terminology mask is used in IC fabrication. Since the process is costly, mask ROM is used when the needed volume is high (hundreds of thousands) and it is absolutely certain that the contents will not change. It is common practice to use UV-EPROM or Flash for the development phase of a project, and only after the code/data have been finalized is the mask version of the product ordered. The main advantage of mask ROM is its cost, since it is significantly cheaper than other kinds of ROM, but if an error in the data/code is found, the entire batch must be thrown away. Many manufacturers of 8051 microcontrollers support the mask ROM version of the 8051. It must be noted that all ROM memories have 8 bits for data pins; therefore, the organization is x8.

RAM (random access memory)

RAM memory is called *volatile* memory since cutting off the power to the IC will result in the loss of data. Sometimes RAM is also referred to as RAWM (read and write memory), in contrast to ROM, which cannot be written to. There are three types of RAM: static RAM (SRAM), NV-RAM (nonvolatile RAM), and dynamic RAM (DRAM). Each is explained separately.

SRAM (static RAM)

Storage cells in static RAM memory are made of flip-flops and therefore do not require refreshing in order to keep their data. This is in contrast to DRAM, discussed below. The problem with the use of flip-flops for storage cells is that each cell requires at least 6 transistors to build, and the cell holds only 1 bit of data. In recent years, the cells have been made of 4 transistors, which still is too many. The use of 4-transistor cells plus the use of CMOS technology has given birth to a high-capacity SRAM, but its capacity is far below DRAM. Table 14-4 shows some examples of SRAM. Figure 14-2 shows the pin diagram for an SRAM chip. In Figure 14-2, notice that WE is write enable, and OE is output enable, for read and write signals, respectively.

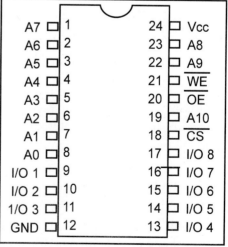

Figure 14-2. 2Kx8 SRAM Pins

Table 14-4: Some SRAM and NV-RAM Chips

SRAM

Part No.	Capacity	Org.	Speed	Pins	V_PP
6116P-1	16K	2Kx8	100 ns	24	CMOS
6116P-2	16K	2Kx8	120 ns	24	CMOS
6116P-3	16K	2Kx8	150 ns	24	CMOS
6116LP-1	16K	2Kx8	100 ns	24	Low power CMOS
6116LP-2	16K	2Kx8	120 ns	24	Low power CMOS
6116LP-3	16K	2Kx8	150 ns	24	Low Power CMOS
6264P-10	64K	8Kx8	100 ns	28	CMOS
6264LP-70	64K	8Kx8	70 ns	28	Low power CMOS
6264LP-12	64K	8Kx8	120 ns	28	Low power CMOS
62256LP-10	256K	32Kx8	100 ns	28	Low Power CMOS
62256LP-12	256K	32Kx8	120 ns	28	Low Power CMOS

NV-RAM from Dallas Semiconductor

Part No.	Capacity	Org.	Speed	Pins	V_PP
DS1220Y-150	16K	2Kx8	150 ns	24	
DS1225AB-150	64K	8Kx8	150 ns	28	
DS1230Y-85	256K	32Kx8	85 ns	28	

NV-RAM (nonvolatile RAM)

Whereas SRAM is volatile, there is a new type of nonvolatile RAM called NV-RAM. Like other RAMs, it allows the CPU to read and write to it, but when the power is turned off the contents are not lost. NV-RAM combines the best of RAM and ROM: the read and write ability of RAM, plus the nonvolatility of ROM. To retain its contents, every NV-RAM chip internally is made of the following components:

1. It uses extremely power-efficient (very, very low power consumption) SRAM cells built out of CMOS.
2. It uses an internal lithium battery as a backup energy source.
3. It uses an intelligent control circuitry. The main job of this control circuitry is to monitor the V_{CC} pin constantly to detect loss of the external power supply. If the power to the V_{CC} pin falls below out-of-tolerance conditions, the control circuitry switches automatically to its internal power source, the lithium battery. In this way, the internal lithium power source is used to retain the NV-RAM contents only when the external power source is off.

It must be emphasized that all three of the components above are incorporated into a single IC chip, and for this reason nonvolatile RAM is a very expensive type of RAM as far as cost per bit is concerned. Offsetting the cost, however, is the fact that it can retain its contents up to ten years after the power has been turned off and allows one to read and write in exactly the same way as SRAM. See Table 14-4 for NV-RAM parts made by Dallas Semiconductor.

Checksum byte ROM

To ensure the integrity of the ROM contents, every system must perform the checksum calculation. The process of checksum will detect any corruption of the contents of ROM. One of the causes of ROM corruption is current surge, either when the system is turned on or during operation. To ensure data integrity in ROM, the checksum process uses what is called a *checksum byte*. The checksum byte is an extra byte that is tagged to the end of a series of bytes of data. To calculate the checksum byte of a series of bytes of data, the following steps can be taken.

1. Add the bytes together and drop the carries.
2. Take the 2's complement of the total sum, and that is the checksum byte, which becomes the last byte of the series.

To perform the checksum operation, add all the bytes, including the checksum byte. The result must be zero. If it is not zero, one or more bytes of data have been changed (corrupted). To clarify these important concepts, see Example 14-4.

Example 14-4

Assume that we have 4 bytes of hexadecimal data: 25H, 62H, 3FH, and 52H.
(a) Find the checksum byte, (b) perform the checksum operation to ensure data integrity, and (c) if the second byte 62H has been changed to 22H, show how checksum detects the error.

Solution:

(a) Find the checksum byte.

```
     25H
  +  62H
  +  3FH
  +  52H
    118H
```
The checksum is calculated by first adding the bytes. The sum is 118H, and dropping the carry, we get 18H. The checksum byte is the 2's complement of 18H, which is E8H.

(b) Perform the checksum operation to ensure data integrity.

```
     25H
  +  62H
  +  3FH
  +  52H
  +  E8H
    200H   (dropping the carries)
```
Adding the series of bytes including the checksum byte must result in zero. This indicates that all the bytes are unchanged and no byte is corrupted.

(c) If the second byte 62H has been changed to 22H, show how checksum detects the error.

```
     25H
  +  22H
  +  3FH
  +  52H
  +  E8H
    1C0H   (dropping the carry, we get C0H.)
```
Adding the series of bytes including the checksum byte shows that the result is not zero, which indicates that one or more bytes have been corrupted.

DRAM (dynamic RAM)

Since the early days of the computer, the need for huge, inexpensive read/write memory has been a major preoccupation of computer designers. In 1970, Intel Corporation introduced the first dynamic RAM (random access memory). Its density (capacity) was 1024 bits and it used a capacitor to store each bit. The use of a capacitor as a means to store data cuts down the number of transistors needed to build the cell; however, it requires constant refreshing due to leakage. This is in contrast to SRAM (static RAM), whose individual cells are made of flip-flops. Since each bit in SRAM uses a single flip-flop and each flip-flop requires 6 transistors, SRAM has much larger memory cells and consequently lower density. The use of capacitors as storage cells in DRAM results in much smaller net memory cell size.

The advantages and disadvantages of DRAM memory can be summarized as follows. The major advantages are high density (capacity), cheaper cost per bit, and lower power consumption per bit. The disadvantage is that it must be refreshed periodically, due to the fact that the capacitor cell loses its charge; furthermore, while it is being refreshed, the data cannot be accessed. This is in contrast to SRAM's flip-flops, which retain data as long as the power is on, which do not need to be refreshed, and whose contents can be accessed at any time. Since 1970, the capacity of DRAM has exploded. After the 1K-bit (1024) chip came the 4K-bit in 1973, and then the 16K chip in 1976. The 1980s saw the introduction of 64K, 256K, and finally 1M and 4M memory chips. The 1990s will see 16M, 64M, 256M, and possibly 1G-bit DRAM chips. Keep in mind that when talking about IC memory chips, the capacity is always assumed to be in bits. Therefore, a 1M chip means 1 megabit and a 256K chip means a 256K-bit memory chip. However, when talking about the memory of a computer system, it is always assumed to be in bytes.

Packaging issue in DRAM

In DRAM there is a problem of packing a large number of cells into a single chip with the normal number of pins assigned to addresses. For example, a 64K-bit chip (64Kx1) must have 16 address lines and 1 data line, requiring 16 pins to send in the address if the conventional method is used. This is in addition to V_{CC} power, ground, and read/write control pins. Using the conventional method of data access, the large number of pins defeats the purpose of high density and small packaging, so dearly cherished by IC designers. Therefore, to reduce the number of pins needed for addresses, multiplexing/demultiplexing is used. The method used is to split the address in half and send in each half of the address through the same pins, thereby requiring fewer address pins. Internally, the DRAM structure is divided into a square of rows and columns. The first half of the address is called the row and the second half is called the column. For example, in the case of DRAM of 64Kx1 organization, the first half of the address is sent in through the 8 pins A0 - A7, and by activating RAS (row address strobe), the internal latches inside DRAM grab the first half of the address. After that, the second half of the address is sent in through the same pins, and by activating CAS (column address strobe), the internal latches inside DRAM latch the second half of the address. This results in using 8 pins for addresses plus RAS and CAS, for a total of 10 pins, instead of the 16 pins that would be required without multiplexing. To access a bit of data from DRAM, both row and column addresses must be provided. For this concept to work, there must be a 2 by 1 multiplexer outside the DRAM circuitry and a demultiplexer inside every DRAM chip. Due to the complexities associated with DRAM interfacing (RAS, CAS, the need for multiplexer and refreshing circuitry), there are DRAM controllers designed to make DRAM interfacing much easier. However, many small microcontroller-based projects that do not require much RAM (usually less than 64K bytes) use SRAM of types EEPROM and NV-RAM, instead of DRAM.

DRAM organization

In the discussion of ROM, we noted that all of them have 8 pins for data. This is not the case for DRAM memory chips, which can have any of the x1, x4, x8, or x16 organizations. See Example 14-5.

In memory chips, the data pins are also called I/O. In some DRAMs there are separate Din and Dout pins. Figure 14-3 shows a 256Kx1 DRAM chip with pins A0 - A8 for address, RAS and CAS, WE (write enable), and data in and data out, as well as power and ground.

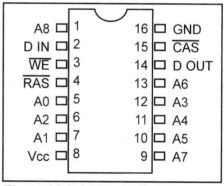

Figure 14-3. 256Kx1 DRAM

Table 14-5: Some Widely Used DRAMs

Part No.	Speed	Capacity	Org.	Pins
4164-15	150 ns	64K	64Kx1	16
41464-8	80 ns	256K	64Kx4	18
41256-15	150 ns	256K	256Kx1	16
41256-6	60 ns	256K	256Kx1	16
414256-10	100 ns	1M	256Kx1	20
511000P-8	80 ns	1M	1Mx1	18
514100-7	70 ns	4M	4Mx1	20

Example 14-5

Discuss the number of pins set aside for addresses in each of the following memory chips.   (a) 16Kx4 DRAM   (b) 16Kx4 SRAM

Solution:

Since 2^{14} = 16K:
(a)   For DRAM we have 7 pins (A0 - A6) for the address pins and 2 pins for RAS and CAS.
(b)   For SRAM we have 14 pins for address and no pins for RAS and CAS since they are associated only with DRAM. In both cases we have 4 pins for the data bus.

Review Questions

1. The speed of semiconductor memory is in the range of
   (a) microseconds (b) milliseconds
   (c) nanoseconds (d) picoseconds

2. Find the organization and chip capacity for each ROM with the indicated number of address and data pins.
   (a) 14 address, 8 data (b) 16 address, 8 data (c) 12 address, 8 data

3. Find the organization and chip capacity for each RAM with the indicated number of address and data pins.
   (a) 11 address, 1 data SRAM   (b) 13 address, 4 data SRAM
   (c) 17 address, 8 data SRAM   (d) 8 address, 4 data DRAM
   (e) 9 address, 1 data DRAM   (f) 9 address, 4 data DRAM

4. Find the capacity and number of pins set aside for address and data for memory chips with following organizations.
   (a) 16Kx4 SRAM   (b) 32Kx8 EPROM   (c) 1Mx1 DRAM
   (d) 256Kx4 SRAM   (e) 64Kx8 EEPROM   (f) 1Mx4 DRAM

5. Which of the following is (are) volatile memory?
   (a) EEPROM   (b) SRAM (c) DRAM   (d) NV-RAM

SECTION 14.2: MEMORY ADDRESS DECODING

In this section we discuss address decoding. The CPU provides the address of the data desired, but it is the job of the decoding circuitry to locate the selected memory block. To explore the concept of decoding circuitry, we look at various methods used in decoding the addresses. In this discussion we use SRAM or ROM for the sake of simplicity.

Memory chips have one or more pins called CS (chip select), which must be activated for the memory's contents to be accessed. Sometimes the chip select is also referred to as chip enable (CE). In connecting a memory chip to the CPU, note the following points.

1. The data bus of the CPU is connected directly to the data pins of the memory chip.
2. Control signals RD (read) and WR (memory write) from the CPU are connected to the OE (output enable) and WE (write enable) pins of the memory chip, respectively.
3. In the case of the address buses, while the lower bits of the addresses from the CPU go directly to the memory chip address pins, the upper ones are used to activate the CS pin of the memory chip. It is the CS pin that along with RD/WR allows the flow of data in or out of the memory chip. No data can be written into or read from the memory chip unless CS is activated.

As can be seen from the data sheets of SRAM and ROM, the CS input of a memory chip is normally active low and is activated by the output of the memory decoder. Normally memories are divided into blocks and the output of the decoder selects a given memory block. There are three ways to generate a memory block selector: (a) using simple logic gates, (b) using the 74LS138, or (c) using programmable logics. Each method is described below with some examples.

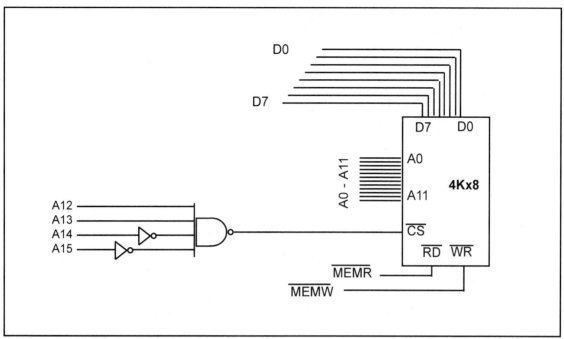

Figure 14-4. Logic Gate as Decoder

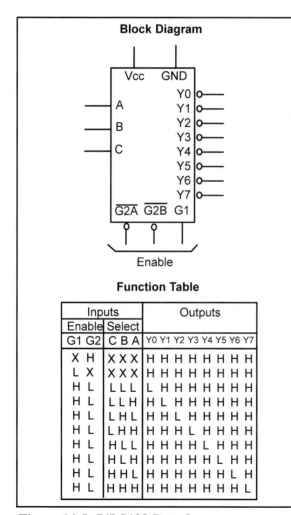

Block Diagram

Vcc   GND

A   Y0
B   Y1
C   Y2
       Y3
       Y4
       Y5
       Y6
       Y7

$\overline{G2A}$ $\overline{G2B}$ G1

Enable

Function Table

Inputs					Outputs							
Enable		Select										
G1	G2	C	B	A	Y0	Y1	Y2	Y3	Y4	Y5	Y6	Y7
X	H	X	X	X	H	H	H	H	H	H	H	H
L	X	X	X	X	H	H	H	H	H	H	H	H
H	L	L	L	L	L	H	H	H	H	H	H	H
H	L	L	L	H	H	L	H	H	H	H	H	H
H	L	L	H	L	H	H	L	H	H	H	H	H
H	L	L	H	H	H	H	H	L	H	H	H	H
H	L	H	L	L	H	H	H	H	L	H	H	H
H	L	H	L	H	H	H	H	H	H	L	H	H
H	L	H	H	L	H	H	H	H	H	H	L	H
H	L	H	H	H	H	H	H	H	H	H	H	L

Figure 14-5. 74LS138 Decoder
(Reprinted by permission of Texas Instruments, Copyright Texas Instruments, 1988)

Simple logic gate address decoder

The simplest method of decoding circuitry is the use of NAND or other gates. The fact that the output of a NAND gate is active low, and that the CS pin is also active low makes them a perfect match. In cases where the CS input is active high, an AND gate must be used. Using a combination of NAND gates and inverters, one can decode any address range. An example of this is shown in Figure 14-4, which shows that A15 - A12 must be 0011 in order to select the chip. This results in the assignment of addresses 3000H to 3FFFH to this memory chip.

Using the 74LS138 3-8 decoder

This is one of the most widely used address decoders. The 3 inputs A, B, and C generate 8 active-low outputs Y0 - Y7. See Figure 14-5. Each Y output is connected to CS of a memory chip, allowing control of 8 memory blocks by a single 74LS138. In the 74LS138, where A, B, and C select which output is activated, there are three additional inputs, G2A, G2B, and G1. G2A and G2B are both active low, and G1 is active high. If any one of the inputs G1, G2A, or G2B is not connected to an address signal (sometimes they are connected to a control signal), they must be activated permanently either by V_{CC} or ground, depending on the activation level. Example 14-6 shows the design and the address range calculation for the 74LS138 decoder.

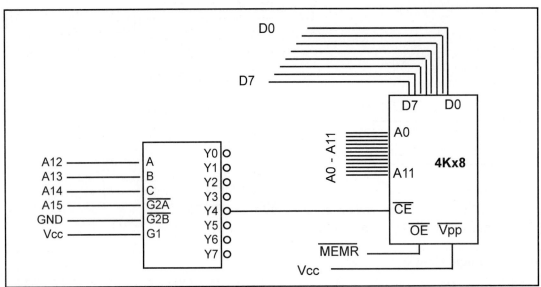

Figure 14-6. 74LS138 as Decoder

Example 14-6

Looking at the design in Figure 14-6, find the address range for the following.
(a) Y4, (b) Y2, and (c) Y7.

Solution:

(a) The address range for Y4 is calculated as follows.

A15	A14	A13	A12	A11	A10	A9	A8	A7	A6	A5	A4	A3	A2	A1	A0
0	1	0	0	0	0	0	0	0	0	0	0	0	0	0	0
0	1	0	0	1	1	1	1	1	1	1	1	1	1	1	1

The above shows that the range for Y4 is 4000H to 4FFFH. In Figure 14-6, notice that A15 must be 0 for the decoder to be activated. Y4 will be selected when A14 A13 A12 = 100 (4 in binary). The remaining A11 - A0 will be 0 for the lowest address and 1 for the highest address.

(b) The address range for Y2 is 2000H to 2FFFH.

A15	A14	A13	A12	A11	A10	A9	A8	A7	A6	A5	A4	A3	A2	A1	A0
0	0	1	0	0	0	0	0	0	0	0	0	0	0	0	0
0	0	1	0	1	1	1	1	1	1	1	1	1	1	1	1

(c) The address range for Y7 is 7000H to 7FFFH.

A15	A14	A13	A12	A11	A10	A9	A8	A7	A6	A5	A4	A3	A2	A1	A0
0	1	1	1	0	0	0	0	0	0	0	0	0	0	0	0
0	1	1	1	1	1	1	1	1	1	1	1	1	1	1	1

Using programmable logic as an address decoder

Other widely used decoders are programmable logic chips such as PAL and GAL chips. One disadvantage of these chips is that one must have access to a PAL/GAL software and burner (programmer), whereas the 74LS138 needs neither of these. The advantage of these chips is that they are much more versatile since they can be programmed for any combination of address ranges. This plus the fact that PALs and GALs have 10 or more inputs (in contrast to 6 in the 74138) means that it can accommodate more address inputs.

Review Questions

1. A given memory block uses addresses 4000H - 7FFFH. How many K bytes is this memory block?
2. The 74138 is a(n) _____ by _____ decoder.
3. In the 74138 give the status of G2A and G2B for the chip to be enabled.
4. In the 74138 give the status of G1 for the chip to be enabled.
5. In Example 14-6, what is the range of address assigned to Y5?

SECTION 14.3: 8031/51 INTERFACING WITH EXTERNAL ROM

As discussed in Chapter 1, the 8031 chip is a ROMless version of the 8051. In other words, it is exactly like any member of the 8051 family such as the 8751 or 89C51 as far as executing the instructions and features are concerned, but it has no on- chip ROM. Therefore, to make the 8031 execute 8051 code, it must be connected to external ROM memory containing the program code. In this section we look at interfacing the 8031 microcontroller with external ROM. Before we discuss this topic, one might wonder why someone would want to use the 8031 when you can buy an 8751, 89C51, or DS5000. The reason is that all these chips have a limited amount of on-chip ROM. Therefore, in many systems where the on-chip ROM of the 8051 is not sufficient, the use of an 8031 is ideal since it allows the program size to be as large as 64K bytes. Although the 8031 chip itself is much cheaper than other family members, an 8031-based system is much more expensive since the ROM containing the program code is connected externally and requires more supporting circuitry, as we explain next. First, we review some of the pins of the 8031/51 used in external memory interfacing. See Figure 14-7.

EA pin

As shown in Chapter 4, for 8751/89C51/DS5000-based systems, we connect the EA pin to V_{CC} to indicate that the program code is stored in the microcontroller's on-chip ROM. To indicate that the program code is stored in external ROM, this pin must be connected to GND. This is the case for the 8031- based system. In fact, there are times when, due to repeated burning and erasing of on-chip ROM, its UV-EPROM is no longer working. In such cases one can also use the 8751 (or 89C51 or any 8051) as the 8031. All we have to do is to connect the EA pin to ground and connect the chip to external ROM containing the program code.

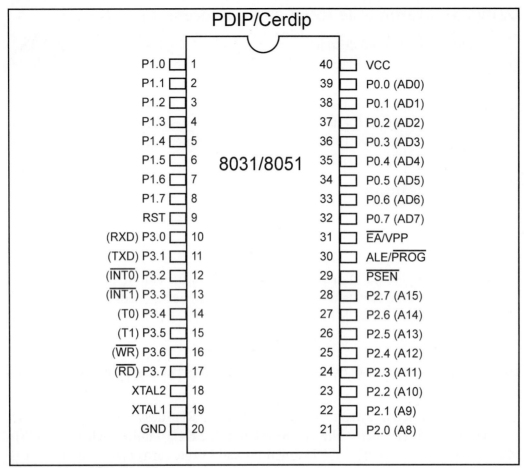

Figure 14-7. 8051 Pin Diagram

P0 and P2 role in providing addresses

Since the PC (program counter) of the 8031/51 is 16-bit, it is capable of accessing up to 64K bytes of program code. In the 8031/51, port 0 and port 2 provide the 16-bit address to access external memory. Of these two ports, P0 provides the lower 8 bit addresses A0 - A7, and P2 provides the upper 8 bit addresses A8 - A15. More importantly, P0 is also used to provide the 8-bit data bus D0 - D7. In other words, pins P0.0 - P0.7 are used for both the address and data paths. This is called address/data multiplexing in chip design. Of course the reason Intel used address/data multiplexing in the 8031/51 is to save pins. How do we know when P0 is used for the data path and when it is used for the address path? This is the job of the ALE (address latch enable) pin. ALE is an output pin for the 8031/51 microcontroller. Therefore, when ALE = 0 the 8031 uses P0 for the data path and when ALE = 1, it is used for the

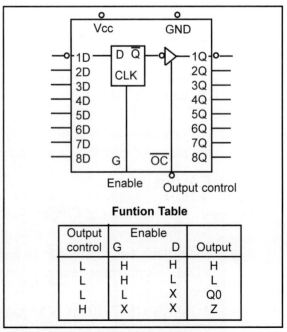

Figure 14-8. 74LS373 D Latch

(Reprinted by permission of Texas Instruments, Copyright Texas Instruments, 1988)

288

address path. As a result, to extract the addresses from the P0 pins we connect P0 to a 74LS373 latch (see Figure 14-8) and use the ALE pin to latch the address as shown in Figure 14-9. This extracting of addresses from P0 is called address/data demultiplexing.

From Figure 14-9, it is important to note that normally ALE = 0, and P0 is used as a data bus, sending data out or bringing data in. Whenever the 8031/51 wants to use P0 as an address bus, it puts the addresses A0 - A7 on the P0 pins and activates ALE = 1 to indicate that P0 has the addresses.

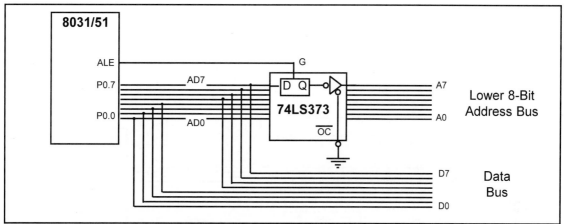

Figure 14-9. Address/Data Multiplexing

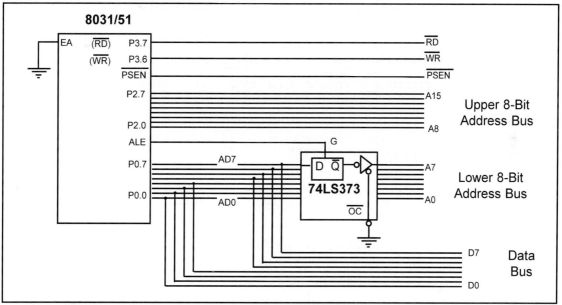

Figure 14-10. Data, Address, and Control Buses for the 8031
(For reset and crystal connection, see Chapter 4.)

PSEN

Another important signal for the 8031/51 is the PSEN (program store enable) signal. PSEN is an output signal for the 8031/51 microcontroller and must be connected to the OE pin of a ROM containing the program code. In other words, to access external ROM containing program code, the 8031/51 uses the PSEN signal. It is important to emphasize the role of EA and PSEN when connecting the 8031/51 to external ROM. When the EA pin is connected to GND, the 8031/51 fetches opcode from external ROM by using PSEN. Notice in Figure 14-11 the connection of the PSEN pin to the OE pin of ROM. In systems based on the 8751/89C51/DS5000 where EA is connected to VCC, these chips do not activate the PSEN pin. This indicates that the on-chip ROM contains program code.

In systems where the external ROM contains the program code, burning the program into ROM leaves the microcontroller chip untouched. This is preferable in some applications due to flexibility. In such applications updating the software is done via the serial or parallel ports of the IBM PC. This is especially the case during software development and this method is widely used in many 8051-based trainers and emulators.

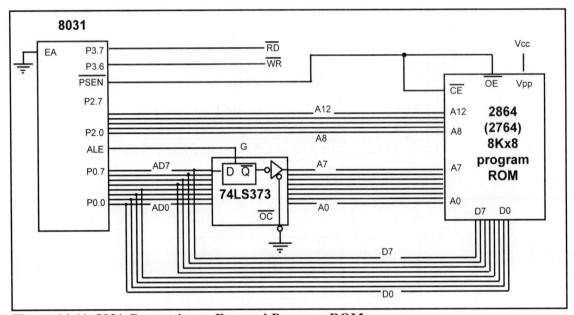

Figure 14-11. 8031 Connection to External Program ROM

On-chip and off-chip code ROM

In all our examples of 8051-based systems so far, we either used the on-chip ROM or the off-chip ROM for the program code. There are times that we want to use both of them. Is this possible? The answer is yes. For example, in an 8751 (or 89C51) system we could use the on-chip ROM for the boot code and an external ROM (using NV-RAM) will contain the user's program. In this way, the system boot code resides on-chip and the user's programs are downloaded into off-chip NV-RAM. In such a system we still have EA = V_{CC}, meaning that upon reset the 8051 executes the on-chip program first, then when it reaches the end of the on-chip ROM it switches to external ROM for the rest of the program code. Many 8051 trainers are designed using this method. Again, notice that this is done automatically by the 8051. For example, in an 8751 (89C51) system with both on-chip and off-chip ROM code where EA = V_{CC}, the controller fetches opcodes starting at address 0000, then goes on to address 0FFF (the last location of on-chip ROM). Then the program counter generates address 1000H and is automatically directed to the external ROM containing the program code. See Examples 14-7 and 14-8. Figure 14-12 shows the memory configuration.

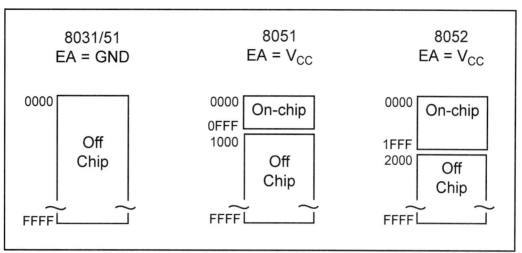

Figure 14-12. On-chip and Off-chip Program Code Access

Example 14-7

Discuss the program ROM space allocation for each of the following cases.
(a) EA = 0 for the 8751(89C51) chip.
(b) EA = V_{CC} with both on-chip and off-chip ROM for the 8751.
(c) EA = V_{CC} with both on-chip and off-chip ROM for the 8752.

Solution:

(a)   When EA=0, the EA pin is strapped to GND, and all program fetches are
       directed to external memory regardless of whether or not the 8751 has some
       on-chip ROM for program code. This external ROM can be as high as 64K bytes
       with address space of 0000 - FFFFH. In this case an 8751 (89C51) is the same
       as the 8031 system.

(b)   With the 8751 (89C51) system where EA=V_{CC}, it fetches the program code of
       addresses 0000 - 0FFFH from on-chip ROM since it has 4K bytes of on-chip
       program ROM and any fetches from addresses 1000H - FFFFH are directed to
       external ROM.

(c)   With the 8752 (89C52) system where EA=V_{CC}, it fetches the program code of
       addresses 0000 - 1FFFH from on-chip ROM since it has 8K bytes of on-chip
       program ROM and any fetches from addresses 2000H-FFFFH are directed to
       external ROM.

Example 14-8

Discuss the role of the PSEN pin in accessing on-chip and off-chip program codes.

Solution:

In the process of fetching the internal on-chip program code the PSEN pin is not used
and is never activated. However, PSEN is used for all external program fetches. In
Figure 14-11, notice that PSEN is also used to activate the CE pin of the program ROM.

Review Questions

1. If EA = GND, indicate from what source the program code is fetched.
2. If EA = V_{CC}, indicate from what source the program code is fetched.
3. Which port of the 8051 is used for address/data multiplexing?
4. Which port of the 8051 provides D0 - D7?
5. Which port of the 8051 provides A0 - A7?
6. Which port of the 8051 provides A8 - A15?
7. True or false. In accessing externally stored program code, the PSEN signal is always activated.

SECTION 14.4: 8051 DATA MEMORY SPACE

So far in this book all our discussion about memory space has involved program code. We have stated that the program counter in the 8051 is 16-bit and therefore can access up to 64K bytes of program code. In some examples in many of the previous chapters we have placed data in the code space and used the instruction "MOVC A, @A+DPTR" to get the data. The MOVC instruction, where C stands for code, indicates that data is located in the code space of the 8051. In the 8051 family there is also a separate data memory space. In this section we describe the data memory space of the 8051 and show how to access it.

Data memory space

In addition to its code space, the 8051 family also has 64K bytes of data memory space. In other words, the 8051 has 128K bytes of address space where 64K bytes are set aside for program code and the other 64K bytes are set aside for data. Program space is accessed using the program counter (PC) to locate and fetch instructions, but the data memory space is accessed using the DPTR register and an instruction called MOVX, where X stands for external. (That means that the data memory space must be implemented externally.)

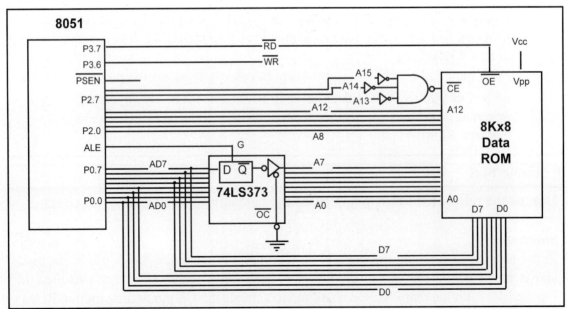

Figure 14-13. 8051 Connection to External Data ROM

292

External ROM for data

To connect the 8031/51 to external ROM containing data, we use RD (pin P3.7). See Figure 14-13. Notice the role of signals PSEN and RD. For the ROM containing the program code, PSEN is used to fetch the code. For the ROM containing data, the RD signal is used to fetch the data. To access the external data memory space we must use the instruction MOVX as described next.

MOVX instruction

MOVX is a widely used instruction allowing access to external data memory space. This is true regardless of which member of the 8051 family is used. To bring externally stored data into the CPU, we use the instruction "MOVX A, @DPTR". This instruction will read the byte of data pointed to by register DPTR and store it in the accumulator. In applications where a large data space is needed, the look-up table method is widely used. See Examples 14-9 and 14-10 for the use of MOVX.

Example 14-9

An external ROM uses the 8051 data space to store the look-up table (starting at 1000H) for DAC data. Write a program to read 30 bytes of these data and send it to P1.

Solution:

```
MYXDATA   EQU   1000H
COUNT     EQU   30
          . . .
          MOV   DPTR,#MYXDATA   ;pointer to external data
          MOV   R2,#COUNT       ;counter
AGAIN:    MOVX  A,@DPTR         ;get byte from external MEM
          MOV   P1,A            ;issue it to P1
          INC   DPTR            ;next location
          DJNZ  R2,AGAIN        ;until all are read
```

Example 14-10

External data ROM has a look-up table for the squares of numbers 0 - 9. Since the internal RAM of the 8031/51 has a shorter access time, write a program to copy the table elements into internal RAM starting at address 30H. The look-up table address starts at address 0 of external ROM.

Solution:

```
TABLE     EQU   000H
RAMTBLE   EQU   30H
COUNT     EQU   10
          . . .
          MOV   DPTR,#TABLE     ;pointer to external data
          MOV   R5,#COUNT       ;counter
          MOV   R0,#RAMTBLE     ;pointer to internal RAM
BACK:     MOVX  A,@DPTR         ;get byte from external MEM
          MOV   @R0,A           ;store it in internal RAM
          INC   DPTR            ;next data location
          INC   R0              ;next RAM location
          DJNZ  R5,BACK         ;until all are read
```

Contrast Example 14-10 with Example 5-8. In that example the table elements are stored in the program code space of the 8051 and we used the instruction "MOVC" to access each element. Although both "MOVC A,@A+DPTR" and "MOVX A,@DPTR" look very similar, one is used to get data in the code space and the other is used to get data in the data space of the microcontroller.

Example 14-11

Show the design of an 8031-based system with 8K bytes of program ROM and 8K bytes of data ROM.

Solution:

Figure 14-14 shows the design. Notice the role of PSEN and RD in each ROM. For program ROM, PSEN is used to activate both OE and CE. For data ROM, we use RD to activate OE, while CE is activated by a simple decoder.

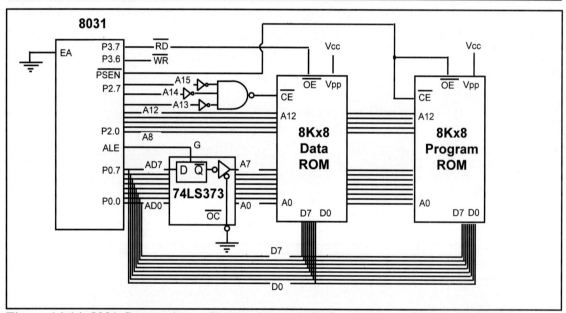

Figure 14-14. 8031 Connection to External Data ROM and External Program ROM

From the discussion so far, we conclude that while we can use internal RAM and registers located inside the CPU for storage of data, any additional memory space for read/write data must be implemented externally. This is discussed further next.

External data RAM

To connect the 8051 to an external SRAM, we must use both RD (P3.7) and WR (P3.6). This is shown in Figure 14-15.

MOVX instruction for external RAM data

In writing data to external data RAM, we use the instruction "MOVX @DPTR,A" where the contents of register A are written to external RAM whose address is pointed to by the DPTR register. This has many applications, especially where we are collecting a large number of bytes of data. In such applications, as we collect data we must store them in NV-RAM so that when power is lost we do not lose the data. See Example 14-12 and Figure 14-15.

Example 14-12

(a) Write a program to read 200 bytes of data from P1 and save the data in external RAM starting at RAM location 5000H.

(b) What is the address space allocated to data RAM in Figure 14-15?

Solution:

(a)
```
RAMDATA     EQU   5000H
COUNT       EQU   200

            MOV   DPTR,#RAMDATA   ;pointer to external NV-RAM
            MOV   R3,#COUNT       ;counter
AGAIN:      MOV   A,P1            ;read data from P1
            MOVX  @DPTR,A         ;save it external NV-RAM
            ACALL DELAY           ;wait before next sample
            INC   DPTR            ;next data location
            DJNZ  R3,AGAIN        ;until all are read
HERE:       SJMP  HERE            ;stay here when finished
```

(b) The data address space is 8000H to BFFFH.

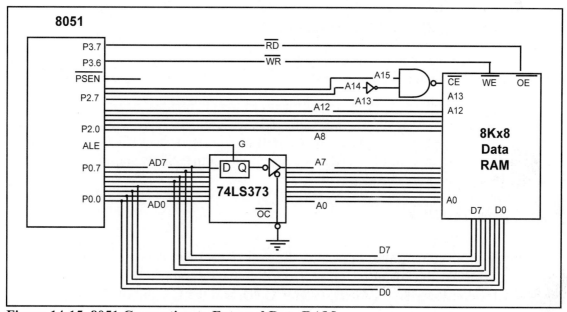

Figure 14-15. 8051 Connection to External Data RAM

A single external ROM for code and data

Assume that we have an 8031-based system connected to a single 64Kx8 (27512) external ROM chip. This single external ROM chip is used for both program code and data storage. For example, the space 0000 - 7FFFH is allocated to program code, and address space D000H - FFFFH is set aside for data. In accessing the data, we use the MOVX instruction. How do we connect the PSEN and RD signals to such a ROM? Notice in the previous discussion that PSEN is used to access the external code space and the RD signal is used to access the external data space. To allow a single ROM chip to provide both program code space and data space, we use an AND gate to signal the OE pin of the ROM chip as shown in Figure 14-16.

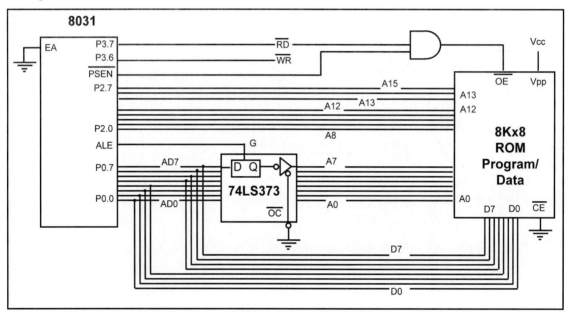

Figure 14-16: A Single ROM for Both Program and Data

8031 system with ROM and RAM

There are times that we need program ROM, data ROM, and data RAM in a system. This is shown in Example 14-13.

Example 14-13
Assume that we need an 8031 system with 16KB of program space, 16KB of data ROM starting at 0000, and 16K of NV-RAM starting at 8000H. Show the design using a 74LS138 for the address decoder.
Solution:
The solution is diagrammed in Figure 14-17. Notice that there is no need for a decoder for program ROM, but we need a 74LS138 decoder for data ROM and RAM. Also notice that G1 = Vcc, G2A = GND, G2B = GND, and the C input of the 74LS138 is also grounded since we use Y0 - Y3 only.

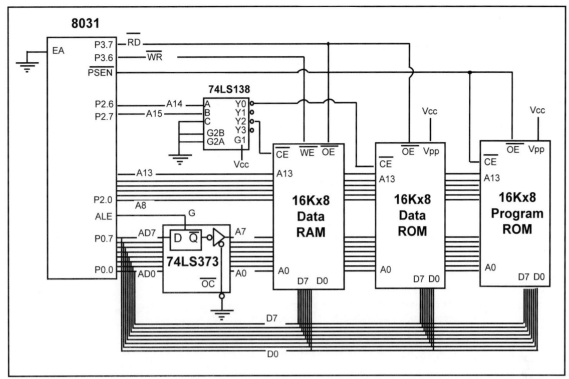

Figure 14-17. 8031 Connection to External Program ROM, Data RAM, and Data ROM

Interfacing to Large External Memory

In some applications we need a large amount (256K bytes, for example) of memory to store data. However, the 8051 can support only 64K bytes of external data memory since DPTR is 16-bit. To solve this problem, we connect A0 - A15 of the 8051 directly to the external memory's A0 - A15 pins, and use some of the P1 pins to access the 64K bytes blocks inside the single 256Kx8 memory chip. This is shown in Example 14-14, and illustrated in Figure 14-18.

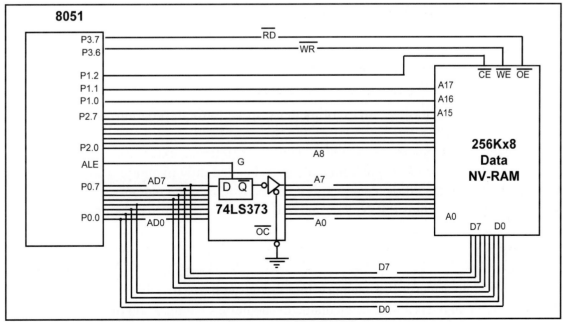

Figure 14-18. 8051 Accessing 256Kx8 External NV-RAM

	Example 14-14

In a certain application, we need 256K bytes of NV-RAM to store data collected by an 8051 microcontroller. (a) Show the connection of an 8051 to a single 256Kx8 NV-RAM chip. (b) Show how various blocks of this single chip are accessed.

Solution:

(a) The 256Kx8 NV-RAM has 18 address pins (A0 - A17) and 8 data lines. As shown in Figure 14-18, A0 - A15 go directly to the memory chip while A16 and A17 are controlled by P1.0 and P1.1, respectively. Also notice that chip select of external RAM is connected to P1.2 of the 8051.

(b) The 256K bytes of memory are divided into four blocks, and each block is accessed as follows:

Chip select P1.2	A17 P1.1	A16 P1.0	Block address space
0	0	0	00000H-0FFFFH
0	0	1	10000H-1FFFFH
0	1	0	20000H-2FFFFH
0	1	1	30000H-3FFFFH
1	x	x	External RAM disabled

For example, to access the 20000H - 2FFFFH address space we need the following:

```
CLR   P1.2          ;enable external RAM
MOV   DPTR,#0        ;start of 64K memory block
CLR   P1.0          ;A16=0
SETB  P1.1          ;A17=1 for 20000H block
MOV   A,SBUF        ;get data from serial port
MOVX  @DPTR,A       ;save data in block 20000H addr.
INC   DPTR          ;next location
...
```

Review Questions

1. The 8051 has a total of ___ bytes of memory space for both program code and data.
2. All the data memory space of the 8051 is _____ (internal, external).
3. True or false. In the 8051, program code must be read-only memory.
4. True or false. In the 8051, data memory can be read or write memory.
5. Explain the role of pins PSEN, RD, and WR in accessing external memory.

SUMMARY

This chapter described memory interfacing with 8031/51-based systems. We began with an overview of semiconductor memories. Comparison of types of memories was given in terms of their capacity, organization, and access time.

ROM (read-only memory) is non-volatile memory typically used to store programs. The relative advantages of various types of ROM were described in this chapter, including PROM, EPROM, UV-EPROM, EEPROM, flash memory EPROM, and mask ROM.

RAM (random-access memory) is typically used to store data or programs. The relative advantages of its various types were discussed, including SRAM, NV-RAM, check-sum byte RAM, and DRAM.

Address decoding techniques using simple logic gates, decoders, and programmable logic were covered. Finally, RAM and ROM memories were interfaced with 8031 systems, and programs were written to access code and data stored on these external memories.

PROBLEMS

SECTION 14.1: SEMICONDUCTOR MEMORY

1. What is the difference between a 4M memory chip and 4M of computer memory as far as capacity is concerned?
2. True or false. The more address pins, the more memory locations are inside the chip. Assume that the number of data pins is fixed.
3. True or false. The more data pins, the more each location inside the chip will hold.
4. True or false. The more data pins, the higher the capacity of the memory chip, assuming a fixed number of address pins.
5. True or false. The more data pins and address pins, the greater the capacity of the memory chip.
6. The speed of a memory chip is referred to as its _____.
7. True or false. The price of memory chips varies according to capacity and speed.
8. The main advantage of EEPROM over UV-EPROM is _____.
9. True or false. SRAM has a larger cell size than DRAM.
10. Which of the following, EPROM, DRAM, and SRAM, must be refreshed periodically?
11. Which memory is used for PC cache?
12. Which of the following, SRAM, UV-EPROM, NV-RAM, and DRAM, is volatile memory?
13. RAS and CAS are associated with which memory?
    (a) EPROM (b) SRAM (c) DRAM (d) all of the above
14. Which memory needs an external multiplexer?
    (a) EPROM (b) SRAM (c) DRAM (d) all of the above
15. Find the organization and capacity of memory chips with the following pins.
    (a) EEPROM A0 - A14, D0 - D7 (b) UV-EPROM A0 - A12, D0 - D7
    (c) SRAM A0 - A11, D0 - D7   (d) SRAM A0 - A12, D0 - D7
    (e) DRAM A0 - A10, D0   (f) SRAM A0 - A12, D0
    (g) EEPROM A0 - A11, D0 - D7 (h) UV-EPROM A0 - A10, D0 - D7
    (i) DRAM A0 - A8, D0 - D3   (j) DRAM A0 - A7, D0 - D7

16. Find the capacity, address, and data pins for the following memory organizations.
    (a) 16Kx8 ROM (b) 32Kx8 ROM
    (c) 64Kx8 SRAM (d) 256Kx8 EEPROM
    (e) 64Kx8 ROM (f) 64Kx4 DRAM
    (g) 1Mx8 SRAM (h) 4Mx4 DRAM
    (i) 64Kx8 NV-RAM
17. Find the checksum byte for these bytes: 34H, 54H, 7FH, 11H, E6H, 99H
18. For each of the following sets of data (the last byte is the checksum byte) verify if the data is corrupted.
    (a) 29H, 1CH, 16H, 38H, and 6DH   (b) 29H, 1CH, 16H, 30H, and 6DH

SECTION 14.2: MEMORY ADDRESS DECODING

19. Find the address range of the memory design in the diagram.
20. Using NAND gates and inverters, design decoding circuitry for the address range 2000H - 2FFFH.
21. Find the address range for Y0, Y3, and Y6 of the 74LS138 for the diagrammed design.
22. Using the 74138, design the memory decoding circuitry in which the memory block controlled by Y0 is in the range 0000H to 1FFFH. Indicate the size of the memory block controlled by each Y.
23. Find the address range for Y3, Y6, and Y7 in Problem 22.
24. Using the 74138, design memory decoding circuitry in which the memory block controlled by Y0 is in the 0000H to 3FFFH space. Indicate the size of the memory block controlled by each Y.
25. Find the address range for Y1, Y2, and Y3 in Problem 24.

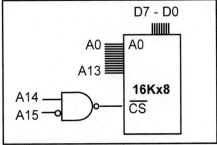

Diagram for Problem 19

Diagram for Problem 21

SECTION 14.3: 8031/51 INTERFACING WITH EXTERNAL ROM

26. In a certain 8031 system, the starting address is 0000H and it has only 16K bytes of program memory. What is the ending address of this system?
27. When the 8031 CPU is powered up, at what address does it expect to see the first opcode?
28. In an 8031/51 microcontroller, RD and WR are pins ____ and ____, respectively. They belong to port ____. Which bits of this port?
29. The 8051 supports a maximum of _____ K bytes of program memory space.
30. True or false. For any member of the 8051 family, if EA = Gnd it fetches program code from external ROM.

31. True or false. For any member of the 8051 family, if EA = Vcc it fetches program code from internal (on-chip) ROM.
32. For which of the following must we have external memory for program code?
    (a) 8751   (b) 89C51   (c) 8031   (d) 8052
33. For which of the following is external memory for program code optional?
    (a) 8751   (b) 89C51   (c) 8031   (d) 8052
34. In the 8051 which port provides the A0 - A7 address bits?
35. In the 8051 which port provides the A8 - A15 address bits?
36. In the 8051 which port provides the D0 - D7 data bits?
37. Explain the difference between ALE = 0 and ALE = 1.
38. RD is pin _____ of P3, and WR is pin _____ of P3. What about PSEN?
39. Which of the following signals must be used in fetching program code from external ROM?
    (a) RD   (b) WR   (c) PSEN
40. For the 8031-based system with external program ROM, when the microcontroller is powered up, it expects to find the first instruction at address _____ of program ROM. Is this internal or external ROM?
41. In an 8051 with 16K bytes of on-chip program ROM, explain what happens if EA = Vcc.
42. True or false. For the 8051 the program code must be read-only memory. In other words, the memory code space of the 8051 is read-only memory.
43. Indicate when PSEN is used. Is it used in accessing on-chip code ROM or external (off-chip) code ROM?

SECTION 14.4: 8051 DATA MEMORY SPACE

44. Indicate when RD and WR are used. Are they used in accessing external data memory?
45. The 8051 supports a maximum of _____ K bytes of data memory space.
46. Which of the following signals must be used in fetching data from external data ROM?
    (a) RD   (b) WR   (c) PSEN   (d) both (a) and (b)
47. For each of the following, indicate if it is active low or active high.
    (a) PSEN   (b) RD   (c) WR
48. True or false. For the 8051, the data memory space can belong to ROM or RAM.
49. Explain the difference between the MOVX and MOVC instructions.
50. Write a program to transfer 20 bytes of data from external data ROM to internal RAM. The external data ROM adddress is 2000H, and internal RAM starts at 60H.
51. Write a program to transfer 30 bytes of data from internal data RAM to external RAM. The external data RAM adddress is 6000H, and internal RAM starts at 40H.
52. Write a program to transfer 100 bytes of data from external data ROM to external data RAM. The external data ROM adddress is 3000H, and the external data RAM starts at 8000H.

ANSWERS TO REVIEW QUESTIONS

SECTION 14.1: SEMICONDUCTOR MEMORY

1. c
2. (a) 16Kx8, 128K bits   (b) 64Kx8, 512K   (c) 4Kx8, 32K
3. (a) 2Kx1, 2K bits   (b) 8Kx4, 32K   (c) 128Kx8, 1M
    (d) 64Kx4, 256K   (e) 256Kx1, 256K   (f) 256Kx4, 1M
4. (a) 64K bits, 14 addr, and 4 data   (b) 256K,15 addr, and 8 data
    (c) 1M, 10 addr, and 1 data   (d) 1M, 18 addr, and 4 data
    (e) 512K, 16, and 8 data   (f) 4M,10 addr, and 4 data
5. b, c

SECTION 14.2: MEMORY ADDRESS DECODING

1. 16K bytes   2. 3,8   3. Both must be low
4. G1 must be high
5. 5000H-5FFFH

SECTION 14.3: 8031/51 INTERFACING WITH EXTERNAL ROM

1. From external ROM (that is off-chip)
2. From internal ROM (that is on-chip)
3. P0
4. P0
5. P0
6. P2
7. True

SECTION 14.4: 8051 DATA MEMORY SPACE

1. 128K   2. External   3. True   4. True
5. Only PSEN is used to access external ROM containing program code, but when accessing external data memory we must use RD and WR signals. In other words RD and WR are only for external data memory and they are never used for external program ROM.

302

CHAPTER 15

8031/51 INTERFACING WITH THE 8255

OBJECTIVES

Upon completion of this chapter, you will be able to:

>> Describe how to expand the I/O ports of the 8031/51 by connecting it to an 8255 chip

>> List the 3 ports of the 8255 and describe their features

>> Explain the use of the control register of the 8255 in selecting a mode

>> Define the modes of the 8255

>> Define the term *memory-mapped I/O* and describe its application

>> Program the 8255 as a simple I/O port for connection with devices such as stepper motors, LCDs, and ADC devices

>> Interface the 8051 with external devices such as stepper motors, LCDs, and ADC devices via the 8255

>> Diagram the 8031 interface with external program ROM and the 8255

>> Explain how address aliases are used in address decoding techniques

>> Describe the hand-shaking feature of the 8255 and its application in printer interfacing

As demonstrated in Chapter 14, in the process of connecting the 8031/51 to external memory, two ports, P0 and P2, are lost. In this chapter we show how to expand the I/O ports of the 8031/51 by connecting it to an 8255 chip. In Section 15.1, the connection of the 8031/51 to an 8255 is shown. In Section 15.2, the 8255 is programmed as a simple I/O port for connection with devices such as LCDs, stepper motors, and ADC. In Section 15.3, more advanced I/O modes of the 8255 are explored, in addition to its hand-shaking capability.

SECTION 15.1: PROGRAMMING THE 8255

In this section we study the 8255 chip, one of the most widely used I/O chips. We first describe its features and then show the connection between the 8031/51 and 8255 chips.

8255 features

The 8255 is a 40-pin DIP chip (see Figure 15-1). It has three separately accessible ports. The ports are each 8-bit, and are named A, B, and C. The individual ports of the 8255 can be programmed to be input or output, and can be changed dynamically. In addition, 8255 ports have hand-shaking capability, thereby allowing interface with devices that also have hand-shaking signals, such as printers. The handshaking capability of the 8255 is discussed in Section 15.3.

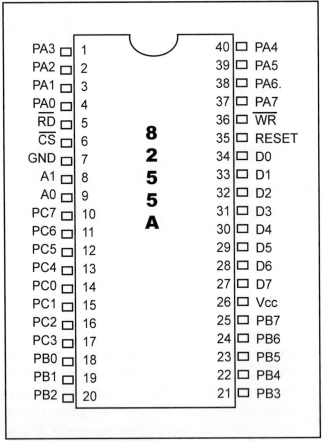

Figure 15-1. 8255 Chip

(Reprinted by permission of Intel Corporation,Copyright Intel Corp. 1983)

PA0 - PA7

The 8-bit port A can be programmed as all input, or as all output, or all bits as bidirectional input/output.

PB0 - PB7

The 8-bit port B can be programmed as all input or as all output. Port B cannot be used as a bidirectional port.

PC0 - PC7

This 8-bit port C can be all input or all output. It can also be split into two parts, CU (upper bits PC4 - PC7) and CL (lower bits PC0 - PC3). Each can be used for input or output. In addition, any of bits PC0 to PC7 can be programmed individually.

$\overline{RD}$ and $\overline{WR}$

These two active-low control signals are inputs to the 8255. The $\overline{RD}$ and $\overline{WR}$ signals from the 8031/51 are connected to these inputs.

D0 - D7 data pin

The data pins of the 8255 are connected to the data pins of the micro-controller allowing it to send data back and forth between the controller and the 8255 chip.

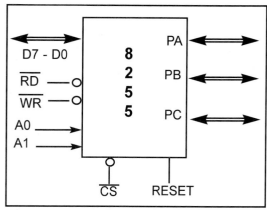

Figure 15-2. 8255 Block Diagram

RESET

This is an active-high signal input into the 8255 used to clear the control register. When RESET is activated, all ports are initialized as input ports. In many designs this pin is connected to the RESET output of the system bus or grounded to make it inactive. Like all IC input pins, it should not be left unconnected.

Table 15-1: 8255 Port Selection

$\overline{CS}$	A1	A0	Selection
0	0	0	Port A
0	0	1	Port B
0	1	0	Port C
0	1	1	Control register
1	x	x	8255 is not selected

A0, A1, and $\overline{CS}$

While $\overline{CS}$ (chip select) selects the entire chip, it is A0 and A1 that select specific ports. These three pins are used to access ports A, B, C, or the control register according to Table 15-1. Note that $\overline{CS}$ is active-low.

Mode selection of the 8255

While ports A, B, and C are used to input or output data, it is the control register that must be programmed to select the operation mode of the three ports. The ports of the 8255 can be programmed in any of the following modes.

1. Mode 0, simple I/O mode. In this mode, any of the ports A, B, CL, and CU can be programmed as input or output. In this mode, all bits are out or all are in. In other words, there is no such thing as single-bit control as we have seen in P0 - P3 of the 8051. Since the vast majority of applications involving the 8255 use this simple I/O mode, we will concentrate on this mode in this chapter.

2. Mode 1. In this mode, ports A and B can be used as input or output ports with handshaking capabilities. Handshaking signals are provided by the bits of port C. The details of this mode are discussed in the third section of this chapter.

3. Mode 2. In this mode, port A can be used as a bidirectional I/O port with hand-shaking capabilities whose signals are provided by port C. Port B can be used either in simple I/O mode or handshaking mode 1. This mode will not be explored further in this book.

4. BSR (bit set/reset) mode. In this mode, only the individual bits of port C can be programmed. This mode is discussed further in the third section of this chapter.

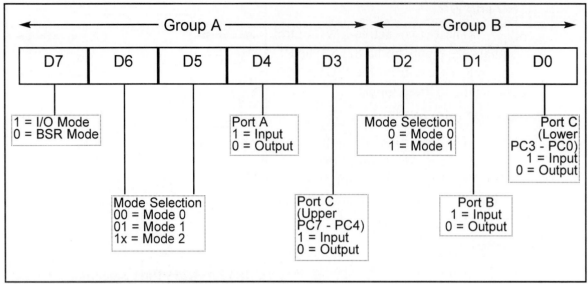

Figure 15-3. 8255 Control Word Format (I/O Mode)

(Reprinted by permission of Intel Corporation, Copyright Intel Corp., 1983)

Simple I/O programming

Intel calls mode 0 the *basic input/output mode*. The more commonly used term is *simple I/O*. In this mode, any of ports A, B, or C can be programmed as input or output. It must be noted that in this mode, a given port cannot be both an input and output port at the same time.

Example 15-1
Find the control word of the 8255 for the following configurations:

Find the control word of the 8255 for the following configurations:
(a) All the ports of A, B, and C are output ports (mode 0).
(b) PA = in, PB = out, PCL = out, and PCH = out.

Solution:

From Figure 15-3 we have:
(a) 1000 0000 = 80H   (b) 1001 0000 = 90H

Connecting the 8031/51 to an 8255

The 8255 chip is programmed in any of the 4 modes mentioned earlier by sending a byte (Intel calls it a control word) to the control register of the 8255. We must first find the port addresses assigned to each of ports A, B, C, and the control register. This is called *mapping* the I/O port.

As can be seen from Figure 15-4, the 8255 is connected to an 8031/51 as if it is RAM memory. Notice the use of $\overline{RD}$ and $\overline{WR}$ signals. This method of connecting an I/O chip to a CPU is called *memory mapped I/O*, since it is mapped into memory space. In other words, we use memory space to access I/O devices. For this reason we use instructions such as MOVX to access the 8255. In Chapter 14 we used MOVX to access RAM and ROM. For an 8255 connected to the 8031/51 we must also use the MOVX instruction to communicate with it. This is shown in Example 15-2.

Example 15-2

For Figure 15-4,
(a) Find the I/O port addresses assigned to ports A, B, C, and the control register.
(b) Program the 8255 for ports A, B, and C to be output ports.
(c) Write a program to send 55H and AAH to all ports continuously.

Solution:

(a)   The base address for the 8255 is as follows:

A15	A14	A13	A12	A11	A10	A9	A8	A7	A6	A5	A4	A3	A2	A1	A0	
x	1	x	x	x	x	x	x	x	x	x	x	x	x	0	0	= 4000H PA
x	1	x	x	x	x	x	x	x	x	x	x	x	x	0	1	= 4001H PB
x	1	x	x	x	x	x	x	x	x	x	x	x	x	1	0	= 4002H PC
x	1	x	x	x	x	x	x	x	x	x	x	x	x	1	1	= 4003H CR

(b)   The control byte (word) for all ports as output is 80H as seen in Example 15-1.
(c)

```
              MOV   A,#80H          ;control word (ports output)
              MOV   DPTR,#4003H;    ;load control reg port addr
              MOVX  @DPTR,A         ;issue control word
              MOV   A,#55H          ;A=55H
AGAIN:        MOV   DPTR,#4000H     ;PA address
              MOVX  @DPTR,A         ;toggle PA bits
              INC   DPTR            ;PB address
              MOVX  @DPTR,A         ;toggle PB bits
              INC   DPTR            ;PC address
              MOVX  @DPTR,A         ;toggle PC bits
              CPL   A               ;toggle bits in reg A
              ACALL DELAY           ;wait
              SJMP  AGAIN           ;continue
```

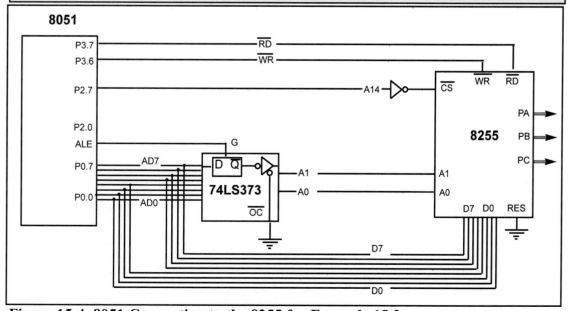

Figure 15-4. 8051 Connection to the 8255 for Example 15-2

Example 15-3

For Figure 15-5,
(a) Find the I/O port addresses assigned to ports A, B, C, and the control register.
(b) Find the control byte for PA = in, PB = out, PC = out.
(c) Write a program to get data from PA and send it to both ports B and C.

Solution:

(a) Assuming all the unused bits are 0s, the base port address for 8255 is 1000H.

Therefore we have: 1000H   PA
                    1001H   PB
                    1002H   PC
                    1003H   Control register

(b) The control word is 10010000, or 90H.

(c)
```
        MOV  A,#90H          ;(PA=IN,PB=OUT,PC=OUT)
        MOV  DPTR,#1003H;    ;load control reg port addr
        MOVX @DPTR,A         ;issue control word
        MOV  DPTR,#1000H     ;PA address
        MOVX A,@DPTR         ;get data from PA
        INC  DPTR            ;PB address
        MOVX @DPTR,A         ;send the data to PB
        INC  DPTR            ;PC address
        MOVX @DPTR,A         ;send it also to PC
```

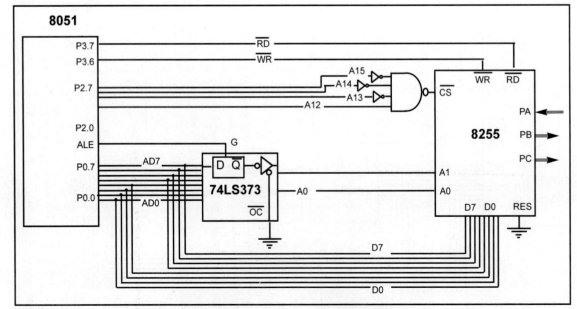

Figure 15-5. 8051 Connection to the 8255 for Example 15-3

For the program in Example 15-3, it is recommended that you use the EQU directive for port addresses as shown next.

```
APORT      EQU   1000H
BPORT      EQU   1001H
CPORT      EQU   1002H
CNTPORT    EQU   1003H

MOV  A,#90H              ;(PA=IN,PB=OUT,PC=OUT)
MOV  DPTR,#CNTPORT       ;load cntr reg port addr
MOVX @DPTR,A             ;issue control word
MOV  DPTR,#APORT         ;PA address
MOVX A,@DPTR             ;get data from PA
INC  DPTR               ;PB address
MOVX @DPTR,A             ;send the data to PB
INC  DPTR               ;PC address
MOVX @DPTR,A             ;send it also to PC
```

or, see the following, also using EQU:

```
CONTRBYT   EQU   90H          ;(PA=IN,PB=OUT,PC=OUT)
BAS8255P   EQU   1000H        ;base address for 8255 chip

MOV  A,#CONTRBYT
MOV  DPTR,#BAS8255P+3        ;load c port addr
MOVX @DPTR,A                 ;issue control word
MOV  DPTR,#BAS8255P          ;PA address
. . .
```

Notice in both Examples 15-2 and 15-3 that we used the DPTR register since the base address assigned to 8255 was 16-bit. If the base address for the 8255 is 8-bit, we can use the instructions "MOVX A, @R0" and "MOVX @R0,A" where R0 (or R1) holds the 8-bit port address of the port. See Example 15-4. Notice in Example 15-4 that we used a simple logic gate to do the address decoding for the 8255. For multiple 8255s in a system, we can use a 74LS138, as shown in Example 15-5.

Address aliases

In Examples 15-4 and 15-5 we decode the A0 - A7 address bits; however, in Examples 15-3 and 15-2 we decoded only a portion of the upper addresses of A8 - A15. This partial address decoding leads to what is called *address aliases*. In other words, the same physical port has different addresses; thus, the same port is known by different names. In Examples 15-2 and 15-3 we could have changed all x's to various combinations of 1s and 0s to come up with different addresses, yet they would all refer to the same physical port. In your hardware reference documentation, make sure that all address aliases are documented, so that the user knows what addresses are available if he/she wants to expand the system.

Example 15-4

For Figure 15-6,

(a) Find the I/O port addresses assigned to ports A, B, C, and the control register.

(b) Find the control byte for PA = out, PB = in, PC0 - PC3 = in, and PC4 - PC7 = out.

(c) Write a program to get data from PB and send it to PA. In addition, data from PCL is sent out to the PCU.

Solution:

(a)   The port addresses are as follows:

$\overline{CS}$	A1	A0	Address	Port
0010 00	0	0	20H	Port A
0010 00	0	1	21H	Port B
0010 00	1	0	22H	Port C
0010 00	1	1	23H	Control Reg

(b)   The control word is 1000 0011 or 83H.

(c)
```
CONTRBYT   EQU   83H     ;A=OUT,PB=IN,PCL=IN,PCU=OUT
APORT      EQU   20H
BPORT      EQU   21H
CPORT      EQU   22H
CNTPORT    EQU   23H

           ...
           MOV   A,#CONTRBYT;PA=OUT,PB=IN,PCL=IN,PCU=OUT
           MOV   R0,#CNTPORT;LOAD CONTROL REG ADDRESS
           MOVX  @R0,A     ;ISSUE CONTROL WORD
           MOV   R0,#BPORT ;LOAD PB ADDRESS
           MOVX  A,@R0     ;READ PB
           DEC   R0        ;POINT TO PA(20H)
           MOVX  @R0,A     ;SEND IT TO PA
           MOV   R0,#CPORT ;LOAD PC ADDRESS
           MOVX  A,@R0     ;READ PCL
           ANL   A,#0FH    ;MASK UPPER NIBBLE
           SWAP  A         ;SWAP LOW AND HIGH NIBBLES
           MOVX  @R0,A     ;SEND TO PCU
```

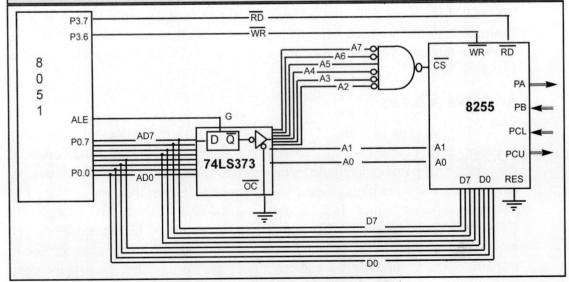

Figure 15-6. 8051 Connection to the 8255 for Example 15-4

Example 15-5

Find the base address for the 8255 in Figure 15-7.

Solution:

G1	$\overline{G2B}$	$\overline{G2A}$	C	B	A			Address
A7	A6	A5	A4	A3	A2	A1	A0	
1	0	0	0	1	0	0	0	88H

8031 system with 8255

In the 8031-based system where external program ROM is an absolute must, the use of an 8255 is most welcome. This is due to the fact that in the process of connecting the 8031 to external program ROM, we lose the two ports P0 and P2, leaving only P1. Therefore, connecting an 8255 is the best way to gain some extra ports. This is shown in Figure 15-8.

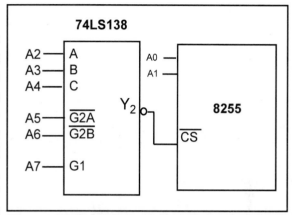

Figure 15-7. 8255 Decoding Using 74LS138

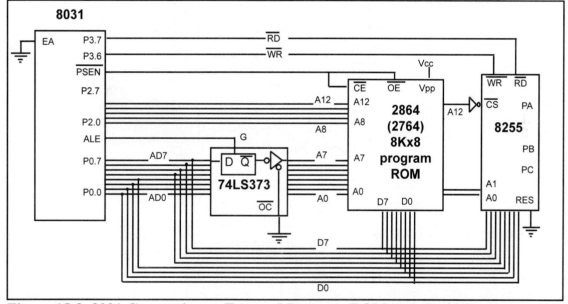

Figure 15-8. 8031 Connection to External Program ROM and the 8255

Review Questions

1. Find the control byte if all ports are inputs.
2. Find the control byte for PC = in, PB = out, and PA = out.
3. True or false. To avoid aliases, we must decode addresses A0 - A15.
4. Can 86H be the base address for port A of the 8255?
5. Why do we use the MOVX instruction to access the ports of the 8255?

SECTION 15.2: 8255 INTERFACING

Chapter 12 detailed real-world interfacing to LCDs, sensors, and ADC devices. In this section we show how to interface the 8255 to LCDs, stepper motors, and ADC devices, then program it using 8051 instructions.

Stepper motor connection to the 8255

Chapter 13 detailed the interface of a stepper motor to the 8051. Here we show stepper motor connection to the 8255 and programming. See Figure 15-9.

```
        MOV   A,#80H      ;control word for PA=out
        MOV   R1,#CRPORT  ;control reg port address
        MOVX  @R1,A       ;configure PA=out
        MOV   R1,#APORT   ;load PA address
        MOV   A,#66H      ;A=66H,stepper motor sequence
AGAIN:  MOVX  @R1,A       ;issue motor sequences to PA
        RR    A           ;rotate sequence for clockwise
        ACALL DELAY       ;wait
        SJMP  AGAIN
```

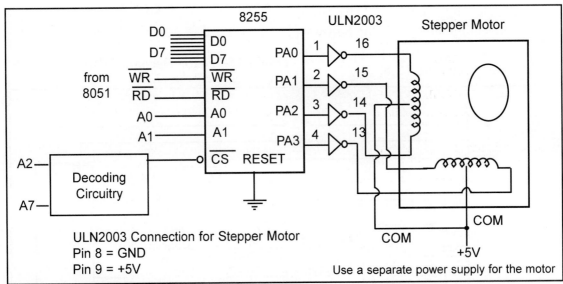

Figure 15-9. 8255 Connection to Stepper Motor

LCD connection to the 8255

Program 15-1 shows how to issue commands and data to an LCD connected to an 8255. See Figure 15-10. In Program 15-1, we must put a long delay before issuing any information (command or data) to the LCD. A better way is to check the busy flag before issuing anything to the LCD. This was discussed in Chapter 12. Program 15-2 is a repeat of Program 15-1 with the checking of the busy flag. Notice that no DELAY is used the main program in Program 15-2.

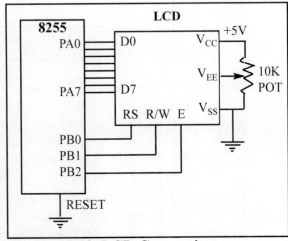

Figure 15-10. LCD Connection

```
;Writing commands and data to LCD without checking busy flag
;Assume PA of 8255 connected to D0-D7 of LCD and
;PB0=RS, PB1=R/W, PB2=E for LCD's control pins connection
            MOV   A,#80H             ;all 8255 ports as output
            MOV   R0,#CNTPORT        ;load control reg address
            MOVX  @R0,A              ;issue control word
            MOV   A,#38H             ;LCD:2 lines, 5X7 matrix
            ACALL CMDWRT             ;write command to LCD
            ACALL DELAY              ;wait before next issue(2 ms)
            MOV   A,#0EH             ;LCD command for cursor on
            ACALL CMDWRT             ;write command to LCD
            ACALL DELAY              ;wait before next issue
            MOV   A,#01H             ;clear LCD
            ACALL CMDWRT             ;write command to LCD
            ACALL DELAY              ;wait before next issue
            MOV   A,#06              ;shift cursor right command
            ACALL CMDWRT             ;write command to LCD
            ACALL DELAY              ;wait before next issue
            .....                    ;etc. for all LCD commands
            MOV   A,#'N'             ;display data (letter N)
            ACALL DATAWRT            ;send data to LCD display
            ACALL DELAY              ;wait before next issue
            MOV   A,#'O'             ;display data (letter O)
            ACALL DATAWRT            ;send data to LCD display
            ACALL DELAY              ;wait before next issue
            ....                     ;etc. for other data
;Command write subroutine, writes instruction commands to LCD
CMDWRT:     MOV   R0,#APORT          ;load port A address
            MOVX  @R0,A              ;issue info to LCD data pins
            MOV   R0,#BPORT          ;load port B address
            MOV   A,#00000100B       ;RS=0,R/W=0,E=1 for H-TO-L
            MOVX  @R0,A              ;activate LCD pins RS,R/W,E
            NOP                      ;make E pin pulse wide enough
            NOP
            MOV   A,#00000000B       ;RS=0,R/W=0,E=0 for H-TO-L
            MOVX  @R0,A              ;latch in data pin info
            RET
;Data write subroutine, write data to be displayed to LCD
DATAWRT:    MOV   R0,#APORT          ;load port A address
            MOVX  @R0,A              ;issue info to LCD data pins
            MOV   R0,#BPORT          ;load port B address
            MOV   A,#00000101B       ;RS=1,R/W=0,E=1 for H-TO-L
            MOVX  @R0,A              ;activate LCD pins RS,R/W,E
            NOP                      ;make E pin pulse wide enough
            NOP
            MOV   A,#00000001B       ;RS=1,R/W=0,E=0 for H-TO-L
            MOVX  @R0,A              ;latch in LCD's data pin info
            RET
```

Program 15-1.

```
;writing commands to the LCD with checking busy flag
;PA of 8255 connected to D0-D7 of LCD and
;PB0=RS, PB1=R/W, PB2=E for 8255 to LCD's control pins connect.
                MOV   A,#80H            ;all 8255 ports as output
                MOV   R0,#CNTPORT       ;load control reg address
                MOVX  @R0,A             ;issue control word
                MOV   A,#38H            ;LCD: 2 LINES, 5X7 matrix
                ACALL NCMDWRT           ;write command to LCD
                MOV   A,#0EH            ;LCD command for cursor on
                ACALL NCMDWRT           ;write command to LCD
                MOV   A,#01H            ;clear LCD
                ACALL NCMDWRT           ;write command to LCD
                MOV   A,#06             ;shift cursor right command
                ACALL NCMDWRT           ;write command to LCD
                .....                   ;etc. for all LCD commands
                MOV   A,#'N'            ;display data (letter N)
                ACALL NDATAWRT          ;send data to LCD display
                MOV   A,#'O'            ;display data (letter O)
                CALL  NDATAWRT          ;send data to LCD display
                ....                    ;etc. for other data
;New command write subroutine with checking busy flag
NCMDWRT:        MOV   R2,A              ;save A value
                MOV   A,#90H            ;PA=IN to read LCD status
                MOV   R0,#CNTPORT       ;load control reg address
                MOVX  @R0,A             ;configure PA=IN, PB=OUT
                MOV   A,#00000110B      ;RS=0,R/W=1,E=1 read command
                MOV   R0,#BPORT         ;load port B address
                MOVX  @R0,A             ;RS=0,R/W=1 for RD and RS pins
                MOV   R0,#APORT         ;load port A address
READY:          MOVX  A,@R0             ;read command reg
                RLC   A                 ;move D7(busy flag) into carry
                JC    READY             ;wait until LCD is ready
                MOV   A,#80H            ;make PA and PB output again
                MOV   R0,#CNTPORT       ;load control port address
                MOVX  @R0,A             ;issue control word to 8255
                MOV   A,R2              ;get back value to LCD
                MOV   R0,#APORT         ;load port A address
                MOVX  @R0,A             ;issue info to LCD'S data pins
                MOV   R0,#BPORT         ;load port B address
                MOV   A,#00000100B      ;RS=0,R/W=0,E=1 for H-TO-L
                MOVX  @R0,A             ;activate RS,R/W,E pins of LCD
                NOP                     ;make E pin pulse wide enough
                NOP
                MOV   A,#00000000B      ;RS=0,R/W=0,E=0 for H-TO-L
                MOVX  @R0,A             ;latch in LCD'S data pin info
                RET
```

Program 15-2. *(continued on following page)*

314

```
;New data write subroutine with checking busy flag
NDATAWRT:
            MOV   R2,A              ;save A value
            MOV   A,#90H            ;PA=IN to read LCD status ,PB=OUT
            MOV   R0,#CNTPORT       ;load control port address
            MOVX  @R0,A             ;configure PA=IN, PB=OUT
            MOV   A,#00000110B      ;RS=0,R/W=1,E=1 to read command reg
            MOV   R0,#BPORT         ;load port B address
            MOVX  @R0,A             ;RS=0,R/W=1 for RD and RS pins
            MOV   R0,#APORT         ;load port A address
READY:      MOVX  A,@R0             ;read command reg
            RLC   A                 ;move D7(busy flag) into carry
            JC    READY             ;wait until LCD is ready
            MOV   A,#80H            ;make PA and PB output again
            MOV   R0,#CNTPORT       ;load control port address
            MOVX  @R0,A             ;issue control word to 8255
            MOV   A,R2              ;get back value to be sent to LCD
            MOV   R0,#APORT         ;load port A address
            MOVX  @R0,A             ;issue info to LCD'S data pins
            MOV   R0,#BPORT         ;load port B address
            MOV   A,#00000101B      ;RS=1,R/W=0,E=1 for H-TO-L
            MOVX  @R0,A             ;activate RS,R/W,E pins of LCD
            NOP                     ;make E pin pulse wide enough
            NOP
            MOV   A,#00000001B      ;RS=1,R/W=0,E=0 for H-TO-L
            MOVX  @R0,A             ;latch in LCD'S data pin info
            RET
```

Program 15-2. *(continued from previous page)*

ADC connection to the 8255

ADC devices were covered in Chapter 12. The following is a program for the ADC connected to the 8255 as shown in Figure 15-11.

```
            MOV   A,#80H       ;control word for PA=OUT,PC=IN
            MOV   R1,#CRPORT   ;control reg port address
            MOVX  @R1,A        ;configure PA=OUT AND PC=IN
BACK:       MOV   R1,#CPORT    ;load port C address
            MOVX  A,@R1        ;read port C to see if ADC is ready
            ANL   A,#00000001B    ;mask all except PC0
            JNZ   BACK         ;keep monitoring PC0 for EOC
            ;end of conversation, now get ADC data
            MOV   R1,#APORT    ;load PA address
            MOVX  A,@R1        ;A=analog data input
```

So far we have discussed the simple I/O mode of the 8255 and showed many applications for it. Next, we discuss more advanced modes of the 8255. In addition to mode 0 (simple I/O mode) discussed in Section 15.1, we also have BSR mode, and mode 1. Each mode is described next.

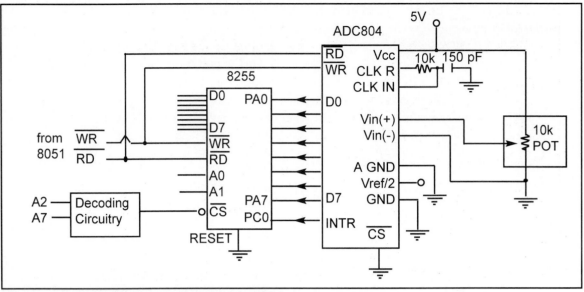

Figure 15-11. 8255 Connection to ADC804

Review Questions

1. Modify the stepper motor program to turn counter-clockwise.
2. True or false. In programming the LCD via an 8255 (without checking the busy flag), port A is always an output port.
3. True or false. In the LCD connection to the 8255, we must have a long delay before issuing the next data if we are not checking the busy flag.

SECTION 15.3: OTHER MODES OF THE 8255

In this section we examine BSR mode and mode 1 of the 8255.

BSR (bit set/reset) mode

A unique feature of port C is that the bits can be controlled individually. BSR mode allows one to set to high or low any of the bits PC0 to PC7, as shown in Figure 15-12. Examples 15-6 and 15-7 show how to use this mode.

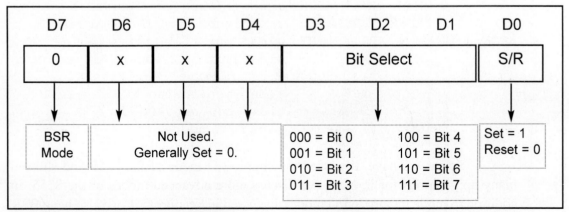

Figure 15-12. BSR Control Word

(Reprinted by permission of Intel Corporation, Copyright Intel Corp. 1983)

316

Example 15-6

Program PC4 of the 8255 to generate a pulse of 50 ms with 50% duty cycle.

Solution:

To program the 8255 in BSR mode, bit D7 of the control word must be low. For PC4 to be high, we need a control word of "0xxx1001". Likewise, for low we would need "0xxx1000" as the control word. The x's are for "don't care" and generally are set to zero.

```
MOV   A,#00001001B    ;control byte for PC4=1
MOV   R1,#CNTPORT     ;load control reg port
MOVX  @R1,A           ;make PC4=1
ACALL DELAY           ;time delay for high pulse
MOV   A,#00001000B    ;control byte for PC4=0
MOVX  @R1,A           ;make PC4=0
ACALL DELAY
```

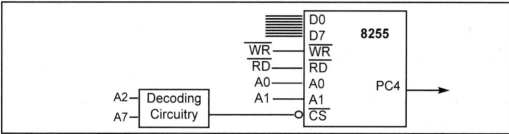

Figure 15-13. Configuration for Examples 15-6, 15-7

Example 15-7

Program the 8255 in Figure 15-13 for the following.
(a) Set PC2 to high.
(b) Use PC6 to generate a square wave of 66% duty cycle continuously.

Solution:
(a)

```
MOV R0,#CNTPORT
MOV A,#0xxx0101              ;control byte
MOVX @R0,A
```

(b)
```
AGAIN:    MOV  A,#00001101B    ;PC6=1
          MOV  R0,#CNTPORT     ;load control port address
          MOVX @R0,A           ;make PC6=1
          ACALL DELAY
          ACALL DELAY
          MOV  A,#00001100B    ;PC6=0
          ACALL DELAY          ;time delay for low pulse
          SJMP AGAIN
```

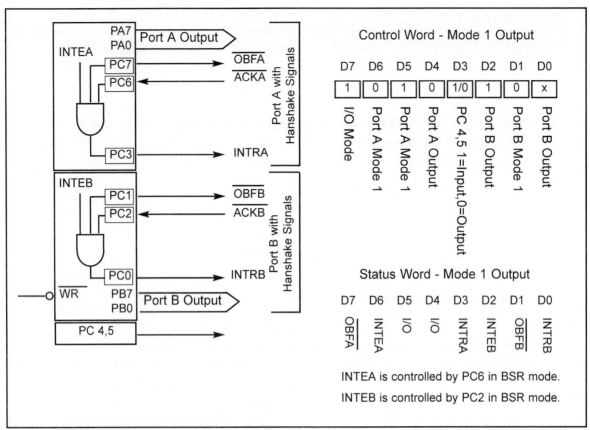

Figure 15-14. 8255A Mode 1 Output Diagram

(Reprinted by permission of Intel Corporation, Copyright Intel Corp. 1983)

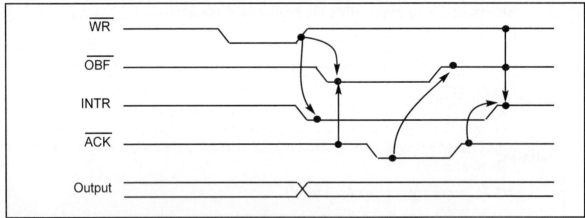

Figure 15-15. Timing Diagram for Mode 1 Output

(Reprinted by permission of Intel Corporation, Copyright Intel Corp. 1983)

8255 in mode 1: I/O with handshaking capability

One of the most powerful features of the 8255 is its ability to handle hand-shaking signals. Handshaking refers to the process of two intelligent devices communicating back and forth. An example of equipment with handshaking signals is the printer. Next we describe the handshaking signals of the 8255 and printers.

Mode 1: outputting data with handshaking signals

As shown in Figure 15-14, A and B can be used as output ports to send data to a device with handshaking signals. The handshaking signals for both ports A and B are provided by the bits of port C. Figure 15-15 provides a timing diagram.

The following paragraphs provide the explanation of and reasoning behind handshaking signals only for port A, but in concept they are exactly the same as for port B.

$\overline{OBF}$a (output buffer full for port A)

This is an active-low signal going out of PC7 to indicate the fact that the CPU has written a byte of data into port A. $\overline{OBF}$a must be connected to STROBE of the receiving equipment (such as a printer) to inform it that it can now read a byte of data from the port A latch.

$\overline{ACK}$a (acknowledge for port A)

This is an active-low input signal received at PC6 of the 8255. Through $\overline{ACK}$, the 8255 knows that the data at port A has been picked up by the receiving device. When the receiving device picks up the data at port A, it must inform the 8255 though the $\overline{ACK}$ signal. The 8255 in turn makes $\overline{OBF}$a high, to indicate that the data at the port is now old data. $\overline{OBF}$a will not go low until the CPU writes a new byte of data to port A.

INTRa (interrupt request for port A)

This is an active-high signal coming out of PC3. The $\overline{ACK}$ signal is a signal of limited duration. When it goes active (low) it makes $\overline{OBF}$a inactive, stays low for a small amount of time and then goes back to high (inactive). It is the rising edge of $\overline{ACK}$ that activates INTRa by making it high. This high signal on INTRa can be used to get the attention of the CPU. The CPU is informed through INTRa that the printer has received the last byte and is ready to receive another one. INTRa interrupts the CPU in whatever it is doing and forces it to write the next byte to port A to be printed. It is important to note that INTRa is set to 1 only if INTEa, $\overline{OBF}$a, and $\overline{ACK}$a are all high. It is reset to zero when the CPU writes a byte to port A.

INTEa (interrupt enable for port A)

The 8255 can disable INTRa to prevent it from interrupting the CPU. This is the function of INTEa. INTEa is an internal flip-flop designed to mask (disable) INTRa. INTEa can be set or reset through port C in BSR mode since the INTEa flip-flop is controlled through PC6. INTEb is controlled by PC2 in BSR mode.

Status word

The 8255 enables monitoring of the status of signals INTR, OBF, and INTE for both ports A and B. This is done by reading port C into the accumulator and testing the bits. This feature allows the implementation of polling instead of a hardware interrupt.

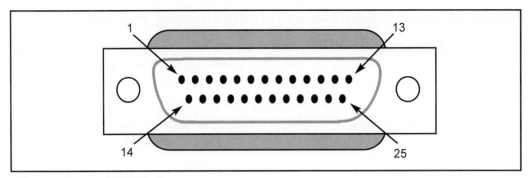

Figure 15-16. DB-25 Connector

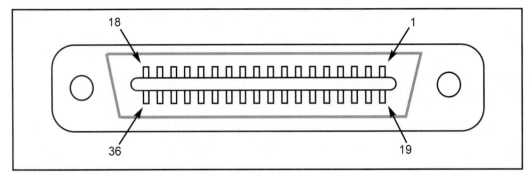

Figure 15-17. 36-Pin Centronics Connector

Printer signal

To understand handshaking with the 8255, we give an overview of printer operation, handshaking signals. The following enumerates the steps of communicating with a printer. Table 15-2 provides a list of signals for Centronics printers.

1. A byte of data is presented to the data bus of the printer.
2. The printer is informed of the presence of a byte of data to be printed by activating its $\overline{\text{Strobe}}$ input signal.
3. Whenever the printer receives the data it informs the sender by activating an output signal called $\overline{\text{ACK}}$ (acknowledge).
4. Signal $\overline{\text{ACK}}$ initiates the process of providing another byte of data to the printer.

As we can see from the steps above, merely presenting a byte of data to the printer is not enough. The printer must be informed of the presence of the data. At the time the data is sent, the printer might be busy or out of paper, so the printer must inform the sender whenever it finally picks up the data from its data bus. Figures 15-16 and 15-17 show the DB-25 and Centronics sides of the printer cable, respectively. Connection of the 8031/51 with the printer and programming are left to the reader to explore.

Table 15-2: DB-25 Printer Pins

Pin	Description
1	Strobe
2	Data bit 0
3	Data bit 1
4	Data bit 2
5	Data bit 3
6	Data bit 4
7	Data bit 5
8	Data bit 6
9	Data bit 7
10	$\overline{\text{ACK}}$ (acknowledge)
11	Busy
12	Out of paper
13	Select
14	Auto feed
15	$\overline{\text{Error}}$
16	Initialize printer
17	Select input
18 - 25	Ground

Reprinted by permission from"IBM Technical Reference" c. 1988 by International Business Machines Corporation)

Table 15-3: Centronics Printer Specifications

Serial	Return	Signal	Direction	Description
1	19	STROBE	IN	STROBE pulse to read data in. Pulse width must be more than 0.5 μs at receiving terminal. The signal level is normally "high"; read-in of data is performed at the "low" level of this signal.
2	20	DATA1	IN	These signals represent information of the 1st to 8th bits of parallel data, respectively. Each signal is at "high" level when data is logical "1", and "low" when logical "0".
3	21	DATA2	IN	" "
4	22	DATA3	IN	" "
5	23	DATA4	IN	" "
6	24	DATA5	IN	" "
7	25	DATA6	IN	" "
8	26	DATA7	IN	" "
9	27	DATA8	IN	" "
10	28	ACKNLG	OUT	Approximately 0.5 μs pulse; "low" indicates data has been received and printer is ready for data.
11	29	BUSY	OUT	A "high" signal indicates that the printer cannot receive data. The signal becomes "high" in the following cases: (1) during data entry, (2) during printing operation, (3) in "off-line" status, (4) during printer error status.
12	30	PE	OUT	A "high" signal indicates that printer is out of paper.
13	--	SLCT	OUT	Indicates that the printer is in the state selected.
14	--	AUTOFEEDXT	IN	When this signal is at "low" level, the paper is fed automatically one line after printing. (The signal level can be fixed to "low" with DIP SW pin 2-3 provided on the control circuit board.)
15	--	NC	--	Not used.
16	--	0V	--	Logic GND level.
17	--	CHASISGND	--	Printer chassis GND. In the printer, chassis GND and the logic GND are isolated from each other.
18	--	NC	--	Not used.
19 - 30	--	GND	--	"Twisted-pair return" signal; GND level.
31	--	INIT	IN	When this signal becomes "low" the printer controller is reset to its initial state and the print buffer is cleared. Normally at "high" level; its pulse width must be more than 50 μs at receiving terminal.
32	--	ERROR	OUT	The level of this signal becomes "low" when printer is in "paper end", "off-line", and "error" state.
33	--	GND	--	Same as with pin numbers 19 to 30.
34	--	NC	--	Not used.
35	--		--	Pulled up to +5V dc through 4.7 K ohms resistance.
36	--	SLCTIN	IN	Data entry to the printer is possible only when the level of this signal is "low". (Internal fixing can be carried out with DIP SW 1 - 8. The condition at the time of shipment is set "low" for this signal.)

(Reprinted by permission from "IBM Technical Reference Options and Adapters"
c. 1981 by International Business Machines Corporation)

Review Questions

1. True or false. In the 8255, only port C is bit-addressable.
2. To set only PC7 high, byte _____ is sent to the _____ port.
3. Find the control word for PA = output with handshaking, PB = output with handshaking. Unused bits of PC = don't care.

SUMMARY

This chapter described how to expand the I/O ports of the 8031/51 by connecting it to an 8255 chip. The 8255 could then be programmed as a simple I/O port to control interfaced devices such as stepper motors, LCDs, and ADC devices.

The 8255 has 3 ports. Ports A and B are 8-bit bidirectional ports. The 8-bit port C can be programmed as one 8-bit port or two 4-bit ports, and it is also bit-programmable. The various operation modes of the 8255 are selected by the control register. Various address decoding techniques were demonstrated to program the 8255 control register, select ports, and set the control register. In addition, numerous program examples were given of 8051/31 instructions to control devices such as LCDs, stepper motors, and ADC devices that were interfaced via the 8255.

Printer and 8255 handshaking signals were discussed in the last section of this chapter.

PROBLEMS

SECTION 15.1: PROGRAMMING THE 8255

1. Find the control byte to set all the ports as simple input.
2. Find the control byte to set all the ports as simple output.
3. Find the control byte to set PA = input, and both ports B and C as outputs.
4. Find the control byte to set PA = output, and both ports B and C as inputs.
5. Find the control byte to set PA = input, PB = input, and PC = output.
6. Show the design of the 8255 connection to the 8051 where port A has the address 2000H. Then program the 8255 to get data from port B and send it to both ports A and C.
7. Show the design of the 8255 connection to the 8051 where port A has the address 800H. Then program the 8255 to get data from port C and send it to both ports A and B.
8. Show the design of the 8255 connection to the 8051 where port A has the address 20H. Then program the 8255 to get data from port B and send it to both ports A and C.
9. Show the design of the 8255 connection to the 8051 where port A has the address 88H. Then program the 8255 to get data from port C and send it to both ports A and B.
10. Using the 74LS138 as an address decoder, show the connection to address A0 - A7 of the 8051 where Y0 is assigned the 8255 base address of 80H.
11. In Problem 10, find the base address of the 8255 for the following.
    (a) Y1   (b) Y2   (c) Y5

12. Using the 74LS138 as an address decoder, show the connection to address A0 - A7 of the 8051 where Y0 is assigned the 8255 base address of C0H. Use any other simple gates you need.

13. In Problem 12, find the base address of the 8255 for the following.
    (a) Y3   (b) Y5   (c) Y7

14. Using a 74LS138 as an address decoder, show the connection to address A0 - A7 of the 8051 where Y0 is assigned to the 8255 base address of 00H.

15. In Problem 14, find the base address of the 8255 for the following.
    (a) Y1   (b) Y2   (c) Y7

16. How many pins of the 8255 are used for ports, and how are they categorized?

17. What is the function of data pins D0 - D7 in the 8255?

18. True or false. All three ports, A, B, and C, are bit-addressable.

19. True or false. All three ports, A, B, and C, can be programmed for simple I/O.

20. True or false. In simple I/O programming of port A of the 8255, we can use PA0 - PA3 for output and PA4 - PA7 for an input port.

21. Show the decoding circuitry for the 8255 if we want port A to have address 68H. Use NAND and inverter gates.

22. Which of the following port addresses cannot be assigned to port A of the 8255, and why?
    (a) 32H   (b) 45H   (c) 89H   (d) BAH

23. If 91H is the control word, indicate which port is input and which is output.

24. Find the control word if PA = input, PB = input, and PC0 - PC7 = output.

25. Write a program to monitor PA for a temperature of 100. If it is equal, it should be saved in register R3. Also send AAH to port B and 55H to port C.

ANSWERS TO REVIEW QUESTIONS

SECTION 15.1: PROGRAMMING THE 8255

1. 9BH   2. 89H   3. True
4. It cannot since we must have A1 = 0 and A0 = 0 for the base address.
5. The MOVX instruction allows access to external memory and 8255 is mapped as memory-mapped I/O.

SECTION 15.2: 8255 INTERFACING

1. "RR A" is changed to "RL A".
2. True   3. True

SECTION 15.3: OTHER MODES OF THE 8255

1. True
2. 0FH to the control register port.
3. A4H

APPENDIX A

8051 INSTRUCTIONS, TIMING AND REGISTERS

OVERVIEW

In the first section of this appendix, we describe the instructions of the 8051 and give their formats with some examples. In many cases, more detailed programming examples will be given to clarify the instructions. These instructions will operate on any 8031, 8032, 8051, or 8052 microcontroller. The first section concludes with a list of machine cycles (clock counts) for each 8051 instruction.

In the second section, a list of all the 8051 registers is provided for ease of reference for the 8051 programmer.

SECTION A.1: THE 8051 INSTRUCTION SET

ACALL target address

Function: Absolute Call
Flags: None

ACALL stands for "absolute call." It calls subroutines with a target address within 2K bytes from the current program counter (PC). See LCALL for more discussion on this.

ADD   A,source byte

Function: ADD
Flags: OV, AC, CY

This adds the source byte to the accumulator (A), and places the result in A. Since register A is one byte in size, the source operands must also be one byte.

The ADD instruction is used for both signed and unsigned numbers. Each one is discussed separately.

Unsigned addition

In the addition of unsigned numbers, the status of CY, AC, and OV may change. The most important of these flags is CY. It becomes 1 when there is a carry from D7 out in 8-bit (D0 - D7) operations.

Example:

```
        MOV    A,#45H           ;A=45H
        ADD    A,#4FH           ;A=94H  (45H+4FH=94H)
                                ;CY=0,AC=1
```

Example:

```
        MOV    A,#0FEH          ;A=FEH
        MOV    R3,#75H          ;R3=75H
        ADD    A,R3             ;A=FE+75=73H
                                ;CY=1,AC=1
```

Example:

```
        MOV    A,#25H           ;A=25H
        ADD    A,#42H           ;A=67H  (25H+42H=67H)
                                ;CY=0,AC=0
```

Addressing modes

The following addressing modes are supported for the ADD instruction:
1. Immediate: `ADD A,#data`   Example: `ADD A,#25H`
2. Register: `ADD A,Rn`   Example: `ADD A,R3`
3. Direct: `ADD A,direct`
   Example: `ADD A,30H ;add to A data in RAM loc. 30H`
4. Register-indirect: `ADD A,@Ri where i=0 or i=1 only`
   Examples:
   `ADD A,@R0 ;add to A data pointed to by R0`
   `ADD A,@R1 ;add to A data pointed to by R1`

In the following example, the contents of RAM locations 50H to 5FH are added together, and the sum is saved in RAM locations 70H and 71H.

```
        CLR  A              ;A=0
        MOV  R0,#50H        ;source pointer
        MOV  R2,#16         ;counter
        MOV  R3,#0          ;clear R3
A_1:    ADD  A,@R0          ;ADD to A from source
        JNC  B_1            ;IF CY=0 go to next byte
        INC  R3             ;otherwise keep carries
B_1:    INC  R0            ;next location
        DJNZ R2,A_1        ;repeat for all bytes
        MOV  70H,A         ;save low byte of sum
        MOV  71H,R3        ;save high byte of sum
```

Notice in all the above examples that we ignored the status of the OV flag. Although ADD instructions do affect OV, it is in the context of signed numbers that the OV flag has any significance. This is discussed next.

Signed addition and negative numbers

In the addition of signed numbers, special attention should be given to the overflow flag (OV) since this indicates if there is an error in the result of the addition. There are two rules for setting OV in signed number operation. The overflow flag is set to 1:
1. If there is a carry from D6 to D7 and no carry from D7 out.
2. If there is a carry from D7 out and no carry from D6 to D7.
   Notice that if there is a carry both from D7 out and from D6 to D7, OV = 0.

Example:
```
        MOV   A,#+8        ;A=0000 1000
        MOV   R1,#+4       ;R1=0000 0100
        ADD   A,R1         ;A=0000 1100 OV=0,CY=0
```
Notice that D7 = 0 since the result is positive and OV = 0 since there is neither a carry from D6 to D7 nor any carry beyond D7. Since OV = 0, the result is correct [(+8) + (+4) = (+12)].

Example:

```
MOV    A,#+66    ;A=0100 0010
MOV    R4,#+69   ;R4=0100 0101
ADD    A,R4      ;A=1000 0111 = -121
                 ;(INCORRECT) CY=0, D7=1, OV=1
```

In the above example, the correct result is +135 [(+66) + (+69) = (+135)], but the result was –121. OV = 1 is an indication of this error. Notice that D7 = 1 since the result is negative; OV = 1 since there is a carry from D6 to D7 and CY=0.

Example:

```
MOV    A,#-12    ;A=1111 0100
MOV    R3,#+18   ;R3=0001 0010
ADD    A,R3      ;A=0000 0110 (+6) correct
                 ;D7=0,OV=0, and CY=1
```

Notice above that the result is correct (OV = 0), since there is a carry from D6 to D7 and a carry from D7 out.

Example:

```
MOV    A,#-30    ;A=1110 0010
MOV    R0,#+14   ;R0=0000 1110
ADD    A,R0      ;A=1111 0000 (-16, CORRECT)
                 ;D7=1,OV=0, CY=0
```

OV = 0 since there is no carry from D7 out nor any carry from D6 to D7.

Example:

```
MOV    A,#-126   ;A=1000 0010
MOV    R7,#-127  ;R7=1000 0001
ADD    A,R7      ;A=0000 0011 (+3, WRONG)
                 ;D7=0, OV=1
```

CY = 1 since there is a carry from D7 out but no carry from D6 to D7.

From the above discussion we conclude that while CY is important in any addition, OV is extremely important in signed number addition since it is used to indicate whether or not the result is valid. As we will see in instruction "DA A", the AC flag is used in the addition of BCD numbers. OV is also used in DIV and MUL instructions as well. See the description of these two instructions for further details.

ADDC A,source byte

Function: Add with carry

Flags: OV, AC, CY

This will add the source byte to A, in addition to the CY flag (A = A + byte + CY). If CY = 1 prior to this instruction, CY is also added to A. If CY = 0 prior to the instruction, source is added to destination plus 0. This is used in multibyte additions. In the addition of 25F2H to 3189H, for example, we use the ADDC instruction as shown below.

Example:

```
CLR   C              ;CY=0
MOV   A,#89H         ;A=89H
ADDC  A,#0F2H        ;A=89H+F2H+0=17BH, A=7B, CY=1
MOV   R3,A           ;SAVE A
MOV   A,#31H
ADDC  A,#25H         ;A=31H+25H+1=57H
```

Therefore the result is:

```
 25F2H
+3189H
 577BH
```

The addressing modes for ADDC are the same as for "ADD A,byte".

AJMP target address

Function: Absolute jump

Flag: None

AJMP stands for "absolute jump." It transfers program execution to the target address unconditionally. The target address for this instruction must be within 2K bytes of program memory. See LJMP for more discussion on this.

ANL   dest-byte,source-byte

Function: Logical AND for byte variables

Flags: None affected

This performs a logical AND on the operands, bit by bit, storing the result in the destination. Notice that both the source and destination values are byte-size only.

A	B	A AND B
0	0	0
0	1	0
1	0	0
1	1	1

Example:

```
MOV   A,#39H   ;A=39H
ANL   A,#09H   ;A=39H ANDed with 09

39    0011 1001
09    0000 1001
09    0000 1001
```

Example:

```
MOV    A,#32H    ;A=32H    32    0011 0010
MOV    R4,#50H   ;R4=50H   50    0101 0000
ANL    A,R4      ;(A=10H)  10    0001 0000
```

For the ANL instruction there are a total of six addressing modes. In four of them, the accumulator must be the destination. They are as follows:

1. Immediate: ANL A,#data   Example: ANL A,#25H
2. Register: ANL A,Rn   Example: ANL A,R3
3. Direct: ANL A,direct
   Example: ANL A,30H   ;AND A with data in RAM location 30H
4. Register-indirect:
   Example: ANL A,@R0   ;AND A with data pointed to by R0

In the next two addressing modes the destination is a direct address (a RAM location or one of the SFR registers) while the source is either A or immediate data.

5. ANL direct,#data
   Example: Assume that RAM location 32H has the value 67H. Find its content after execution of the following code.

```
ANL 32H,#44H
44H   0100 0100
67H   0110 0111
44H   0100 0101    Therefore, it has 44H.
```

Or look at these examples:

```
ANL P1,#11111110B    ;mask P1.0(D0 of Port 1)
ANL P1,#01111111B    ;mask P1.7(D7 of Port 1)
ANL P1,#11110111B    ;mask P1.3(D3 of Port 1)
ANL P1,#11111100B    ;mask P1.0 and P1.1
```

The above instructions clear (mask) certain bits of the output port of P1.

6. ANL direct,A
   Example: Find the contents of register B after the following:

```
MOV B,#44H     ;B=44H
MOV A,#67H     ;A=67H
ANL 0F0H,A     ;A AND B(B is located at RAM F0H)
               ;after the operation B=44H
```

Note: We cannot use this to mask bits of input ports! For example, "ANL A,P1" is incorrect!

ANL C,source-bit

Function: Logical AND for bit variable
Flag: CY

In this instruction the carry flag bit is ANDed with a source bit and the result is placed in carry. Therefore, if source bit = 0, CY is cleared; otherwise, the CY flag remains unchanged.

Example: Write code to clear the accumulator if bits P2.1 and P2.2 are both high; otherwise, make A=FFH.

```
        MOV  A,#0FFH    ;A=FFH
        MOV  C,P2.1     ;copy bit P2.1 to carry flag
        ANL  C,P2.2     ;and then
        JNC  B_1        ;jump if one of them is low
        CLR  A
B_1:
```

Another variation of this instruction involves the ANDing of the CY flag bit with the complement of the source bit. Its format is "ANL C,/bit". See the following example.

Example: Clear A if P2.1 is high and P2.2 is low; otherwise, make A=FFH.
```
        MOV  A,#0FFH
        MOV  C,P2.1     ;get a copy of P2.1 bit
        ANL  C,/P2.2    ;AND P2.1 with complement of P2.2
        JNC  B_1
        CLR  A
B_1:
```

CJNE dest-byte,source-byte,target

Function: Compare and jump if not equal
Flag: CY

The magnitudes of the source byte and destination byte are compared. If they are not equal, it jumps to the target address.

Example: Keep monitoring P1 indefinitely for the value of 99H. Get out only when P1 has the value 99H.
```
        MOV  P1,0FFH    ;make P1 an input port
BACK:   MOV  A,P1       ;read P1
        CJNE A,#99,BACK ;keep monitoring
```

Notice that CJNE jumps only for the not-equal value. To find out if it is greater or less after the comparison, we must check the CY flag. Notice also that the CJNE instruction affects the CY flag only, and after the jump to target address the carry flag indicates which value is greater, as shown here.

| Dest < Source CY = 1 |
| Dest > Source CY = 0 |
| |

In the following example, P1 is read and compared with value 65. Then:
(1) If P1 is equal to 65, the accumulator keeps the result,
(2) If P1 has a value less than 65, R2 has the result, and finally,
(3) If P1 has a value greater than 65, it is kept by R3.

At the end of the program, A will contain the equal value, or R2 the smaller value, or R3 the greater value.

Example:

```
        MOV  A,P1              ;READ P1
        CJNE A,#65,NEXT        ;IS IT 65?
        SJMP EXIT              ;YES, A KEEPS IT, EXIT
NEXT:   JNC  OVER              ;NO
        MOV  R2,A              ;SAVE THE SMALLER IN R2
        SJMP EXIT              ;AND EXIT
OVER:   MOV  R3,A              ;SAVE THE LARGER IN R3
EXIT:
```

This instruction supports four addressing modes. In two of them, A is the destination.

1. Immediate: CJNE A,#data,target
      Example: CJNE A,#96,NEXT   ;JUMP IF A IS NOT 96
2. Direct: CJNE A,direct,target
      Example: CJNE A,40H,NEXT   ;JUMP IF A NOT =
                                       ;WITH THE VALUE HELD BY RAM LOC. 40H

Notice the absence of the "#" sign in the above instruction. This indicates RAM location 40H. Notice in this mode that we can test the value at an input port. This is a widely used application of this instruction. See the following:

```
        MOV  P1,#0FF          ;P1 is an input port
        MOV  A,#100           ;A = 100
HERE:   CJNE A,P1,HERE        ;WAIT HERE TIL P1 = 100
```

In the third addressing mode, any register, R0 - R7, can be the destination.

3. Register: CJNE Rn,#data,target
      Example: CJNE R5,#70,NEXT   ;jump if R5 is not 70

In the fourth addressing mode, any RAM location can be the destination. The RAM location is held by register R0 or R1.

4. Register-indirect: CJNE @Ri,#data,target
      Example: CJNE @R1,#80,NEXT   ;jump if RAM
                                        ;location whose address is held by R1
                                        ;is not equal to 80

Notice that the target address can be no more than 128 bytes backward or 127 bytes forward, since it is a 2-byte instruction. For more on this see SJMP.

CLR   A

Function: Clear accumulator
Flag: None are affected
This instruction clears register A. All bits of the accumulator are set to 0.

Example:

```
        CLR   A
        MOV   R0,A        ;clear R0
        MOV   R2,A        ;clear R2
        MOV   P1,A        ;clear port 1
```

Example: The following simple program clears RAM locations 40H - 4FH.

```
        MOV   R3,#16          ;count
        MOV   R0,#40H         ;starting RAM location
        CLR   A
BACK:   MOV   @R0,A           ;clear RAM location
        INC   R0             ;next location
        DJNZ  R3,BACK         ;repeat for all locations
```

CLR   bit

Function: Clear bit
This instruction clears a single bit. The bit can be the carry flag, or any bit-addressable location in the 8051. Here are some examples of its format:

```
CLR   C           ;CY=0
CLR   P2.4        ;CLEAR P2.5 (P2.5=0)
CLR   P1.7        ;CLEAR P1.7 (P1.7=0)
CLR   ACC.7       ;CLEAR D7 OF ACCUMULATOR  (ACC.7=0)
```

CPL   A

Function: Complement accumulator
Flags: None are affected
This complements the contents of register A, the accumulator. The result is the 1's complement of the accumulator. That is: 0s become 1s and 1s become 0s.

Example:

```
        MOV A,#55H        ;A=01010101
AGAIN:  CPL A             ;complement reg. A
        MOV P1,A          ;toggle bits of P1
        SJMP AGAIN        ;continuously
```

Function: Decimal-adjust accumulator after addition

Flags: CY

This instruction is used after addition of BCD numbers to convert the result back to BCD. The data is adjusted in the following 2 possible cases.

1. It adds 6 to the lower 4 bits of A if it is greater than 9 or if AC = 1.

2. It also adds 6 to the upper 4 bits of A if it is greater than 9 or if CY = 1.

Example:

```
MOV    A,#47H    ;A=0100 0111
ADD    A,#38H    ;A=47H+38H=7FH, invalid BCD
DA     A         ;A=1000 0101=85H, valid BCD


  47H
+ 38H
  7FH    (invalid BCD)
+  6H    (after DA A)
  85H    (valid BCD)
```

In the above example, since the lower nibble was greater than 9, DA added 6 to A. If the lower nibble is less than 9 but AC = 1, it also adds 6 to the lower nibble. See the following example.

Example:

```
MOV    A,#29H       ;A=0010 1001
ADD    A,#18H       ;A=0100 0001 INCORRECT
DA     A         ;A=0100 0111 = 47H VALID BCD


  29H
+ 18H
  41H    (incorrect result in BCD)
+  6H
  47H    correct result in BCD
```

The same thing can happen for the upper nibble. See the following example.

Example:

```
MOV    A,#52H    ;A=0101 0010
ADD    A,#91H    ;A=1110 0011 INVALID BCD
DA     A         ;A=0100 0011 AND CY=1


  52H
+ 91H
  E3H            (invalid BCD)
+  6            (after DA A, adding to upper nibble)
 143H            valid BCD
```

Similarly, if the upper nibble is less than 9 and CY = 1, it must be corrected. See the following example.

Example:

```
MOV   A,#94H       ;A=1001 0100
ADD   A,#91H       ;A=0010 0101 INCORRECT
DA    A            ;A=1000 0101, VALID BCD
                   ;FOR 85,CY=1
```

It is possible that 6 is added to both the high and low nibbles. See the following example.

Example:

```
MOV   A,#54H       ;A=0101 0100
ADD   A,#87H       ;A=1101 1011 INVALID BCD
DA    A          ;A=0100 0001, CY=1 (BCD 141)
```

DEC   byte

Function: Decrement
Flags: None

This instruction subtracts 1 from the byte operand. Note that CY (carry/borrow) is unchanged even if a value 00 is decremented and becomes FF.

This instruction supports four addressing modes.

1. Accumulator: DEC A   Example: DEC A
2. Register: DEC Rn   Example: DEC R1 or DEC R3
3. Direct: DEC direct
   Example: DEC 40H   ;dec byte in RAM location 40H
4. Register-indirect: DEC @Ri   ;where i=0 or 1 only
   Example: DEC @R0   ;decr. byte pointed to by R0

DIV   AB

Function: Divide
Flags: CY and OV

This instruction divides a byte accumulator by the byte in register B. It is assumed that both registers A and B contain an unsigned byte. After the division, the quotient will be in register A and the remainder in register B. If you divide by zero (that is setting register B = 0 before the execution of "DIV AB"), the values in register A and B are undefined and the OV flag is set to high to indicate an invalid result. Notice that CY is always 0 in this instruction.

Example:

```
MOV A,#35
MOV B,#10
DIV AB            ;A=3  and B=5
```

Example:
```
        MOV A,#97H
        MOV B,#12H
        DIV AB          ;A=8 and B=7
```

Notice in this instruction that the carry and OV flags are both cleared, unless we divide A by 0, in which case the result is invalid and OV=1 to indicate the invalid condition.

DJNZ byte, target

Function: Decrement and jump if not zero
Flags: None
In this instruction a byte is decremented, and if the result is not zero it will jump to target address.

Example: Count from 1 to 20 and send the count to P1.
```
        CLR A                   ;A=0
        MOV R2,#20              ;R2=20 counter
BACK:   INC A
        MOV P1,A
        DJNZ R2,BACK    ;repeat if R2 not = zero
```

The following two formats are supported by this instruction.

1. Register: DJNZ Rn,target (where n=0 to 7)
   Example: `DJNZ R3,HERE`
2. Direct: DJNZ direct,target

Notice that the target address can be no more than 128 bytes backward or 127 bytes forward, since it is a 2-byte instruction. For more on this see SJMP.

INC   byte

Function: Increment
Flags: None

This instruction adds 1 to the register or memory location specified by the operand. Note that CY is not affected even if value FF is incremented to 00. This instruction supports four addressing modes.

1. Accumulator: INC A   Example: INC A
2. Register: INC Rn   Example: INC R1 or INC R5
3. Direct: INC direct   Example:
                   `INC 30H   ;incr. byte in RAM loc. 30H`
4. Register -indirect: INC @Ri (i = 0 or 1)   Example:
                   `INC @R0   ;incr. byte pointed to by R0`

INC   DPTR

Function:	Increment data pointer	
Flags:	None	

This instruction increments the 16-bit register DPTR (data pointer) by 1. Notice that DPTR is the only 16-bit register that can be incremented. Also notice that there is no decrement version of this instruction.

Example:

```
MOV  DPTR,#16FFH    ;DPTR=16FFH
INC  DPTR           ;now DPTR=1700H
```

JB   bit,target   also: JNB   bit,target

Function:	Jump if bit set	Jump if bit not set
Flags:	None	

These instructions are used to monitor a given bit and jump to a target address if a given bit is high or low. In the case of JB, if the bit is high it will jump while for JNB if the bit is low it will jump. The given bit can be any of the bit-addressable bits of RAM, ports, or registers of the 8051.

Example: Monitor bit P1.5 continously. When it becomes low, send 55H to P2.

```
        SETB P1.5      ;make P1.5 an input bit
HERE:   JB P1.5,HERE   ;stay here as long as P1.5=1
        MOV P2,#55H    ;since P1.5=0 send 55H to P2
```

Example: See if register A has an even number. If so, make it odd.

```
        JB   ACC.0,NEXT    ;jump if it is odd
        INC A             ;it is even, make it odd
NEXT:   ...
```

Example: Monitor bit P1.4 continously. When it becomes high, send 55H to P2.

```
        SETB P1.4      ;make P1.4 an input bit
HERE:   JNB P1.4,HERE  ;stay here as long as P1.4=0
        MOV P2,#55H    ;since P1.4=1 send 55H to P2
```

Example: See if register A has an even number. If not, make it even.

```
        JNB  ACC.0,NEXT    ;jump if D0 is 0 (even)
        INC  A             ;D0=1, make it even
NEXT:   ...
```

JBC   bit,target

      Function: Jump if bit is set and clear bit
      Flags: None

If the desired bit is high it will jump to the target address while at the same time the bit is cleared to zero.

Example: The following instruction will jump to label NEXT if D7 of register A is high; at the same time D7 is cleared to zero.

```
        JBC ACC.7,NEXT
        MOV P1,A
        . . .
NEXT:
```

Notice that the target address can be no more than 128 bytes backward or 127 bytes forward since it is a 2-byte instruction. For more on this see SJMP.

JC   target

      Function: Jump if CY=1.
      Flags: None

This instruction examines the CY flag; if it is high, it will jump to the target address.

JMP   @A+DPTR

      Function: Jump indirect
      Flags: None

The JMP instruction is an unconditional jump to a target address. The target address is provided by the total sum of register A and the DPTR register. Since this is not a widely used instruction we will bypass further discussion of it.

JNB   bit,target

See JB and JNB.

JNC   target

      Function: Jump if no carry (CY=0)
      Flags: None

This instruction examines the CY flag, and if it is zero it will jump to the target address.

Example: Find the total sum of the bytes F6H, 98H, and 8AH. Save the carries in register R3.

```
            CLR   A              ;A=0
            MOV   R3,A           ;R3=0
            ADD   A,#0F6H
            JNC   OVER1
            INC   R3
OVER1:      ADD   A,#98H
            JNC   OVER2
            INC   R3
OVER2:      ADD   A,#8AH
            JNC   OVER3
            INC   R3
OVER3:
```

Notice that this is a 2-byte instruction and the target address cannot be farther than –128 to +127 bytes from the program counter. See J condition for more on this.

JNZ   target

Function:	Jump if accumulator is not zero
Flags:	None

This instruction jumps if register A has a value other than zero.

Example: Search RAM locations 40H - 4FH to find how many of them have the value 0.

```
            MOV   R5,16          ;set counter
            MOV   R3,#0          ;R3 holds number of 0s
            MOV   R1,#40H        ;address
BACK:       MOV   A,@R1          ;bring data to reg A
            JNZ   OVER
            INC   R3
OVER:       INC   R1             ;point to next location
            DJNZ  R5,BACK        ;repeat for all locations
```

The above program will bring the data into the accumulator and if it is zero, it increments counter R3. Notice that this is a 2-byte instruction; therefore, the target address cannot be more than –128 to +127 bytes away from the program counter. See J condition for further discussion on this.

JZ   target

Function: Jump if A = zero
Flags: None

This instruction examines the contents of the accumulator and jumps if it has value 0.

Example: A string of bytes of data is stored in RAM locations starting at address 50H. The end of the string is indicated by the value 0. Send the values to P1 one by one with a delay in between each.

```
        MOV  R0,#50H    ;address
BACK:   MOV  A,@R0      ;bring the value into reg A
        JZ   EXIT       ;end of string, exit
        MOV  P1,A       ;send it to P1
        ACALL DELAY
        INC  R0         ;point to next
        SJMP BACK
EXIT:   ...
```

Notice that this is a 2-byte instruction; therefore, the target address cannot be more than –128 to +127 bytes away from the program counter. See J condition for further discussion on this.

J condition   target

Function: Conditional jump

In this type of jump, control is transferred to a target address if certain conditions are met. The target address cannot be more than –128 to +127 bytes away from the current PC.

JC	Jump carry	jump if CY = 1
JNC	Jump no carry	jump if CY = 0
JZ	Jump zero	jump if register A = 0
JNZ	Jump no zero	jump if register A is not 0
JNB bit	Jump no bit	jump if bit = 0
JB bit	Jump bit	jump if bit = 1
JBC bit	Jump bit clear bit	jump if bit = 1 and clear bit
DJNZ Rn,...	Decrement and jump if not zero	
CJNE A,#val,...	Compare A with value and jump if not equal	

Notice that all "J condition" instructions are short jumps, meaning that the target address cannot be more than –128 bytes backward or +127 bytes forward of the PC of the instruction following the jump (see SJMP). What happens if a programmer needs to use a "J condition" to go to a target address beyond the –128 to +127 range? The solution is to use the "J condition" along with the unconditional LJMP instruction, as shown below.

```
              ORG   100H
              ADD   A,R0
              JNC   NEXT
              LJMP  OVER        ;target more than 128 bytes away
NEXT:         ...
              ORG 300H
OVER:         ADD   A,R2
```

LCALL   16-bit addr   also: ACALL   11-bit addr

Function: Transfers control to a subroutine
Flags: None

There are two types of CALLs: ACALL and LCALL. In ACALL, the target address is within 2K bytes of the current PC (program counter). To reach the target address in the 64K bytes maximum ROM space of the 8051, we must use LCALL. If calling a subroutine, the PC register (which has the address of the instruction after the ACALL) is pushed onto the stack and the stack pointer (SP) is incremented by 2. Then the program counter is loaded with the new address and control is transferred to the subroutine. At the end of the procedure, when RET is executed, PC is popped off the stack, which returns control to the instruction after the CALL.

Notice that LCALL is a 3-byte instruction, in which one byte is the opcode, and the other two bytes are the 16-bit address of the target subroutine. ACALL is a 2-byte instruction, in which 5 bits are used for the opcode and the remaining 11 bits are used for the target subroutine address. An 11-bit address limits the range to 2K bytes.

LJMP 16-bit addr   also: SJMP 8-bit addr

Function: Transfers control unconditionally to a new address.

In the 8051 there are two unconditional jumps: LJMP (long jump) and SJMP (short jump). Each is described next.

1. LJMP (long jump): This is a 3-byte instruction. The first byte is the opcode and the next two bytes are the target address. As a result, LJMP is used to jump to any address location within the 64K byte code space of the 8051. Notice that the difference between LJMP and LCALL is that the CALL instruction will return and continue execution with the instruction following the CALL, whereas JMP will not return.

2. SJMP (short jump): This is a 2-byte instruction. The first byte is the opcode and the second byte is the signed number displacement, which is added to the PC (program counter) of the instruction following the SJMP to get the target address. Therefore, in this jump the target address must be within –128 to +127 bytes of the PC (program counter) of the instruction after the SJMP since a sin-

---

gle byte of address can take values of +127 to –128. This address is often referred to as a *relative address* since the target address is –128 to +127 bytes relative to the program counter (PC). In this Appendix, we have used the term target address in place of relative address only for the sake of simpilicity.

Example:

Line 2 of the code below shows 803E as the object code for "SJMP OVER", which is a forward jump instruction. The 80H, located at address 100H, is the opcode for the SJMP, and 3EH, located at address 101H, is the relative address. The address is relative to the next address location, which is 102H. Adding 102H + F8H = 140H gives the target address of the "OVER" label.

```
LOC   OBJ        LINE
0100             1                    ORG 100H
0100  803E       2                    SJMP OVER
0140             3                    ORG 140H
0140  7A0A       4        OVER:       MOV R2,#10
0142  7B64       5        AGAIN:      MOV R3,#100
0144  00         6        BACK:       NOP
0145  00         7                    NOP
0146  DBFC       8                    DJNZ R3,BACK
0148  80F8       9                    SJMP AGAIN
```

Example:

Line 9 of the code above shows 80F8 for "SJMP AGAIN", which is a backward jump instruction. The 80H, located at address 148H, is the opcode for the SJMP, and F8H, located at address 149H, is the relative address. The address is relative to next address location, which is 14AH. Therefore, adding 14AH + F8H = 142H gives the target address of the "AGAIN" label.

If the target address is beyond the –128 to +127 byte range, the assembler gives an error. All the conditional jumps are short jumps as discussed next.

MOV dest-byte,source-byte

Function:	Move byte variable
Flags:	None

This copies a byte from the source location to the destination. There are fifteen possible combinations for this instruction. They are as follows:

(a) Register A as the destination. This can have the following 4 formats.
1. MOV A,#data   Example: MOV A,#25H   ; (A=25H)
2. MOV A,Rn   Example: MOV A,R3
3. MOV A,direct   Example: MOV A,30H ;A= data in 30H
4. MOV A,@Ri (i=0 or 1)
         Examples: MOV A,@R0 ;A = data pointed to by R0
                      MOV A,@R1 ;A = data pointed to by R1
     Notice that "MOV A,A" is invalid.

342

(b) Register A is the source. The destination can take the following forms.
    5. MOV Rn,A
    6. MOV direct,A
    7. MOV @Ri,A

(c) Rn is the destination.
    8. MOV Rn,#immediate
    9. MOV Rn,A
    10. MOV Rn,direct

(d) The destination is a direct address.
    11. MOV direct,#data
    12. MOV direct,@Ri
    13. MOV direct,A
    14. MOV direct, Rn

(d) Destination is an indirect address held by R0 or R1.
    15. MOV @Ri,#data
    16. MOV @Ri,A
    17. MOV @Ri,direct

MOV dest-bit,source-bit

    Function: Move bit data

This MOV instruction copies the source bit to the destination bit. In this instruction one of the operands must be the CY flag. Look at the following examples.

```
MOV  P1.2,C     ;copy carry bit to port bit P1.2
MOV  C,P2.5     ;copy port bit P2.5 to carry bit
```

MOV DPTR,#16-bit value

    Function: Load data pointer
    Flags: None

This instruction loads the 16-bit DPTR (data pointer) register with a 16-bit immediate value.

Examples:

```
MOV  DPTR,#456FH      ;DPTR=456FH
MOV  DPTR,#MYDATA     ;load 16-bit address
                      ;assigned to MYDATA
```

Function: Move code byte
Flags: None

This instruction moves a byte of data located in program (code) ROM into register A. This allows us to put strings of data, such as look-up table elements, in the code space and read them into the CPU. The address of the desired byte in the code space (on-chip ROM) is formed by adding the original value of the accumulator to the 16-bit DPTR register.

Example: Assume that an ASCII character string is stored in the on-chip ROM program memory starting at address 200H. Write a program to bring each character into the CPU and send it to P1 (port 1).

```
        ORG   100H
        MOV   DPTR,#200H     ;load data pointer
B1:     CLR   A              ;A=0
        MOVC  A,@A+DPTR      ;move data at A+DPTR into A
        JZ    EXIT           ;exit if last (null) char
        MOV   P1,A           ;send character to P1
        INC   DPTR           ;next character
        SJMP  B1             ;continue
EXIT:   ..

        ORG   200H
DATA:   DB    "The earth is but one country and "
        DB    "mankind its citizens","Baha'u'llah",0
        END
```

In the program above first A = 0 and then it is added to DPTR to form the address of the desired byte. After the MOVC instruction, register A has the character. Notice that the DPTR is incremented to point to the next character in the DATA table.

Example: Look-up table SQUR has the squares of values between 0 and 9, and register R3 has the values of 0 to 9. Write a program to fetch the squares from the look-up table.

```
        MOV   DPTR,#SQUR     ;load pointer for table
        MOV   A,R3
        MOVC  A,@A+DPTR

        ORG   100H
SQUR:   DB    0,1,4,9,16,25,36,49,64,81
```

Notice that the MOVC instruction transfers data from the internal ROM space of the 8051 into register A. This internal ROM space belongs to program (code) on-chip ROM of the 8051. To access off-chip memory, that is, memories connected externally, we use the MOVX instruction. See MOVX for further discussion.

MOVC   A,@A+PC

Function:	Move code byte
Flags:	None

This instruction moves a byte of data located in the program (code) area to A. The address of the desired byte of data is formed by adding the program counter (PC) register to the original value of the accumulator. Contrast this instruction with "MOVC A, @A+DPTR". Here the PC is used to generate the data address instead of DPTR.

Example: Look-up table SQUR has the squares of values between 0 and 9, and register R3 has the values of 0 to 9. Write a program to fetch the squares from the table. Use the "MOVC A, @A+PC" instruction (this is a rewrite of example of the previous instruction "MOVC A, @A+DPTR").

```
        MOV  A,R3
        INC  A
        MOVC A,@A+PC
        RET
SQUR:   DB   0,1,4,9,16,25,36,49,64,81
```

The following should be noted concerning the above code.

(a) The program counter, which is pointing to instruction RET, is added to register A to form the address of the desired data. In other words, the PC is incremented to the address of the next instruction before it is added to the original value of the accumulator.
(b) The role of "INC A" should be emphasized. We need instruction "INC A" to bypass the single byte of opcode belonging to the RET instruction.
(c) This method is preferable over "MOVC A, @A+DPTR" if we do not want to divide the program code space into two separate areas of code and data. As a result, we do not waste valuable on-chip code space located between the last byte of program (code) and the beginning of the data space where the look-up table is located.

Function:	Move external
Flags:	None

This instruction transfers data between external memory and register A. As discussed in Chapter 14, the 8051 has 64K bytes of data space in addition to the 64K bytes of code space. This data space must be connected externally. This instruction allows us to access externally connected memeory. The address of external memory being accessed can be 16-bit or 8-bit as explained below.

(a) The 16-bit external memory address is held by the DPTR register.

```
MOVX A,@DPTR
```

This moves into the accumulator a byte from external memory whose address is pointed to by DPTR. In other words, this brings data into the CPU (register A) from the off-chip memory of the 8051.

```
MOVX  @DPTR,A
```

This moves the contents of the accumulator to the external memory location whose address is held by DPTR. In other words, this takes data from inside the CPU (register A) to memory outside of the 8051.

(b) The 8-bit address of external memory is held by R0 or R1.

```
MOVX A,@Ri     ;where i = 0 or 1
```

This moves to the accumulator a byte from external memory whose 8-bit address is pointed to by R0 (or R1 in MOVX A,@R1).

```
MOVX @Ri,A
```

This moves a byte from register A to an external memory location whose 8-bit address is held by R0 (or R1 in MOVX @R1,A)

The 16-bit address version of this instruction is widely used to access external memory while the 8-bit version is used to access external I/O ports.

MUL AB

Function:	Multiply $A \times B$
Flags:	OV, CY.

This multiplies an unsigned byte in A by a unsigned byte in register B and the result is placed in A and B where A has the lower byte and B has the higher byte.

Example:
```
            MOV  A,#5
            MOV  B,#7
            MUL  AB      ;A=35=23H,  B=00
```
Example:
```
            MOV  A,#10
            MOV  B,#15
            MUL  AB      ;A=150=96H,  B=00
```

This instruction always clears the CY flag; however, OV is changed according to the product. If the product is greater than FFH, OV=1; otherwise, it is cleared (OV=0).

Example:
```
            MOV  A,#25H
            MOV  B,#78H
            MUL  AB    ;A=58H,  B=11H   CY=0  and  OV=1
                       ;(25H  x  78H  =  1158H)
```

Example:
```
            MOV  A,#100
            MOV  B,#200
            MUL  AB    ;A=20H,  B=4EH,  OV=1  and  CY=0
                       ;(100  x  200  =  20,000  =  4E20H)
```

NOP

Function:	No operation
Flags:	None

This performs no operation and execution continues with the next instruction. It is sometimes used for timing delays to waste clock cycles. This instruction only updates the PC (program counter) to point to the next instruction following NOP.

ORL dest-byte,source-byte

Function:	Logical-OR for byte variable
Flags:	None

This performs a logical OR on the byte operands, bit by bit, and stores the result in the destination.

A	B	A OR B
0	0	0
0	1	1
1	0	1
1	1	1

Example:
```
        MOV     A,#39H    ;A=39H
        ORL     A,#09H    ;A=39H OR 09 (A=39H)

        39H     0011 1001
        09H     0000 1001
        39      0011 1001
```
Example:
```
        MOV     A,#32H    ;A=32H
        MOV     R4,#50H   ;R4=50H
        ORL     A,R4      ; (A=72H)

        32H     0011 0010
        50H     0101 0000
        72H     0111 0010
```

For the ORL instruction there are a total of six addressing modes. In four of them the accumulator must be the destination. They are as follows:

1. Immediate: ORL A,#data   Example: ORL A,#25H
2. Register: ORL A,Rx   Example: ORL A,R3
3. Direct: ORL A,direct
      Example: ORL A,30H   ;OR A with data located in RAM 30H
4. Register-Indirect: ORL A,@Rn
      Example: ORL A,@R0   ;OR A with data pointed to by R0

In the next two addressing modes the destination is a direct address (a RAM location or one of the SFR registers), while the source is either A or immediate data as shown below:

5. ORL direct,#data
      Example: Assuming that RAM location 32H has the value 67H, find the content of A after the following:

```
      ORL 32H,#44H  ;OR 44H with contents of RAM loc. 32H
      MOV A,32H     ;move content of RAM loc. 32H to A

      44H   0100 0100
      67H   0110 0111
      67H   0110 0111  Therefore A will have 67H.
```

6. ORL direct,A
      Example: Find the content of B after following:
```
      MOV B,#44H     ;B=44H
      MOV A,#67H     ;A=67H
      ORL 0F0H,A     ;OR A and B (B is at RAM F0H)
      ;After the operation B=67H.
```
Note: This cannot be used to modify data at input pins.

348

ORL   C,source-bit

Function: Logical-OR for bit variables
Flags: CY

In this instruction the carry flag bit is ORed with a source bit and the result is placed in the carry flag. Therefore, if the source bit is 1, CY is set; otherwise, the CY flag remains unchanged.

Example: Set the carry flag if either P1.5 or ACC.2 is high.

```
MOV  C,P1.5      ;get P1.5 status
ORL  C,ACC.2
```

Example: Write a program to clear A if P1.2 or P2.2 is high. Otherwise, make A=FFH.

```
        MOV  A,#FFH
        MOV  C,P1.2
        ORL  C,P2.2
        JNC  OVER
        CLR  A
OVER:
```

Another variation of this instruction involves ORing CY with the complement of the source bit. Its format is "ORL C,/bit". See the following example.

Example: Clear A if P2.1 is high or P2.2 is low. Otherwise, make A=FFH.

```
        MOV  A,#0FFH
        MOV  C,P2.1      ;get a copy of P2.1 bit
        ORL  C,/P2.2     ;OR P2.1 with complement of P2.2
        JNC  OVER
        CLR  A
OVER:
```

POP   direct

Function: Pop from the stack
Flags: None

This copies the byte pointed to by SP (stack pointer) to the location whose direct address is indicated, and decrements SP by 1. Notice that this instruction supports only direct addressing mode. Therefore, instructions such as "POP A" or "POP R3" are illegal. Instead we must write "POP 0E0H" where E0H is the RAM address belonging to register A and "POP 03" where 03 is the RAM address of R3 of bank 0.

PUSH direct

Function:	Push onto the stack
Flags:	None

This copies the indicated byte onto the stack and increments SP by 1. Notice that this instruction supports only direct addressing mode. Therefore, instructions such as "PUSH A" or "PUSH R3" are illegal. Instead, we must write "PUSH 0E0H" where E0H is the RAM address belonging to register A and "PUSH 03" where 03 is the RAM address of R3 of bank 0.

RET

Function:	Return from subroutine
Flags:	None

This instruction is used to return from a subroutine previously entered by instructions LCALL or ACALL. The top two bytes of the stack are popped into the program counter (PC) and program execution continues at this new address. After popping the top two bytes of the stack into the program counter, the stack pointer (SP) is decremented by 2.

RETI

Function:	Return from interrupt
Flags:	None

This is used at the end of an interrupt service routine (interrupt handler). The top two bytes of the stack are popped into the program counter and program execution continues at this new address. After popping the top two bytes of the stack into the program counter (PC), the stack pointer (SP) is decremented by 2.

Notice that while the RET instruction is used at the end of a subroutine associated with the ACALL and LCALL instructions, IRET must be used for the interrupt service subroutines.

RL   A

Function:	Rotate left the accumulator
Flags:	None

This rotates the bits of A left. The bits rotated out of A are rotated back into A at the opposite end.

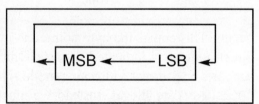

Example:
```
MOV A,#69H      ;A=01101001
RL  A           ;Now A=11010010
RL  A           ;Now A=10100101
```

RLC   A

Function: Rotate A left through carry
Flags: CY

This rotates the bits of the
accumulator left. The bits rotated out
of register A are rotated into CY, and
the CY bit is rotated into the opposite
end of the accumulator.

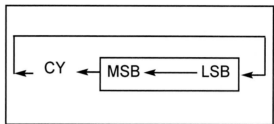

Example:
```
CLR C           ;CY=0
MOV A,#99H      ;A=10011001
RLC A           ;Now A=00110010 and CY=1
RLC A           ;Now A=01100101 and CY=0
```

RR   A

Function: Rotate A right
Flags: None

This rotates the bits of register A
right. The bits rotated out of A are rotated
back into A at the opposite end.

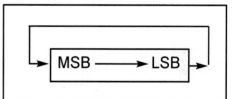

Example:
```
MOV A,#66H      ;A=01100110
RR  A           ;Now A=00110011
RR  A           ;Now A=10011001
```

RRC   A

Function: Rotate A right through carry
Flags: CY

This rotates the bits of the
accumulator right. The bits rotated
out of register A are rotated into CY
and the CY bit is rotated into the
opposite end of the accumulator.

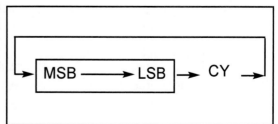

Example:
```
SETB C              ;CY=1
MOV A,#99H          ;A=10011001
RRC A               ;Now A=11001100 and CY=1
SETB C              ;CY=1
RRC A               ;Now A=11100110 and CY=0
```

SETB bit

Function: Set bit

This sets high the indicated bit. The bit can be the carry or any directly addressable bit of a port, register, or RAM location.

Examples:
```
SETB P1.3           ;P1.3=1
SETB P2.6           ;P2.6=1
SETB ACC.6          ;ACC.6=1
SETB 05             ;set high D5 of RAM loc. 20H
SETB C              ;Set Carry Flag CY=1
```

SJMP

See LJMP & SJMP.

SUBB A,source-byte

Function: Subtract with borrow
Flags: OV, AC,CY

This subtracts the source-byte and the carry flag from the accumulator and puts the result in the accumulator. The steps for subtraction performed by the internal hardware of the CPU are as follows:

1. Take the 2's complement of the source byte.
2. Add this to register A .
3. Invert the carry.

This instruction sets the carry flag according to the following:

	CY	
dest > source	0	the result is positive
dest=source	0	the result is 0
dest < source	1	the result is negative in 2's comp

Notice that there is no SUB instruction in the 8051. Therefore, we perform the SUB instruction by making CY = 0 and then using SUBB: A = (A − byte − CY).

Example:

```
MOV  A,#45H
CLR  C
SUBB A,#23H      ;45H-23H-0=22H
```

Addressing Modes

The following four addressing modes are supported for the SUBB.

1. Immediate: SUBB A,#data
   Example: SUBB A,#25H   ;A=A-25H-CY
2. Register   SUBB A,Rn   Example:
   Example: SUBB A,R3   ;A=A-R3-CY
3. Direct: SUBB A,direct
   Example: SUBB A,30H   ;A - data at (30H) - CY
4. Register-indirect: SUBB A,@Rn
   Example: SUBB A,@R0   ;A - data at (R0) - CY

SWAP A

Function: Swap nibbles within the accumulator
Flags: None

The SWAP instruction interchanges the lower nibble (D0 - D3) with the upper nibble (D4 - D7) inside register A.

Example:

```
MOV A,#59H      ;A=59H (0101 1001 in binary)
SWAP A          ;A=95H (1001 0101 in binary)
```

XCH   A,Byte

Function: Exchange A with a byte variable
Flags: None

This instruction swaps the contents of register A and the source byte. The source byte can be any register or RAM location.

Example:

```
MOV A,#65H      ;A=65H
MOV R2,#97H     ;R2=97H
XCH A,R2        ;now A=97H and R2=65H
```

For the "XCH A, byte" instruction there are a total of three addressing modes. They are as follows:

1. Register: XCH A,Rn   Example: XCH A,R3
2. Direct: XCH A,direct
   Example: XCH A,40H   ;exchange A with data in RAM loc. 40H
3. Register-Indirect XCH A,@Rn
   Examples: XCH A,@R0   ;XCH A with data pointed to by R0
                XCH A,@R1   ;XCH A with data pointed to by R1

XCHD A,@Ri

Function:	Exchange digits
Flags:	None

The XCHD instruction exchanges only the lower nibble of A with the lower nibble of the RAM location pointed to by Ri while leaving the upper nibbles in both places intact.

Example: Assuming RAM location 40H has the value 97H, find its content after the following instructions.

```
;40H= (97H)
MOV A,#12H      ;A=12H (0001 0010 binary)
MOV R1,#40H     ;R1=40H, load pointer
XCHD A,@R1      ;exchange the lower nibble of
                ;A and RAM location 40H
```

After execution of the XCHD instruction, we have A = 17H and RAM location 40H has 92H.

XRL   dest-byte,source-byte

Function:	Logical exclusive-OR for byte variables
Flags:	None

This performs a logical exclusive-OR on the operands, bit by bit, storing the result in the destination.

A	B	A XOR B
0	0	0
0	1	1
1	0	1
1	1	0

Example:
```
MOV    A,#39H    ;A=39H
XRL    A,#09H    ;A=39H ORed with 09

39H    0011 1001
09H    0000 1001
30     0011 0000
```

Example:

```
MOV    A,#32H    ;A=32H
MOV    R4,#50H   ;R4=50H
XRL    A,R4      ;(A=62H)

32H    0011 0010
50H    0101 0000
62H    0110 0010
```

For the XRL instruction there are total of six addressing modes. In four of them the accumulator must be the destination. They are as follows:

1. Immediate: XRL A,#data   Example: XRL A,#25H
2. Register: XRL A,Rn   Example: XRL A,R3
3. Direct: XRL A,direct
   Example: XRL A,30H   ;XRL A with data in RAM location 30H
4. Register-indirect: XRL A,@Rn
   Example: XRL A,@R0   ;XRL A with data pointed to by R0

In the next two addressing modes the destination is a direct address (a RAM location or one of the SFR registers) while the source is either A or immediate data as shown below:

5. XRL direct,#data
   Example: Assume that RAM location 32H has the value 67H.
   Find the content of A after execution of the following code.

```
XRL 32H,#44H
MOV A,32H ;move content of RAM loc. 32H to A

44H   0100 0100
67H   0110 0111
23H   0010 0011 Therefore A will have 23H.
```

6. XRL direct,A

Example: Find the contents of B after following:

```
MOV B,#44H     ;B=44H
MOV A,#67H     ;A=67H
XRL 0F0H,A     ;OR register A and B
;(register B is located at RAM location F0H)
;after the operation B=23H
```

Note: We cannot use this instruction to exclusive-OR the data at the input port.

Table A-1: 8051 Instruction Set Summary

Mnemonic		Byte	Machine Cycle
Arithmetic Operations			
ADD	A,Rn	1	1
ADD	A,direct	2	1
ADD	A,@Ri	1	1
ADD	A,#data	2	1
ADDC	A,Rn	1	1
ADDC	A,direct	2	1
ADDC	A,@Ri	1	1
ADDC	A,#data	2	1
SUBB	A,Rn	1	1
SUBB	A,direct	2	1
SUBB	A,@Ri	1	1
SUBB	A,#data	2	1
INC	A	1	1
INC	Rn	1	1
INC	direct	2	1
INC	@Ri	1	1
DEC	A	1	1
DEC	Rn	1	1
DEC	direct	2	1
DEC	@Ri	1	1
INC	DPTR	1	2
MUL	AB	1	4
DIV	AB	1	4
DA	A	1	1
Logical Operations			
ANL	A,Rn	1	1
ANL	A,direct	2	1
ANL	A,@Ri	1	1
ANL	A,#data	2	1
ANL	direct,A	2	1
ANL	direct,#data	3	2
ORL	A,Rn	1	1
ORL	A,direct	2	1
ORL	A,@Ri	1	1
ORL	A,#data	2	1
ORL	direct,A	2	1
ORL	direct,#data	3	2
XRL	A,Rn	1	1
XRL	A,direct	2	1
XRL	A,@Ri	1	1
XRL	A,#data	2	1
XRL	direct,A	2	1
XRL	direct,#data	3	2
CLR	A	1	1
CPL	A	1	1
RL	A	1	1
RLC	A	1	1
RR	A	1	1
RRC	A	1	1
SWAP	A	1	1

Mnemonic		Byte	Machine Cycle
Data Transfer			
MOV	A,Rn	1	1
MOV	A,direct	2	1
MOV	A,@Ri	1	1
MOV	A,#data	2	1
MOV	Rn,A	1	1
MOV	Rn,direct	2	2
MOV	Rn,#data	2	1
MOV	direct,A	2	1
MOV	direct,Rn	2	2
MOV	direct,direct	3	2
MOV	direct,@Ri	2	2
MOV	direct,#data	3	2
MOV	@Ri,A	1	1
MOV	@Ri,direct	2	2
MOV	@Ri,#data	2	1
MOV	DPTR,#data16	3	2
MOVX	A,@Ri	1	2
MOVX	A,@DPTR	1	2
MOVX	@Ri,A	1	2
MOV	@DPTR,A	1	2
PUSH	direct	2	2
POP	direct	2	2
XCH	A,Rn	1	1
XCH	A,direct	2	1
XCH	A,@Ri	1	1
XCHD	A,@Ri	1	1
Boolean Variable Manipulation			
CLR	C	1	1
CLR	bit	2	1
SETB	C	1	1
SETB	bit	2	1
CPL	C	1	1
CPL	bit	2	1
ANL	C,bit	2	2
ANL	C,/bit	2	2
ORL	C,bit	2	2
ORL	C,/bit	2	2
MOV	C,bit	2	1
MOV	bit,C	2	2
JC	rel	2	2
JNC	rel	2	2
JB	bit,rel	3	2
JNB	bit,rel	3	2
JBC	bit,rel	3	2

(continued)

Mnemonic	Byte	Machine Cycle
Program Branching		
ACALL addr11	2	2
LCALL addr16	3	2
RET	1	2
RETI	1	2
AJMP addr11	2	2
LJMP addr16	3	2
SJMP rel	2	2
JMP @A+DPTR	1	2
JZ rel	2	2
(continued)		

Mnemonic	Byte	Machine Cycle
Program Branching *(continued)*		
JNZ rel	2	2
CJNE A,direct,rel	3	2
CJNE A,#data,rel	3	2
CJNE Rn,#data,rel	3	2
CJNE @Ri,#data,rel	3	2
DJNZ Rn,rel	2	2
DJNZ direct,rel	3	2
NOP	1	2

SECTION A.2: 8051 REGISTERS

Table A-2: Special Function Register (SFR) Addresses

Symbol	Name	Address
ACC*	Accumulator	0E0H
B*	B register	0F0H
PSW*	Program status word	0D0H
SP	Stack pointer	81H
DPTR	Data pointer 2 bytes	
DPL	Low byte	82H
DPH	High byte	83H
P0*	Port 0	80H
P1*	Port 1	90H
P2*	Port 2	0A0H
P3*	Port 3	0B0H
IP*	Interrupt priority control	0B8H
IE*	Interrupt enable control	0A8H
TMOD	Timer/counter mode control	89H
TCON*	Timer/counter control	88H
T2CON*	Timer/counter 2 control	0C8H
T2MOD	Timer/counter mode control	0C9H
TH0	Timer/counter 0 high byte	8CH
TL0	Timer/counter 0 low byte	8AH
TH1	Timer/counter 1 high byte	8DH
TL1	Timer/counter 1 low byte	8BH
TH2	Timer/counter 2 high byte	0CDH
TL2	Timer/counter 2 low byte	0CCH
RCAP2H	T/C 2 capture register high byte	0CBH
RCAP2L	T/C 2 capture register low byte	0CAH
SCON*	Serial control	98H
SBUF	Serial data buffer	99H
PCON	Power control	87H

* Bit addressable

```
Byte
address              Bit address

FF  ┌──────────────────────────────┐
F0  │ F7  F6  F5  F4  F3  F2  F1  F0 │ B
    └──────────────────────────────┘
E0  ┌──────────────────────────────┐
    │ E7  E6  E5  E4  E3  E2  E1  E0 │ ACC
    └──────────────────────────────┘
D0  ┌──────────────────────────────┐
    │ D7  D6  D5  D4  D3  D2  D1  D0 │ PSW
    └──────────────────────────────┘
B8  ┌──────────────────────────────┐
    │ --  --  --  BC  BB  BA  B9  B8 │ IP
    └──────────────────────────────┘
B0  ┌──────────────────────────────┐
    │ B7  B6  B5  B4  B3  B2  B1  B0 │ P3
    └──────────────────────────────┘
A8  ┌──────────────────────────────┐
    │ AF  --  --  AC  AB  AA  A9  A8 │ IE
    └──────────────────────────────┘
A0  ┌──────────────────────────────┐
    │ A7  A6  A5  A4  A3  A2  A1  A0 │ P2
    └──────────────────────────────┘

99  ┌──────────────────────────────┐
    │        not bit addressable    │ SBUF
    └──────────────────────────────┘
98  ┌──────────────────────────────┐
    │ 9F  9E  9D  9C  9B  9A  99  98 │ SCON
    └──────────────────────────────┘
90  ┌──────────────────────────────┐
    │ 97  96  95  94  93  92  91  90 │ P1
    └──────────────────────────────┘

8D  │        not bit addressable    │ TH1
8C  │        not bit addressable    │ TH0
8B  │        not bit addressable    │ TL1
8A  │        not bit addressable    │ TL0
89  │        not bit addressable    │ TMOD
88  │ 8F  8E  8D  8C  8B  8A  89  88 │ TCON
87  │        not bit addressable    │ PCON

83  │        not bit addressable    │ DPH
82  │        not bit addressable    │ DPL
81  │        not bit addressable    │ SP
80  │ 87  86  85  84  83  82  81  80 │ P0
          Special Function Registers
```

Figure A-1. SFR RAM Address (Byte and Bit)

7F								
			General purpose RAM					
30								
2F	7F	7E	7D	7C	7B	7A	79	78
2E	77	76	75	74	73	72	71	70
2D	6F	6E	6D	6C	6B	6A	69	68
2C	67	66	65	64	63	62	61	60
2B	5F	5E	5D	5C	5B	5A	59	58
2A	57	56	55	54	53	52	51	50
29	4F	4E	4D	4C	4B	4A	49	48
28	47	46	45	44	43	42	41	40
27	3F	3E	3D	3C	3B	3A	39	38
26	37	36	35	34	33	32	31	30
25	2F	2E	2D	2C	2B	2A	29	28
24	17	26	25	24	23	22	21	20
23	1F	1E	1D	1C	1B	1A	19	18
22	17	16	15	14	13	12	11	10
21	0F	0E	0D	0C	0B	0A	09	08
20	07	06	05	04	03	02	01	00

Bit-addressable locations

1F 18	Bank 3
17 10	Bank 2
0F 08	Bank 1
07 00	Default register bank for R0 - R7

Figure A-2. 128 Bytes of Internal RAM

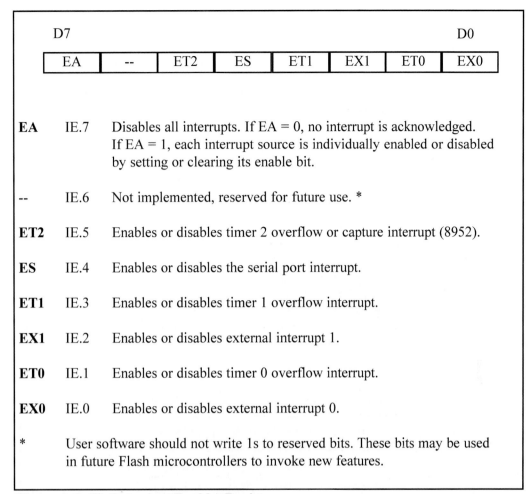

D7							D0
EA	--	ET2	ES	ET1	EX1	ET0	EX0

EA IE.7 Disables all interrupts. If EA = 0, no interrupt is acknowledged. If EA = 1, each interrupt source is individually enabled or disabled by setting or clearing its enable bit.

-- IE.6 Not implemented, reserved for future use. *

ET2 IE.5 Enables or disables timer 2 overflow or capture interrupt (8952).

ES IE.4 Enables or disables the serial port interrupt.

ET1 IE.3 Enables or disables timer 1 overflow interrupt.

EX1 IE.2 Enables or disables external interrupt 1.

ET0 IE.1 Enables or disables timer 0 overflow interrupt.

EX0 IE.0 Enables or disables external interrupt 0.

* User software should not write 1s to reserved bits. These bits may be used in future Flash microcontrollers to invoke new features.

Figure A-3. IE (Interrupt Enable) Register

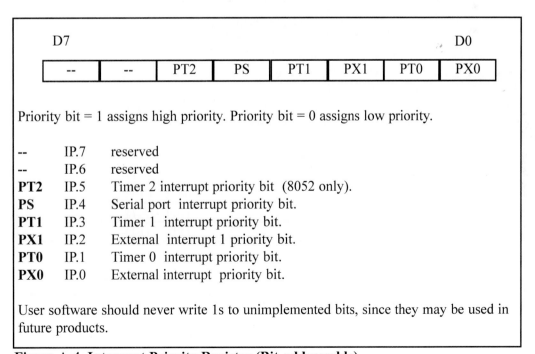

D7							D0
--	--	PT2	PS	PT1	PX1	PT0	PX0

Priority bit = 1 assigns high priority. Priority bit = 0 assigns low priority.

-- IP.7 reserved
-- IP.6 reserved
PT2 IP.5 Timer 2 interrupt priority bit (8052 only).
PS IP.4 Serial port interrupt priority bit.
PT1 IP.3 Timer 1 interrupt priority bit.
PX1 IP.2 External interrupt 1 priority bit.
PT0 IP.1 Timer 0 interrupt priority bit.
PX0 IP.0 External interrupt priority bit.

User software should never write 1s to unimplemented bits, since they may be used in future products.

Figure A-4. Interrupt Priority Register (Bit-addressable)

SMOD	--	--	--	GF1	GF0	PD	IDL

Figure A-5. PCON Register (not bit addressable)

Finding the TH_1 value for various baud rates:

SMOD = 0 (default on reset)

$$TH_1 = 256 - \frac{\text{Crystal frequency}}{384 \times \text{Baud rate}}$$

SMOD = 1

$$TH_1 = 256 - \frac{\text{Crystal frequency}}{192 \times \text{Baud rate}}$$

CY	AC	F0	RS1	RS0	OV	--	P

CY	PSW.7	Carry flag.
AC	PSW.6	Auxiliary carry flag.
--	PSW.5	Available to the user for general purposes.
RS1	PSW.4	Register Bank selector bit 1.
RS0	PSW.3	Register Bank selector bit 0.
OV	PSW.2	Overflow flag.
--	PSW.1	User definable bit.
P	PSW.0	Parity flag. Set/cleared by hardware each instuction cycle to indicate an odd/even number of 1 bits in the accumulator.

RS1	RS0	Register Bank	Address
0	0	0	00H - 07H
0	1	1	08H - 0FH
1	0	2	10H - 17H
1	1	3	18H - 1FH

Figure A-6. Bits of the PSW Register (bit addressable)

SM0	SM1	SM2	REN	TB8	RB8	T1	R1

SM0	SCON.7	Serial port mode specifier
SM1	SCON.6	Serial port mode specifier
SM2	SCON.5	Used for multiprocessor communication. (Make it 0)
REN	SCON.4	Set/cleared by software to enable/disable reception.
TB8	SCON.3	Not widely used.
RB8	SCON.2	Not widely used.
TI	SCON.1	Transmit interrupt flag. Set by hardware at the beginning of the stop bit in mode 1. Must be cleared by software.
RI	SCON.0	Receive interrupt flag. Set by hardware halfway through the stop bit time in mode 1. Must be cleared by software.

Note: Make SM2, TB8, and RB8 = 0.

Figure A-7. SCON Serial Port Control Register (Bit Addressable)

Finding the TH_1 value for various baud rates:

SMOD = 0 (default on reset)

$$TH_1 = 256 - \frac{\text{Crystal frequency}}{384 \times \text{Baud rate}}$$

SMOD = 1

$$TH_1 = 256 - \frac{\text{Crystal frequency}}{192 \times \text{Baud rate}}$$

(MSB)   (LSB)

GATE	C/T	M1	M0	GATE	C/T	M1	M0
Timer 1				Timer 0			

GATE   Gating control when set. Timer/counter is enabled only while the INTx pin is high and the TRx control pin is set. When cleared, the timer is enabled whenever the TRx control bit is set.

C/T   Timer or counter selected cleared for timer operation (input from internal system clock). Set for counter operation (input from Tx input pin).

M1   Mode bit 1

M0   Mode bit 0

M1	M0	Mode	Operating Mode
0	0	0	13-bit timer mode
			8-bit timer/counter THx with TLx as 5-bit prescalar
0	1	1	16-bit timer mode
			16-bit timer/counters THx and TLx are cascaded; there is no prescalar
1	0	2	8-bit auto reload
			8-bit auto reload timer/counter; THx holds a value which is to be reloaded into TLx each time it overflows.
1	1	3	Split timer mode

Figure A-8. TMOD Register (not bit addressable)

```
       D7                                                          D0
     ┌───────┬───────┬───────┬───────┬───────┬───────┬───────┬───────┐
     │  TF1  │  TR1  │  TF0  │  TR0  │  IE1  │  IT1  │  IE0  │  IT0  │
     └───────┴───────┴───────┴───────┴───────┴───────┴───────┴───────┘
```

TF1   TCON.7   Timer 1 overflow flag. Set by hardware when timer/counter 1 overflows. Cleared by hardware as the processor vectors to the interrupt service routine.

TR1   TCON.6   Timer 1 run control bit. Set/cleared by software to turn timer/counter 1 on/off.

TF0   TCON.5   Timer 0 overflow flag. Set by hardware when timer/counter 0 overflows. Cleared by hardware as the processor vectors to the service routine.

TR0   TCON.4   Timer 0 run control bit. Set/cleared by software to turn timer/counter 0 on/off.

IE1   TCON.3   External interrupt 1 edge flag. Set by CPU when the external interrupt edge (H-to-L transition) is detected. Cleared by CPU when the interrupt is processed. *Note:* This flag does not latch low-level triggered interrupts.

IT1   TCON.2   Interrupt 1 type control bit. Set/cleared by software to specify falling edge/low-level triggered external interrupt.

IE0   TCON.1   External interrupt 0 edge flag. Set by CPU when external interrupt (H-to-L transition) edge detected. Cleared by CPU when interrupt is processed. *Note:* This flag does not latch low-level triggered interrupts.

IT0   TCON.0   Interrupt 0 type control bit. Set/cleared by software to specify falling edge/low-level triggered external interrupt.

Figure A-9. TCON (Timer/Counter) Register (Bit-addressable)

APPENDIX B

8051-BASED SYSTEMS: WIRE-WRAPPING AND TESTING

OVERVIEW

This appendix demonstrates how to design and wire-wrap 8051-based systems. In addition, it covers the design and wire-wrap of DS5000-based systems. Finally, this appendix shows how to download programs into the DS5000 using an IBM PC.

Section B.1 of this appendix describes the basics of wire-wrapping. Section B.2 describes wire-wrapping and testing procedures for 8051-based systems, including DS5000-based systems.

SECTION B.1: BASICS OF WIRE-WRAPPING

Note: For this tutorial appendix, you will need the following:
Wire-wrapping tool (Radio Shack part number 276-1570)
30-gauge (30-AWG) wire for wire wrapping
(Thank you Shannon Looper and Greg Boyle for your assistance on this section.)

The following describes the basics of wire wrapping.

1. There are several different types of wire-wrap tools available. The best one is available from Radio Shack for less than $10. The part number for Radio Shack is 276-1570. This tool combines the wrap and unwrap functions in the same end of the tool and includes a separate stripper. We found this to be much easier to use than the tools that combined all these features on one two-ended shaft. There are also wire-wrap guns, which are of course more expensive.

2. Wire-wrapping wire is available prestripped in various lengths or in bulk on a spool. The prestripped wire is usually more expensive and you are restricted to the different wire lengths you can afford to buy. Bulk wire can be cut to any length you wish which allows each wire to be custom fit.

3. There are a few different types of wire-wrap boards available. These are usually called perfboards or wire-wrap boards. These types of boards are sold at many electronics stores (such as Radio Shack). The best type of board has plating around the holes on the bottom of the board. These boards are better because the sockets and pins can be soldered to the board which makes the circuit more mechanically stable.

4. Choose a board that is large enough to accommodate all the parts in your design with room to spare so that the wiring does not become too cluttered. If you wish to expand your project in the future, you should be sure to include enough room on the original board for the complete circuit. Also, if possible, the layout of the IC on the board needs be done such that signals go from left to right just like the schematics.

5. To make the wiring easier and to keep pressure off the pins, install one standoff on each corner of the board. You may also wish to put standoffs on the top of the board to add stability when the board is on its back.

6. For power hook-up, use some type of standard binding post. Solder a few single wire-wrap pins to each power post to make circuit connections (to at least one pin for each IC in the circuit).

7. To further reduce problems with power, each IC must have its own connection to the main power of the board. If your perfboard does not have built-in power buses, run a separate power and ground wire from each IC to the main power. In other words, DO NOT daisy chain (chip-to-chip connection is called daisy chain) power connections, as each connection down the line will have more wire and more resistance to get power through. However, daisy chaining is acceptable for other connections such as data, address, and control buses.

8. You must use wire wrap sockets. These sockets have long square pins whose edges will cut into the wire as it is wrapped around the pin.

9. Wire-wrapping will not work on round legs. If you need to wrap to components, such as capacitors, that have round legs, you must also solder these connections. The best way to connect single components is to install individual wire-wrap pins into the board and then solder the components to the pins. An alternate method is to use an empty IC socket to hold small components such as resistors and wrap to the socket.

10. The wire should be stripped about 1 inch. This will allow for 7 to 10 turns for each connection. The first turn or turn-and-a-half should be insulated. This prevents stripped wire from coming in contact with other pins. This can be accomplished by inserting the wire as far as it will go into the tool before making the connection.

11. Try to keep wire lengths to a minimum. This prevents the circuit from looking like a bird nest. Be neat and use color coding as much as possible. Use only red wires for V_{CC} and black wires for ground connections. Also use different colors for data, address, and control signal connections. These suggestions will make troubleshooting much easier.

12. It is standard practice to connect all power lines first and check them for continuity. This will eliminate trouble later on.

13. It's also a good idea to mark the pin orientation on the bottom of the board. There are plastic templates available with pin numbers preprinted on them specifically for this purpose or you can make your own from paper. Forgetting to reverse pin order when looking at the bottom of the board is a very common problem when wire-wrapping circuits.

14. To prevent damage to your circuit, place a diode (such as IN5338) in reverse bias across the power supply. If the power gets hooked up backwards, the diode will be forward biased and will act as a short, keeping the reversed voltage from your circuit.

15. In digital circuits, there can be a problem with current demand on the power supply. To filter the noise on the power supply, a 100 μF electrolytic capacitor and 0.1 μF monolithic capacitor is connected from V_{CC} to ground, in parallel with each other, at the entry point of the power supply to the board. These two together will filter both the high- and the low-frequency noises. Instead of using two capacitors in parallel, you can use a single 20-100 μF tantalum capacitor. Remember that the long lead is the positive one.

16. To filter the transient current, use a 0.1 μF monolithic capacitor for each IC. Place the 0.1 μF monolithic capacitor between V_{CC} and ground of each IC. Make sure the leads are as short as possible.

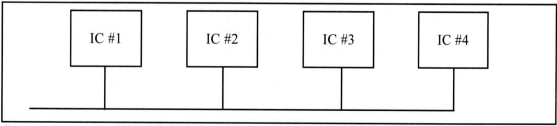

Figure B-1. Daisy Chain Connection (not recommended for power lines)

SECTION B.2: WIRE-WRAPPING AN 8051-BASED SYSTEM

In this section we show the connections for 8051-based systems. First we show an 8751 (89C51) system, then a DS5000 system. If you decide to wire-wrap one of these, make sure that you read Section B.1 on wire-wrapping.

8751/89C51-based system

In systems based on an 8751 or 89C51 microcontroller, you need a ROM burner to burn your program into the microcontroller. In the case of the 8751, you also need an EPROM erasure tool since it uses UV-EPROM. For the 89C51, the ROM burner can erase the Flash ROM in addition to burning a program into it. In burning the 8751, you need to erase its contents first, which takes approximately 20 minutes since it uses UV-EPROM. For the 89C51, this is not required since it has Flash ROM.

8751/89C51 connection

Figure B.2 shows the minimum connection for the 8751 or 89C51-based system. Notice that "EA=VCC" indicates that an 8751 or 89C51 has on-chip ROM for the program. Also notice the P0 connection to pull-up resistors to ensure the availability of P0 for I/O operations. If you need to use a debounce switch for RESET, refer to Figure 4-3 (b).

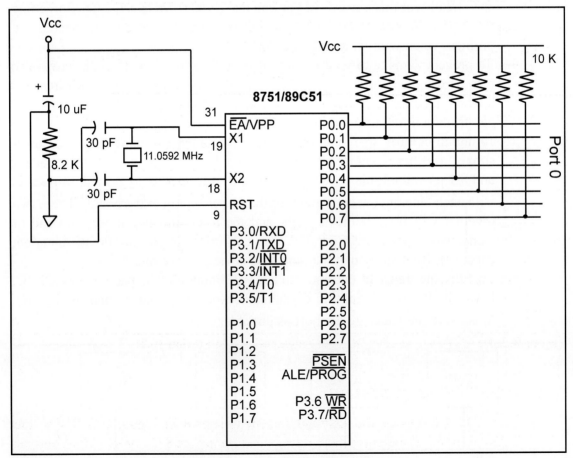

Figure B-2. Minimum Connection For 8751/89C51-Based Systems

Testing an 8751/ 89C51 system

After the 8751, 89C51, or DS5000 system is wire-wrapped, it must be tested. The simplest test is to write a program in which all the bits of ports P0, P1, and P2 are toggled continuously. The following program should do that.

```
            ORG  0
            MOV  A,#55H
OVER:       MOV  P0,A
            MOV  P1,A
            MOV  P2,A
            ACALL DELAY
            CPL  A
            SJMP OVER
;——one second delay(approx. for XTAL = 11.0592 MHz)
DELAY:      MOV  R3,#10
AGAIN:      MOV  R4,#200
BACK:       MOV  R5,#230
HERE:       DJNZ R5,HERE
            DJNZ R4,BACK
            DJNZ R3,AGAIN
            RET
            END
```

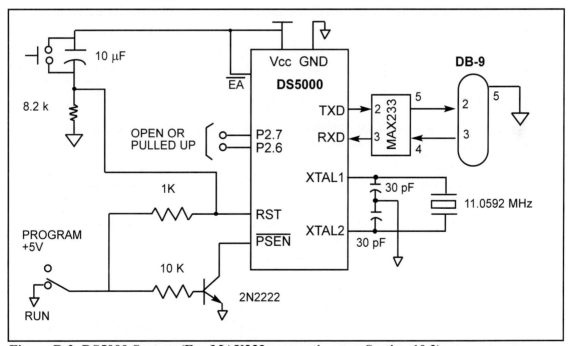

Figure B-3. DS5000 System (For MAX232 connection, see Section 10.2)

DS5000-based system

The design of the DS5000 system is shown in Figure B-3. Using the DS5000 for development is more advantageous than the 8751 or 89C51 system for the following reasons.

1. Using the DS5000 for an 8051 microcontroller allows us to program the chip without any need for a ROM burner.

2. Although the DS5000 chip is more expensive than either the 8751 or the 89C51, the fact that not everyone has access to a ROM burner makes DS5000 an ideal home-development system.

3. Another advantage of DS5000 is that it can be programmed via the COM port of a PC (x86 IBM or compatible PC) while it is in the system. Contrast this with the 8751 or 89C51 system in which you must remove the chip, program it, and install it back in the system every time you want to change the program contents of the on-chip ROM. This results in a much longer development time for the 8751 or 89C51 system compared with the DS5000 system.

4. Since DS5000 on-chip ROM is an NV-RAM, we can selectively change the bytes and test the program. In the case of the 8751 or 89C51, the entire on-chip ROM must be erased before it can be reprogrammed.

5. The DS5000 has two versions: DS5000 and DS5000T. While DS5000 has Timer 0 and Timer 1, just like any 8051, the DS5000T has an additional clock generator which is called a real-time clock (RTC). The on-chip RTC of the DS5000T is a permanently powered timekeeping device which provides the user both the time and date. It keeps track of hundredths of seconds, seconds, minutes, hours, days of the month, months, and years. Notice that the DS5000T keeps the time and date even if the external power and crystal oscillator are disconnected. For projects using the 8751/89C51 or DS5000 chips you must use an external real time clock such as the DS1284 in order to have the time and date since these chips do not have on-chip RTC. Since the DS5000T is a DS5000 with an on-chip RTC, any discussion about the DS5000 also applies to the DS5000T. For this reason, from now on we simply refer to both as the DS5000.

The design of a DS5000 system is shown in Figure B-3. Notice from Figure B-3 that the reset circuitry and serial port connections are the same as in any 8051-based system. However, the extra circuitry needed for programming is a transistor, a switch, and 10 K and 1 K ohm resistors. In fact you can add these components to your 8751/89C51 system and use it as a DS5000 system by simply plugging a DS5000 chip in the socket. The switch allows you to select between the program and run options. We can load our program into the DS5000 by setting the switch to V_{CC}, and run the program by setting it to Gnd, as shown in Figure B-3.

Communicating with the DS5000

After we build our DS5000-based system, we can communicate with it using the HyperTerminal or Terminal software. HyperTerminal comes with MS Windows 95 (and higher), and the Terminal utility comes with all versions of Windows. We first show how to use HyperTerminal to communicate with the DS5000 system and then the use of Terminal is discussed.

Using HyperTerminal with the DS5000

In order to communicate with DS5000 using HyperTerminal the following steps are needed.

1. With the power off, examine the COM2 port on the back of your PC and connect your DS5000 system to it via a serial cable. Notice that your serial cable has a DB-9 connector on both ends. If one side is DB-25, then you need a DB-9-to-DB-25 converter. You might also need a gender bender if the connector is male-male. After you hook up your DS5000 system to your PC, go ahead and power it up.
2. In Windows Accessory, click on Hyperterminal. (If you get a modem installation option, choose "No".)
3. Type a dummy name, and click OK (or Windows will not let you go on).
4. For "Connect Using" select COM2 and click OK. We use COM2 since COM1 is normally used by the mouse.
5. Pick 9600 baud rate, 8 bit data, no parity bit, and 1 stop bit.
6. Change the "Flow Control" to NONE or Xon/Xoff, and click OK. (Definitely do not choose the hardware option).
7. Now you are in Windows HyperTerminal, and when you press the ENTER key the DS5000 will respond with "> ".

If you do not see ">" after pressing the ENTER key go through the above steps one more time.

At this stage, if you do not get ">" after pressing the ENTER key, you need to check your hardware connections, such as the MAX232/233. See the end of this chapter for some troubleshooting tips.

Loading a program into a DS5000 system

After we get the ">" from the DS5000, we are ready to load the program into it and run. First make sure that the file you are loading is in Intel hex format. The Intel hex format is provided by your 8051 Assembler. More about Intel hex format is given at the end of this section.

Here are the steps to load the program into DS5000 after you go through the above 7 steps and get ">" on your screen.

1. At the ">" prompt, enter L (L is for Load).
2. In HyperTerminal, click on the Transfer menu option. Click on Send Text File.
3. Select your file. Example: "A:test.hex"
4. Wait until the loading is complete. The appearance of the ">" prompt indicates that the loading is finished.
5. Now use D to dump the contents of the NV-RAM of the DS5000 onto the screen. Example: >D 00 FF
   The dump will give you Figure B-4, which is the Intel hex file for the test program. (We will soon examine the Intel hex file and compare it with the list file of the test program given in Figure B-5).
6. Change the switch to the run position, press the reset button on the DS5000 system, and the program will execute. Use a logic probe (or scope) to see the P0, P1, and P2 bits toggle "on" and "off" continuously with one second delay in between the "ON" and "OFF" states.

Use the following steps to communicate with DS5000 using the Windows Terminal utility.

1. In the Windows Accessory, click on Terminal, or run terminal.exe in the Windows directory.
2. Click on Settings and choose Communications. Then select COM2.
3. Also select 9600 baud rate, 8 bit data, 1 stop bit, and no parity bit.
4. Change the "Flow Control" to NONE or Xon/Xoff and click OK. (Definitely do not choose the hardware option)
5. Now you are in Windows Terminal. By pressing the enter key, the DS5000 will respond with ">".

The loading of the file using Windows Terminal is the same as for HyperTerminal as shown above.

```
:CC   AAAA TT    DDDDDDDDDDDDDDDDDDDDDDDDDDDDDDDD    SS
:10   0000 00    7455F580F590F5A0110DF480F57B0A7C    10
:0A   0010 00    C8ADE6DDFEDCFADBF622E7
:00   0000 01    FF
```

Figure B-4. Intel Hex File Format for Test Program

Note: In Figure B-4, the spaces between the fields are for clarification and do not exist in the Intel hex file format.

Explaining the Intel hex file

Intel hex file is a widely used file format designed to standardize the loading of executable machine codes into a ROM chip. Therefore, loaders that come with every ROM burner (programmer) support the Intel hex file format. While in many newer Windows-based assemblers, the Intel hex file is produced automatically (by selecting the right setting), in a DOS-based PC you need a utility called OH (object-to-hex). In the DOS environment, the object file is fed into the linker to produce the abs file; then the abs file is fed into the OH utility to create the Intel hex file. While the abs file is used by systems which have a monitor program, the hex file is used only by the loader of an EPROM programmer to load it into the ROM chip. Figure B-4 shows the Intel hex file format for the test program. From the Intel hex file, the loader will know (1) the number of bytes of information, (2) the information, and (3) the starting address where the information must be placed. Each line of the hex file consists of six parts. The following describes each part.

1. ":" Each line starts with a colon.
2. CC, the count byte. This tells the loader how many bytes are in the line. CC can range from 00 to 16 (10 in hex).
3. AAAA is for the address. This is a 16-bit address. The loader places the first byte of data into this memory address.
4. TT is for type. This field is either 00 or 01. If it is 00, it means that there are more lines to come after this line. If it is 01, it means that this is the last line and the loading should stop after this line.

5. DD...D is the real information (data or code). There is a maximum of 16 bytes in this part. The loader places this information into successive memory locations.

6. SS is a single byte. This last byte is the check-sum byte of everything in the line. The check-sum byte is used for error checking. Check-sum bytes are discussed in detail in Chapter 14. Notice that the check-sum byte at the end of each line represents the check-sum byte for everything in that line and not just for the data portion.

DS5000 commands

There are several commands embedded in the DS5000. We showed the use of L (load) and D (dump) commands earlier. A complete list of commands can be obtained from the Web site www.dalsemi.com. The following are a few of these commands.

D <begin> <end>	Dump Intel hex file.
F <byte> <begin> <end>	Fill embedded RAM block with a constant.
L	Load Intel hex file.
T	Trace (echo) incoming Intel hex data.

```
LOC      OBJ     LINE
0000             1                      ORG  0
0000     7455    2                      MOV  A,#55H
0002     F580    3        OVER:         MOV  P0,A
0004     F590    4                      MOV  P1,A
0006     F5A0    5                      MOV  P2,A
0008     110D    6                      ACALL DELAY
000A     F4      7                      CPL  A
000B     80F5    8                      SJMP OVER
                 9        ;----ONE SECOND DELAY
000D     7B0A    10       DELAY:        MOV  R3,#10
000F     7CC8    11       AGAIN:        MOV  R4,#200
0011     ADE6    12       BACK:         MOV  R5,230
0013     DDFE    13       HERE:         DJNZ R5,HERE
0015     DCFA    14                     DJNZ R4,BACK
0017     DBF6    15                     DJNZ R3,AGAIN
0019     22      16                     RET
                 17                     END
```

Figure B-5. List File For Test Program

Example B-1

From Figure B-4, analyze the six parts of line 2.

Solution:

After the colon(:), we have 0A which means that ten bytes of data are in this line. 0010H is the address at which the data starts. Next, 00 means that this is not the last line of the record. Then the data, which is ten bytes, is as follows: C8ADE6DDFEDC-FADBF622. Finally, the last byte, E7, is the check-sum byte.

Example B-2

Verify the check-sum byte for line 2 of figure B-4.

Solution:

0A + 00 + 10 + 00 + C8 + AD + E6 + DD + FE + DC + FA + DB + F6 + 22 = 819H
Dropping the carries (8), gives 19H, and its 2's complement is E7, which is the last byte of line 2.

Example B-3

Compare the data portion of the Intel hex file of Figure B-4 with the opcodes in the list file of the test program given in Figure B-5. Do they match?

Solution:

In the first line of Figure B-4, the data portion starts with 74H, the opcode for the instruction "MOV A, #55H", as shown in the list file of Figure B-5. The last byte of the data in line 2 of Figure B-4 is 22, which is the opcode for the "RET" instruction in the list file of Figure B-5.

Some troubleshooting tips

Running the test program on your wire-wrapped 8051 system should toggle all the I/O bits with a 1-second delay. If your system does not work, follow these steps to find the problem.

1. With the power off, check your wire-wrapping for all pins, especially V_{CC} and GND.
2. Check RST (pin #9) using a oscilloscope. When the system is powered up, pin #9 is low. Upon pressing the debounce switch it goes high. Make sure the debounce switch is connected properly.
3. Observe the XTAL2 pin on the oscilloscope while the power is on. You should see a crude square wave. This indicates that the crystal oscillator is good.
4. If all the above steps pass inspection, check the contents of the on-chip ROM starting at memory location 0000. It must be the same as the opcodes provided by the list file of Figure B-5. Your assembler produces the list file which lists the opcodes and operands on the left side of the assembly instructions. This must match exactly the contents of your on-chip ROM if the proper steps were taken in burning and loading the program into the on-chip ROM.

APPENDIX C

IC TECHNOLOGY AND SYSTEM DESIGN ISSUES

OVERVIEW

This appendix provides an overview of IC technology and 8051 interfacing. In addition, we look at the microcontroller-based system as a whole and examine some general issues in system design.

First, in Section C.1 we provide an overview of IC technology. Then, in Section C.2, the internal details of 8051 I/O ports and interfacing are discussed. Section C.3 examines system design issues.

C.1: OVERVIEW OF IC TECHNOLOGY

In this section we examine IC technology, and discuss some major developments in advanced logic families. Since this is an overview, it is assumed that the reader is familiar with logic families on a level presented in basic digital electronics books.

Transistors

The transistor was invented in 1947 by three scientists at Bell Laboratory. In the 1950s, transistors replaced vacuum tubes in many electronics systems, including computers. It was not until 1959 that the first integrated circuit was successfully fabricated and tested by Jack Kilby of Texas Instruments. Prior to the invention of the IC, the use of transistors, along with other discrete components such as capacitors and resistors, was common in computer design. Early transistors were made of germanium, which was later abandoned in favor of silicon. This was due to the fact that the slightest rise in temperature resulted in massive current flows in germanium-based transistors. In semiconductor terms, it is because the band gap of germanium is much smaller than that of silicon, resulting in a massive flow of electrons from the valence band to the conduction band when the temperature rises even slightly. By the late 1960s and early 1970s, the use of the silicon-based IC was widespread in mainframes and minicomputers. Transistors and ICs at first were based on P-type materials. Later on, due to the fact that the speed of electrons is much higher (about two and a half times) than the speed of holes, N-type devices replaced P-type devices. By the mid-1970s, NPN and NMOS transistors had replaced the slower PNP and PMOS transistors in every sector of the electronics industry, including in the design of microprocessors and computers. Since the early 1980s, CMOS (complementary MOS) has become the dominant technology of IC design. Next we provide an overview of differences between MOS and bipolar transistors. See Figure C-1.

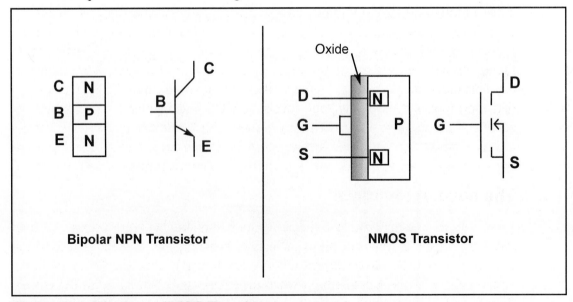

Figure C-1. Bipolar vs. MOS Transistors

MOS vs. bipolar transistors

There are two type of transistors: bipolar and MOS (metal-oxide semiconductor). Both have three leads. In bipolar transistors, the three leads are referred to as the *emitter*, *base*, and *collector*, while in MOS transistors they are named *source*, *gate*, and *drain*. In bipolar transistors, the carrier flows from the emitter to the collector, and the base is used as a flow controller. In MOS transistors, the carrier flows from the source to the drain, and the gate is used as a flow controller. In NPN-type bipolar transistors, the electron carrier leaving the emitter must overcome two voltage barriers before it reaches the collector (see Figure C-1). One is the N-P junction of the emitter-base and the other is the P-N junction of the base-collector. The voltage barrier of the base-collector is the most difficult one for the electrons to overcome (since it is reverse-biased) and it causes the most power dissipation. This led to the design of the unipolar type transistor called MOS. In N-channel MOS transistors, the electrons leave the source reaching the drain without going through any voltage barrier. The absence of any voltage barrier in the path of the carrier is one reason why MOS dissipates much less power than bipolar transistors. The low power dissipation of MOS allows putting millions of transistors on a single IC chip. In today's technology, putting 10 million transistors into an IC is common, and it is all because of MOS technology. Without the MOS transistor, the advent of desktop personal computers would not have been possible, at least not so soon. The bipolar transistors in both the mainframes and minicomputers of the 1960s and 1970s were bulky and required expensive cooling systems and large rooms. MOS transistors do have one major drawback: They are slower than bipolar transistors. This is due partly to the gate capacitance of the MOS transistor. For a MOS to be turned on, the input capacitor of the gate takes time to charge up to the turn-on (threshold) voltage, leading to a longer propagation delay.

Overview of logic families

Logic families are judged according to (1) speed, (2) power dissipation, (3) noise immunity, (4) input/output interface compatibility, and (5) cost. Desirable qualities are high speed, low power dissipation, and high noise immunity (since it prevents the occurrence of false logic signals during switching transition). In interfacing logic families, the more inputs that can be driven by a single output, the better. This means that high-driving-capability outputs are desired. This, plus the fact that the input and output voltage levels of MOS and bipolar transistors are not compatible mean that one must be concerned with the ability of one logic family to drive the other one. In terms of the cost of a given logic family, it is high during the early years of its introduction but it declines as production and use rise.

The case of inverters

As an example of logic gates, we look at a simple inverter. In a one-transistor inverter, the transistor plays the role of a switch, and R is the pull-up resistor. See Figure C-2. However, for this inverter to work most effectively in digital circuits, the R value must be high when the transistor is "on" to limit the current flow from V_{CC} to ground in order to have low power dissipation (P = VI, where V = 5 V). In other words, the lower the I, the lower the power dissipation. On the other hand, when the transistor is "off", R must be a small value to limit the voltage drop across R, thereby making sure that V_{OUT} is close to V_{CC}. This is a contradictory demand on R. This is one reason that logic gate designers use active components (transistors) instead of passive components (resistors) to implement the pull-up resistor R.

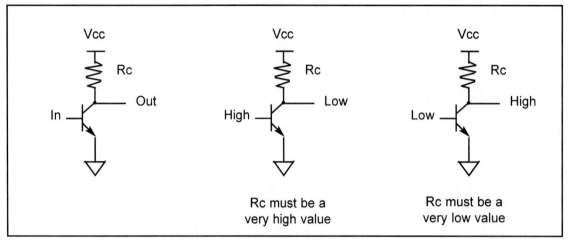

Figure C-2. One-Transistor Inverter with Pull-up Resistor

The case of a TTL inverter with totem pole output is shown in Figure C-3. In Figure C-3, Q3 plays the role of a pull-up resistor.

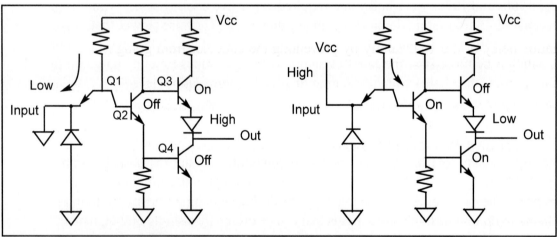

Figure C-3. TTL Inverter with Totem-Pole Output

CMOS inverter

In the case of CMOS-based logic gates, PMOS and NMOS are used to construct a CMOS (complementary MOS) inverter as shown in Figure C-4. In CMOS inverters, when the PMOS transistor is off, it provides a very high impedance path, making leakage current almost zero (about 10 nA); when the PMOS is on, it provides a low resistance on the path of V_{DD} to load. Since the speed of the hole is slower than that of the electron, the PMOS transistor is wider to compensate for this disparity; therefore, PMOS transistors take more space than NMOS transistors in the CMOS gates. At the end of this section we will see an open-collector gate in which the pull-up resistor is provided externally, thereby allowing system designers to choose the value of the pull-up resistor.

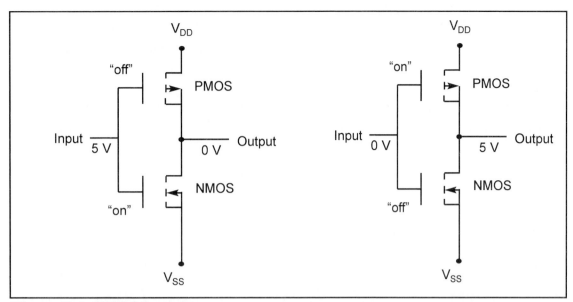

Figure C-4. CMOS Inverter

Input/output characteristics of some logic families

In 1968 the first logic family made of bipolar transistors was marketed. It was commonly referred to as the standard TTL (transistor-transistor logic) family. The first MOS-based logic family, the CD4000/74C series, was marketed in 1970. The addition of the Schottky diode to the base-collector of bipolar transistors in the early 1970s gave rise to the S family. The Schottky diode shortens the propagation delay of the TTL family by preventing the collector from going into what is called deep saturation. Table C-1 lists major characteristics of some logic families. In Table C-1, note that as the CMOS circuit's operating frequency rises, the power dissipation also increases. This is not the case for bipolar-based TTL.

Table C-1: Characteristics of Some Logic Families

Characteristic	STD TTL	LSTTL	ALSTTL	HCMOS
V_{CC}	5 V	5 V	5 V	5 V
V_{IH}	2.0 V	2.0 V	2.0 V	3.15 V
V_{IL}	0.8 V	0.8 V	0.8 V	1.1 V
V_{OH}	2.4 V	2.7 V	2.7 V	3.7 V
V_{OL}	0.4 V	0.5 V	0.4 V	0.4 V
I_{IL}	−1.6 mA	−0.36 mA	−0.2 mA	−1 μA
I_{IH}	40 μA	20 μA	20 μA	1 μA
I_{OL}	16 mA	8 mA	4 mA	4 mA
I_{OH}	−400 μA	−400 μA	−400 μA	4 mA
Propagation delay	10 ns	9.5 ns	4 ns	9 ns
Static power dissipation (F=0)	10 mW	2 mW	1 mW	0.0025 nW
Dynamic power dissipation at F=100 kHz	10 mW	2 mW	1 mW	0.17 mW

History of logic families

Early logic families and microprocessors required both positive and negative power voltages. In the mid-1970s, 5 V V_{CC} became standard. In the late 1970s, advances in IC technology allowed combining the speed and drive of the S family with the lower power of LS to form a new logic family called FAST (Fairchild Advanced Schottky TTL). In 1985, AC/ACT (Advanced CMOS Technology), a much higher speed version of HCMOS, was introduced. With the introduction of FCT (Fast CMOS Technology) in 1986, at last the speed gap between CMOS and TTL was closed. Since FCT is the CMOS version of FAST, it has the low power consumption of CMOS but the speed is comparable with TTL. Table C-2 provides an overview of logic families up to FCT.

Table C-2: Logic Family Overview

Product	Year Introduced	Speed (ns)	Static Supply Current (mA)	High/Low Family Drive (mA)
Std TTL	1968	40	30	–2/32
CD4K/74C	1970	70	0.3	–0.48/6.4
LS/S	1971	18	54	–15/24
HC/HCT	1977	25	0.08	–6/–6
FAST	1978	6.5	90	–15/64
AS	1980	6.2	90	–15/64
ALS	1980	10	27	–15/64
AC/ACT	1985	10	0.08	–24/24
FCT	1986	6.5	1.5	–15/64

Reprinted by permission of Electronic Design Magazine, c. 1991.

Recent advances in logic families

As the speed of high-performance microprocessors reached 25 MHz, it shortened the CPU's cycle time, leaving less time for the path delay. Designers normally allocate no more than 25% of a CPU's cycle time budget to path delay. Following this rule means that there must be a corresponding decline in the propagation delay of logic families used in the address and data path as the system frequency is increased. In recent years, many semiconductor manufacturers have responded to this need by providing logic families that have high speed, low noise, and high drive I/O. Table C-3 provides the characteristics of high-performance logic families introduced in recent years. ACQ/ACTQ are the second-generation advanced CMOS (ACMOS) with much lower noise. While ACQ has the CMOS input level, ACQT is equipped with TTL-level input. The FCTx and FCTx-T are second-generation FCT with much higher speed. The "x" in the FCTx and FCTx-T refers to various speed grades, such as A, B, and C, where the A designation means low speed and C means high speed. For designers who are well versed in using the FAST logic family, the use of FASTr is an ideal choice since it is faster than FAST, has higher driving capability (I_{OL}, I_{OH}), and produces much lower noise than FAST. At the time of this writing, next to ECL and gallium arsenide logic gates, FASTr is the fastest logic family in the market (with the 5 V V_{CC}), but the power consumption is high relative to other logic families, as shown in Table

C-3. The combining of high-speed bipolar TTL and the low power consumption of CMOS has given birth to what is called BICMOS. Although BICMOS seems to be the future trend in IC design, at this time it is expensive due to extra steps required in BICMOS IC fabrication, but in some cases there is no other choice. (For example, Intel's Pentium microprocessor, a BICMOS product, had to use high-speed bipolar transistors to speed up some of the internal functions.) Table C-3 provides advanced logic characteristics. The "x" is for different speeds where A, B, and C are used for designation. A is the slowest one while C is the fastest one. The above data is for the 74244 buffer.

Table C-3: Advanced Logic General Characteristics

Family	Year	Number Suppliers	Tech Base	I/O Level	Speed (ns)	Static Current	I_{OH}/I_{OL}
ACQ	1989	2	CMOS	CMOS/CMOS	6.0	80 μA	−24/24 mA
ACQT	1989	2	CMOS	TTL/CMOS	7.5	80 μA	−24/24 mA
FCTx	1987	3	CMOS	TTL/CMOS	4.1 - 4.8	1.5 mA	−15/64 mA
FCTxT	1990	2	CMOS	TTL/TTL	4.1 - 4.8	1.5 mA	−15/64 mA
FASTr	1990	1	Bipolar	TTL/TTL	3.9	50 mA	−15/64 mA
BCT	1987	2	BICMOS	TTL/TTL	5.5	10 mA	−15/64 mA

Reprinted by permission of Electronic Design Magazine, c. 1991.

Since the late 70s, the use of a +5 V power supply has become standard in all microprocessors and microcontrollers. To reduce power consumption, 3.3 V V_{CC} is being embraced by many designers. The lowering of V_{CC} to 3.3 V has two major advantages: (1) it lowers the power consumption, resulting in prolonging the life of the battery in systems using a battery, and (2) it allows a further reduction of line size (design rule) to submicron dimensions. This reduction results in putting more transistors in a given die size. The decline in the line size is expected to reach 0.1 micron by the year 2002 and transistor density per chip will reach 100 million transistors.

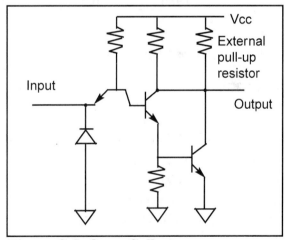

Figure C-5. Open Collector

Open collector and open drain gates

To allow multiple outputs to be connected together, we use open-collector logic gates. In such cases, an external resistor will serve as load. This is shown in Figures C-5 and C-6.

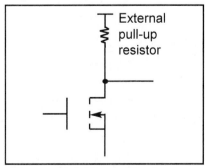

Figure C-6. Open Drain

SECTION C.2: 8051 I/O PORT STRUCTURE AND INTERFACING

In interfacing the 8051 microcontroller with other IC chips or devices, fan-out is the most important issue. To understand the 8051 fan-out we must first understand the port structure of the 8051. This section provides a detailed discussion of the 8051 port structure and its fan-out. It is very critical that we understand the I/O port structure of the 8051 lest we damage it while trying to interface it with an external device.

IC fan-out

In connecting IC chips together, we need to find out how many input pins can be driven by a single output pin. This is a very important issue and involves the discussion of what is called IC fan-out. The IC fan-out must be addressed for both logic "0" and logic "1" outputs. Fan-out for logic low and fan-out for logic high are defined as follows:

$$\text{fan-out (of low)} = \frac{I_{OL}}{I_{IL}} \qquad\qquad \text{fan-out (of high)} = \frac{I_{OH}}{I_{IH}}$$

Of the above two values, the lower number is used to ensure the proper noise margin. Figure C-7 shows the sinking and sourcing of current when ICs are connected together.

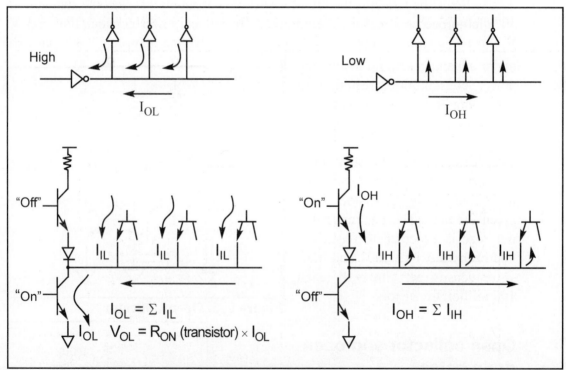

Figure C-7. Current Sinking and Sourcing in TTL

Notice that in Figure C-7, as the number of input pins connected to a single output increases, I_{OL} rises, which causes V_{OL} to rise. If this continues, the rise of V_{OL} makes the noise margin smaller and this results in the occurrence of false logic due to the slightest noise.

Example C-1

Find how many unit loads (UL) can be driven by the output of the LS logic family.

Solution:

The unit load is defined as $I_{IL} = 1.6$ mA and $I_{IH} = 40$ μA. Table C-1 shows $I_{OH} = 400$ μA and $I_{OL} = 8$ mA for the LS family. Therefore, we have

$$\text{fan-out (low)} = \frac{I_{OL}}{I_{IL}} = \frac{8 \text{ mA}}{1.6 \text{ mA}} = 5$$

$$\text{fan-out (high)} = \frac{I_{OH}}{I_{IH}} = \frac{400 \text{ μA}}{40 \text{ μA}} = 10$$

This means that the fan-out is 5. In other words, the LS output must not be connected to more than 5 inputs with unit load characteristics.

74LS244 and 74LS245 buffers/drivers

In cases where the receiver current requirements exceed the driver's capability, we must use buffers/drivers such as the 74LS245 and 74LS244. Figure C-8 shows the internal gates for the 74LS244 and 74LS245. While the 74LS245 is used for bidirectional data buses, the 74LS244 is used for unidirectional address buses.

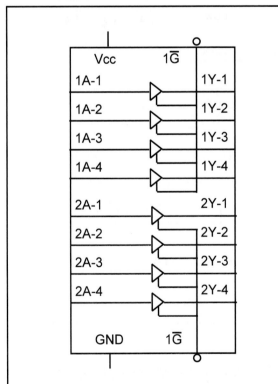

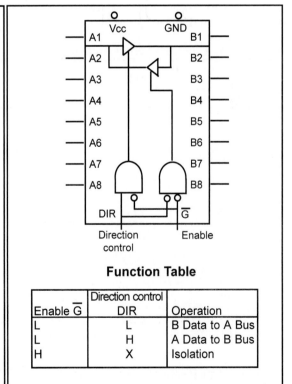

Function Table

Enable $\overline{G}$	Direction control DIR	Operation
L	L	B Data to A Bus
L	H	A Data to B Bus
H	X	Isolation

Figure C-8 (a). 74LS244 Octal Buffer
(Reprinted by permission of Texas Instruments, Copyright Texas Instruments, 1988)

Figure C-8 (b). 74LS245 Bidirectional Buffer
(Reprinted by permission of Texas Instruments, Copyright Texas Instruments, 1988)

Tri-state buffer

Notice that the 74LS244 is simply 8 tri-state buffers in a single chip. As shown in Figure C-9 a tri-state buffer has a single input and single output and the enable control input. By activating the enable, data at the input is transferred to the output. The enable can be an active-low or an active-high. Notice that the enable input for the 74LS244 is an active low whereas the enable input pin for Figure C-9 is active high.

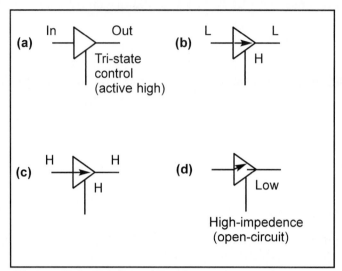

Figure C-9. Tri-State Buffer

74LS245 and 74LS244 fan-out

It must be noted that the output of 74LS245 and 74LS244 can sink and source a much larger amount of current than other LS gates. See Table C-4. That is the reason we use these buffers for driver when a signal is travelling a long distance through a cable or it has to drive many inputs.

Table C-4: Electrical Specifications for Buffers/Drivers

	IOH(mA)	IOL(mA)
74LS244	3	12
74LS245	3	12

After this background on the fan-out, next we discuss the structure of 8051 ports. We first discuss the structure of P1 - P3 since their structure is slightly different from the structure of P0.

P1 - P3 structure and operation

Since all the ports of 8051 are bidirectional they all have the following three components in their structure:
1. D latch
2. Output driver
3. Input buffer

Figure C-10 shows the structure of P1 and its three components. The other ports, P2 and P3, are basically the same except with extra circuitry to allow their dual functions (see Chapter 14). Notice in Figure C-10 that the L1 load is an internal load for P1, P2, and P3. As we will see at the end of this section, that is not the case for P0.

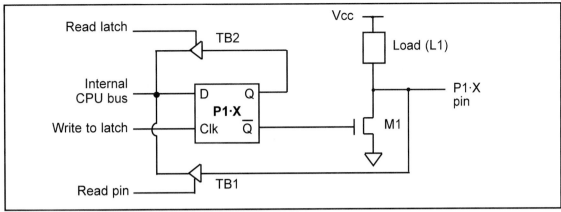

Figure C-10. 8051 Port 1 Structure

Also notice that in Figure C-10, the 8051 ports have both the latch and buffer. Now the question is, in reading the port, are we reading the status of the input pin or are we reading the status of the latch? That is an extremely important question and its answer depends on which instruction we are using. Therefore, when reading the ports there are two possibilities: (1) reading the input pin, or (2) reading the latch. The above distinction is very important and must be understood lest you damage the 8051 port. Each is described next.

Reading the input pin

As we stated in Chapter 4, to make any bits of any port of 8051 an input port, we first must write a 1 (logic high) to that bit. Look at the following sequences of events to see why.

1. As can be seen from Figure C-11, by writing 1 to the port bit it is written to the latch and the D latch has "high" on its Q. Therefore, Q = 1 and $\overline{Q}$ = 0.
2. Since $\overline{Q}$ = 0 and is connected to the transistor M1 gate, the M1 transistor is off.
3. When the M1 transistor is off, it blocks any path to the ground for any signal connected to the input pin and the input signal is directed to the tri-state TB1.
4. When reading the input port in instructions such as "MOV A, P1" we are really reading the data present at the pin. In other words, it is bringing into the CPU the status of the external pin. This instruction activates the read pin of TB1 (tristate buffer 1) and lets data at the pins flow into the CPU's internal bus. Figures C-11 and C-12 show high and low signals at the input, respectively.

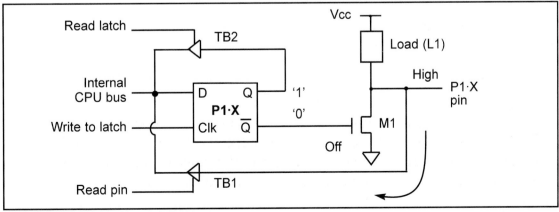

Figure C-11. Reading "High" at Input Pin

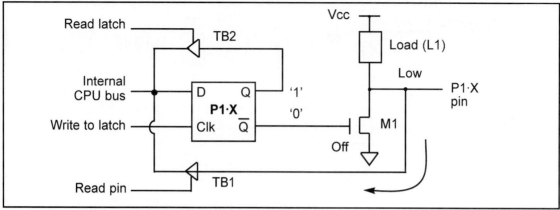

Figure C-12. Reading "Low" at the Input Pin

Writing "0" to the port

The above discussion showed why we must write a "high" to the ports bits in order to make them an input port. What happens if we write a "0" to a port that was configured as an input port? From Figure C-13 we see that if we write a 0 (low) to port bits, then $Q = 0$ and $\overline{Q} = 1$. As a result of $\overline{Q} = 1$, the M1 transistor is "on". If M1 is "on," it provides the path to ground for both L1 and the input pin. Therefore, any attempt to read the input pin will always get the "low" ground signal regardless of the status of the input pin. This can also lead to damage to the port, as explained next.

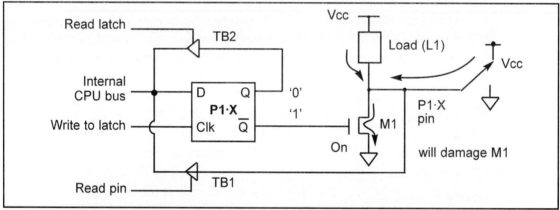

Figure C-13. Never Connect Direct V_{CC} to the 8051 Port Pin

Avoid damaging the port

When connecting a switch to an input port of the 8051 we must be very careful. This is due to the fact that the wrong kind of connection can damage the port. Look at Figure C-13. If a switch with V_{CC} and ground is connected directly to the pin and the M1 transistor is "on" it will sink current from both internal load L1 and external V_{CC}. This can be too much current for M1 and will blow the transistor and, as a result, damage the port bit. There are several ways to avoid this problem. They are shown in Figures C-14, C-15, and C-16.

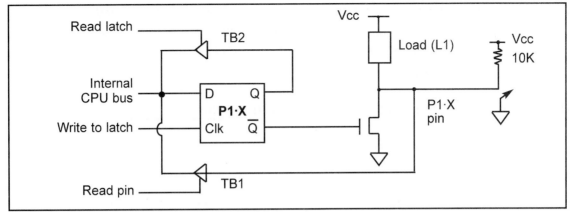

Figure C-14. Input Switch with Pull-Up Resistor

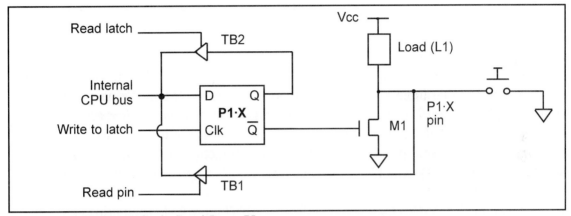

Figure C-15. Input Switch with no V$_{CC}$

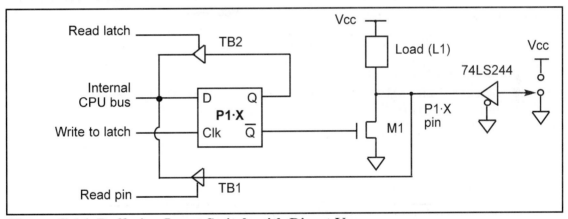

Figure C-16. Buffering Input Switch with Direct V$_{CC}$

1. One way is to have a 10 K-ohm resistor on the V$_{CC}$ path to limit current flow through the M1 transistor. See Figure C-14.
2. The second method is to use a switch with a ground only, and no V$_{CC}$, as shown in Figure C-15. In this method, we read a low when the switch is pressed and we read a high when it is released.
3. Another way is to connect any input switch to a 74LS244 tri-state buffer before it is fed to the 8051 pin. This is shown in Figure C-16.

The above points are extermely important and must be emphasized since many people damage their ports and wonder how it happened. We must also use the right instruction when we want to read the status of an input pin. Table C-5 shows the list of instructions in which reading the port reads the status of the input pin.

Table C-5: Instructions Reading the Status of Input Port

Mnemonics	Examples
MOV A,PX	MOV A,P1
JNB PX.Y,...	JNB P1.2,TARGET
JB PX.Y,...	JB P1.3,TARGET
MOV C,PX.Y	MOV C,P1.4
CJNE A,PX,...	CJNE A,P1,TARGET

Reading latch

Since in reading the port, some instructions read the port and some others read the latch, we next consider the case of reading the port where it reads the internal port latch. "ANL P1,A" is an example of an instruction that reads the latch instead of the input pin. Look at the sequence of actions taking place when an instruction such as "ANL P1,A" is executed.

1. The read latch activates the tri-state buffer of TB2 (Figure C-17) and brings the data from the Q latch into CPU.
2. This data is ANDed with the contents of register A.
3. The result is rewritten to the latch.

After rewriting the result to the latch, there are two possibilities: (1) If Q = 0, then $\overline{Q}$ = 1 and M1 is "on," and the output pin has "0," the same as the status of the Q latch, (2) If Q = 1, then $\overline{Q}$ = 0 and the M1 is "off," and the output pin has "1," the same as the status of the Q latch.

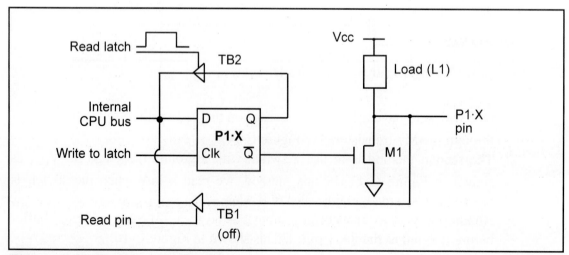

Figure C-17. Reading the Latch

From the above discussion, we conclude that the instruction that reads the latch normally reads a value, performs an operation (possibly changing the value), and rewrites it to the latch. This is often called "read-modify-write." Table C-6 provides a list of read-modify-write instructions. Notice from Table C-6 that all the read-modify-write instructions use the port as the destination operand.

Table C-6: Read-Modify-Write Instructions

Mnemonics	Example
ANL	ANL P1,A
ORL	ORL P1,A
XRL	XRL P1,A
JBC	JBC P1.1,TARGET
CPL	CPL P1.2
INC	INC P1
DEC	DEC P1
DJNZ	DJNZ P1,TARGET
MOV PX.Y,C	MOV P1.2,C
CLR PX.Y	CLR P1.3
SETB PX.Y	SETB P1.4

P0 structure

A major difference between P0 and other ports is that P0 has no internal pull-up resistors. (The reason is to allow it to multiplex address and data. See Chapter 14 for a detailed discussion of address/data muliplexing.) Since P0 has no internal pull-up resistors, it is simply an open-drain as shown in Figure C-18. (Open-drain in MOS is same as open-collector in TTL). Now by writing a "1" to the bit latch, the M1 transistor is "off" and that causes the pin to float. That is the reason why when P0 is used for simple data I/O we must connect it to external pull-up resistors. As can be seen from Figures C-18 and C-19, for a P0 bit to drive an input, there must be a pull-up resistor to source current.

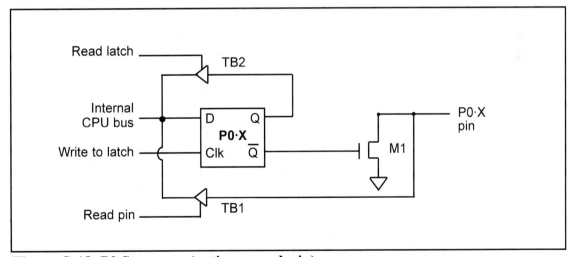

Figure C-18. P0 Structure (notice open drain)

It must be noticed that when P0 is used for address/data multiplexing and it is connected to the 74LS373 to latch the address, there is no need for external pull-up resistors as shown in detail in Chapter 14.

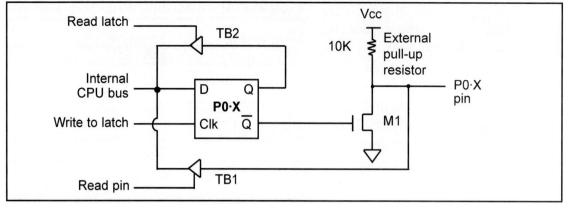

Figure C-19. P0 With External Pull-Up Resistor

8051 fan-out

Now that we are familiar with the port structure of the 8051, we need to examine the fan-out for the 8051 microconctroller. While the early 8051 micro-controllers were based on NMOS IC technology, today's 8051 microcontrollers are all based on CMOS technology. However, note that while the core of the 8051 microcontroller is CMOS, the circuitry driving its pins is all TTL compatible. That is, the 8051 is a CMOS-based product with TTL-compatible pins.

P1, P2, and P3 fan-out

The three ports of P1, P2, and P3 have the same I/O structure, and therefore the same fan-out. Table C-7 provides the I/O characteristics of P1, P2, and P3.

Table C-7: 8051 Fan-out for P1, P2, P3

Pin	Fan-out
IOL	1.6 mA
IOH	60 μA
IIL	50 μA
IIH	650 μA

Note: P1, P2, and P3 can drive up to 4 LS TTL inputs when connected to other IC chips.

Port 0 fan-out

P0 requires external pull-up resistors in order to drive an input since it is an open drain I/O. The value of this resistor decides the fan-out. However, since I_{OL} = 3.2 mA for V_{OL} = 0.45 V, we must make sure that the pull-up resistor connected to each pin of the P0 is no less than 1422 ohms, since (5 V − 0.45 V) / 3.2 mA = 1422 ohms. In applications in which P0 is not connected to an external pull-up resistor, or is used in bus mode connected to a 74LS373 or other chip, it can drive up to 8 LS TTL inputs.

74LS244 driving an output pin

In many cases when an 8051 port is driving multiple inputs, or driving a single input via a long wire or cable (e.g., printer cable), we need to use the 74LS244 as a driver. When driving an off-board circuit, placing the 74LS244 buffer between your 8051 and the circuit is essential since the 8051 lacks sufficient current. See Figure C-20.

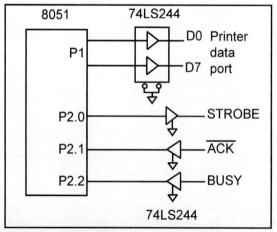

Figure C-20. 8051 Connection to Printer Signals

SECTION C.3: SYSTEM DESIGN ISSUES

In addition to fan-out, the other issues related to system design are power dissipation, ground bounce, V_{CC} bounce, crosstalk, and transmission lines. In this section we provide an overview of these topics.

Power dissipation considerations

Power dissipation of a system is a major concern of system designers, especially for laptop and hand-held systems in which batteries provide the power source. Power dissipation is a function of frequency and voltage as shown below:

$$Q = CV$$

$$\frac{Q}{T} = \frac{CV}{T}$$

since $\quad F = \dfrac{1}{T} \qquad$ and $\quad I = \dfrac{Q}{T}$

$$I = CVF$$

now $\quad P = VI = CV^2F$

In the above equations, the effects of frequency and V_{CC} voltage should be noted. While the power dissipation goes up linearly with frequency, the impact of the power supply voltage is much more pronounced (squared). See Example C-2.

Example C-2

Compare the power consumption of two 8051 systems. One uses 5 V and the other uses 3 V for V_{CC}.

Solution:

Since P = VI, by substituting I = V/R we have $P = V^2/R$. Assuming that R = 1, we have $P = 5^2 = 25$ W and $P = 3^2 = 9$ W. This results in using 16 W less power, which means a 64% power saving. (16/25 × 100) for systems using 3 V for power source.

Dynamic and static currents

There are two major types of currents flowing through an IC: dynamic and static. A dynamic current is I = CVF. It is a function of the frequency under which the component is working. This means that as the frequency goes up, the dynamic current and power dissipation go up. The static current, also called DC, is the current consumption of the component when it is inactive (not selected). The dynamic current dissipation is much higher than the static current consumption. To reduce power consumption, many microcontrollers, including the 8051, have power-saving modes. In the 8051, the power saving modes are called *idle mode* and *power down mode*. Each one is described next.

Idle mode

In idle mode, which is also called *sleep mode*, the core CPU is put to sleep while all on-chip peripherals such as the serial port, timers, and interrupts remain

---

active and continue to function. In this mode, the oscillator continues to provide clock to the serial port, interrupt, and the timers, but no clock is provided to the CPU. Notice that during this mode all the contents of the registers and on-chip RAM remain unchanged.

Power down mode

In the power down mode, the on-chip oscillator is frozen which cuts off frequency to the CPU and peripheral functions, such as serial port, interrupts, and timers. Notice that while this mode brings down power consumption to an absolute minimum, the contents of RAM and the SFR registers are saved and remain unchanged.

Ground bounce

One of the major issues that designers of high-frequency systems must grapple with is ground bounce. Before we define ground bounce, we will discuss lead inductance of IC pins. There is a certain amount of capacitance, resistance, and inductance associated with each pin of the IC. The size of these elements varies depending on many factors such as length, area, and so on.

The inductance of the pins is commonly referred to as *self-inductance* since there is also what is called *mutual inductance*, as we will show below. Of the three components of capacitor, resistor, and inductor, self-inductance is the one that causes the most problems in high-frequency system design since it can result in ground bounce. Ground bounce is caused when a massive amount of current flows through the ground pin caused by many outputs changing from high to low all at the same time. See Figure C-21(a). The voltage is related to the inductance of the ground lead as follows:

$$V = L\frac{di}{dt}$$

As we increase the system frequency, the rate of dynamic current, di/dt, is also increased, resulting in an increase in the inductance voltage L (di/dt) of the ground pin. Since the low state (ground) has a small noise margin, any extra voltage due to the inductance voltage can cause a false signal. To reduce the effect of ground bounce, the following steps must be taken where possible.

1. The V_{CC} and ground pins of the chip must be located in the middle rather than at the opposite ends of the IC chip (the 14-pin TTL logic IC uses pins 14 and 7 for ground and V_{CC}). This is exactly what we see in high-performance logic gates such as Texas Instrument's advanced logic AC11000 and ACT11000 families. For example, the ACT11013 is a 14-pin DIP chip in which pin numbers 4 and 11 are used for the ground and V_{CC}, instead of 7 and 14 as in the traditional TTL family. We can also use the SOIC packages instead of DIP.

2. Another solution is to use as many pins for ground and V_{CC} as possible to reduce the lead length. This is exactly why all high-performance microprocessors and logic families use many pins for V_{CC} and ground instead of the traditional single pin for V_{CC} and single pin for GND. For example, in the case of Intel's Pentium processor there are over 50 pins for ground, and another 50 pins for V_{CC}.

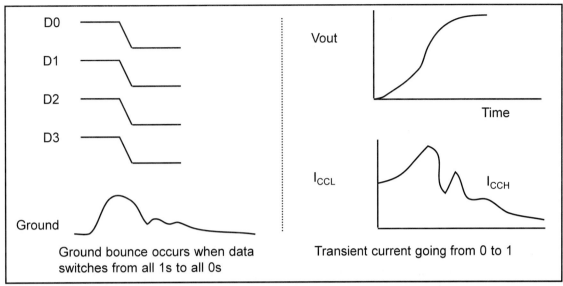

D0

D1

D2

D3

Ground

Ground bounce occurs when data switches from all 1s to all 0s

Vout

Time

I_{CCL}

I_{CCH}

Transient current going from 0 to 1

Figure C-21 (a). Ground Bounce   (b). Transient Current

The above discussion of ground bounce is also applicable to V_{CC} when a large number of outputs changes from the low to the high state; this is referred to as *V_{CC} bounce*. However, the effect of V_{CC} bounce is not as severe as ground bounce since the high ("1") state has a wider noise margin than the low ("0") state.

Filtering the transient currents using decoupling capacitors

In the TTL family, the change of the output from low to high can cause what is called *transient current*. In a totem-pole output in which the output is low, Q4 is on and saturated, whereas Q3 is off. By changing the output from the low to the high state, Q3 becomes on and Q4 becomes off. This means that there is a time that both transistors are on and drawing current from V_{CC}. The amount of current depends on the R_{ON} values of the two transistors, which in turn depend on the internal parameters of the transistors. However, the net effect of this is a large amount of current in the form of a spike for the output current, as shown in Figure C-21(b). To filter the transient current, a 0.01 μF or 0.1 μF ceramic disk capacitor can be placed between the V_{CC} and ground for each TTL IC. However, the lead for this capacitor should be as small as possible since a long lead results in a large self-inductance and that results in a spike on the V_{CC} line [V = L (di/dt)]. This spike is called V_{CC} bounce. The ceramic capacitor for each IC is referred to as a *decoupling capacitor*. There is also a bulk decoupling capacitor, as described next.

Bulk decoupling capacitor

If many IC chips change state at the same time, the combined currents drawn from the board's V_{CC} power supply can be massive and may cause a fluctuation of V_{CC} on the board where all the ICs are mounted. To eliminate this, a relatively large decoupling tantalum capacitor is placed between the V_{CC} and ground lines. The size and location of this tantalum capacitor varies depending on the number of ICs on the board and the amount of current drawn by each IC, but it is common to have a single 22 μF to 47 μF capacitor for each of the 16 devices, placed between the V_{CC} and ground lines.

Crosstalk

Crosstalk is due to mutual inductance. See Figure C-22. Previously, we discussed self-inductance, which is inherent in a piece of conductor. *Mutual inductance* is caused by two electric lines running parallel to each other. The mutual inductance is a function of l, the length of two conductors running in parallel, and d, the distance between them, and the medium material placed between them. The effect of crosstalk can be reduced by increasing the distance between the parallel or adjacent lines (in printed circuit boards, they will be traces). In many cases, such as printer and disk drive cables, there is a dedicated ground for each signal. Placing ground lines (traces) between signal lines reduces the effect of crosstalk. This method is used even in some ACT logic families where there are a V_{CC} and GND pin next to each other. Crosstalk is also called *EMI* (electromagnetic interference). This is in contrast to *ESI* (electrostatic interference), which is caused by capacitive coupling between two adjacent conductors.

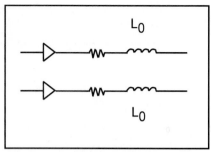

Figure C-22. Crosstalk (EMI)

Transmission line ringing

The square wave used in digital circuits is in reality made of a single fundamental pulse and many harmonics of various amplitudes. When this signal travels on the line, not all the harmonics respond in the same way to the capacitance, inductance, and resistance of the line. This causes what is called *ringing*, which depends on the thickness and the length of the line driver, among other factors. To reduce the effect of ringing, the line drivers are terminated by putting a resistor at the end of the line. See Figure C-23. There are three major methods of line driver termination: parallel, serial, and Thevenin.

In serial termination, resistors of 30 - 50 ohms are used to terminate the line. The parallel and Thevenin methods are used in cases where there is a need to match the impedance of the line with the load impedance. This requires a detailed analysis of the signal traces and load impedance, which is beyond the scope of this book. In high-frequency systems, wire traces on the printed circuit board (PCB) behave like transmission lines, causing ringing. The severity of this ringing depends on the speed and the logic family used. Table C-8 provides the length of the traces, beyond which the traces must be looked at as transmission lines.

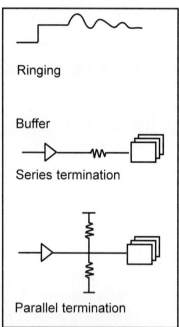

Figure C-23. Reducing Transmission Line Ringing

Table C-8. Line Length Beyond Which Traces Behave Like Transmission Lines

Logic Family	Line Length (in.)
LS	25
S, AS	11
F, ACT	8
AS, ECL	6
FCT, FCTA	5

(Reprinted by permission of Integrated Device Technology, c. IDT 1991)

APPENDIX D

FLOWCHARTS AND PSEUDOCODE

OVERVIEW

This appendix provides an introduction to writing flowcharts and pseudocode.

Flowcharts

If you have taken any previous programming courses, you are probably familiar with flowcharting. Flowcharts use graphic symbols to represent different types of program operations. These symbols are connected together into a flowchart to show the flow of execution of a program. Figure D-1 shows some of the more commonly used symbols. Flowchart templates are available to help you draw the symbols quickly and neatly.

Pseudocode

Flowcharting has been standard practice in industry for decades. However, some find limitations in using flowcharts, such as the fact that you can't write much in the little boxes, and it is hard to get the "big picture" of what the program does without getting bogged down in the details. An alternative to using flowcharts is pseudocode, which involves writing brief descriptions of the flow of the code. Figures D-2 through D-6 show flowcharts and pseudocode for commonly used control structures.

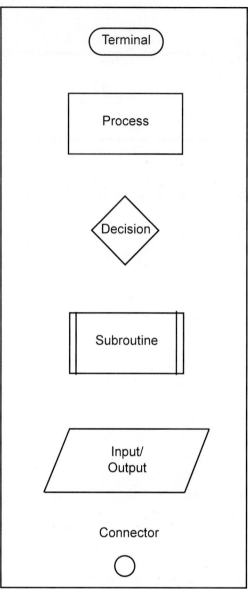

Figure D-1. Commonly Used Flowchart Symbols

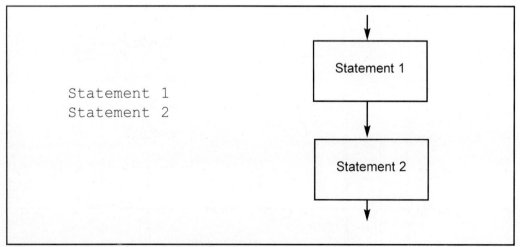

```
Statement 1
Statement 2
```

Figure D-2. SEQUENCE Pseudocode versus Flowchart

Structured programming uses three basic types of program control structures: sequence, control, and iteration. Sequence is simply executing instructions one after another. Figure D-2 shows how sequence can be represented in pseudocode and flowcharts.

Figures D-3 and D-4 show two control programming structures: IF-THEN-ELSE and IF-THEN in both pseudocode and flowcharts.

Note in Figures D-2 through D-6 that "statement" can indicate one statement or a group of statements.

Figures D-5 and D-6 show two iteration control structures: REPEAT UNTIL and WHILE DO. Both structures execute a statement or group of statements repeatedly. The difference between them is that the REPEAT UNTIL structure always executes the statement(s) at least once, and checks the condition after each iteration, whereas the WHILE DO may not execute the statement at all since the condition is checked at the beginning of each iteration.

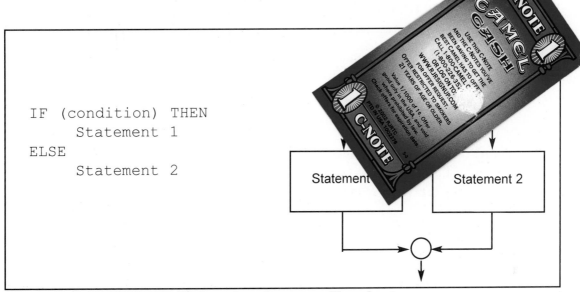

```
IF (condition) THEN
      Statement 1
ELSE
      Statement 2
```

Figure D-3. IF THEN ELSE Pseudocode versus Flowchart

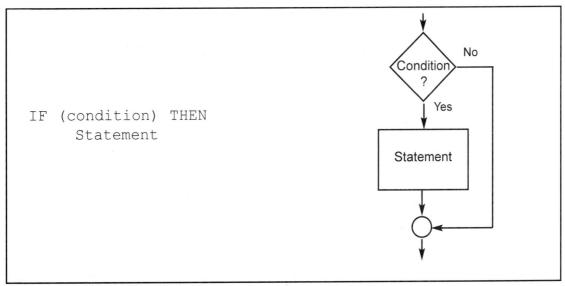

```
IF (condition) THEN
      Statement
```

Figure D-4. IF THEN Pseudocode versus Flowchart

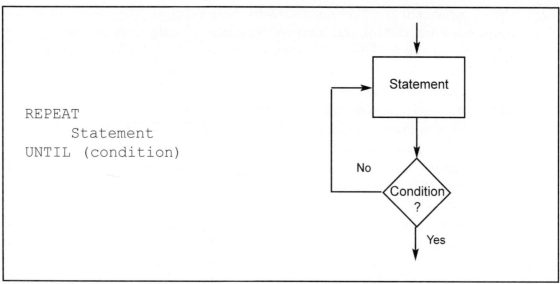

```
REPEAT
      Statement
UNTIL (condition)
```

Figure D-5.REPEAT UNTIL Pseudocode versus Flowchart

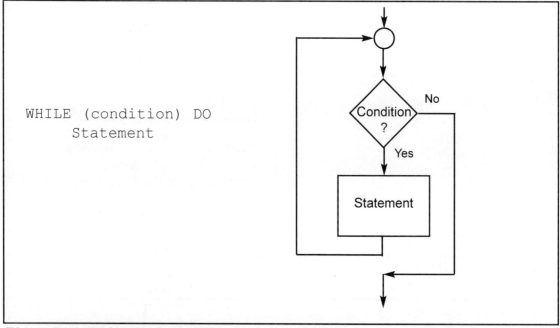

```
WHILE (condition) DO
        Statement
```

Figure D-6. WHILE DO Pseudocode versus Flowchart

Program D-1 finds the sum of a series of bytes. Compare the flowchart versus the pseudocode for Program D-1. In this example, more program details are given than one usually finds. For example, this shows steps for initializing and decrementing counters. Another programmer may not include these steps in the flowchart or pseudocode. It is important to remember that the purpose of flowcharts or pseudocode is to show the flow of the program and what the program does, not the specific Assembly language instructions that accomplish the program's objectives. Notice also that the pseudocode gives the same information in a much more compact form than the flowchart. It is important to note that sometimes pseudocode is written in layers, so that the outer level or layer shows the flow of the program and subsequent levels show more details of how the program accomplishes its assigned tasks.

```
Count = 5
Address = 40H
Repeat
    Add next byte
    Increment address
    Decrement counter
Until Count = 0

Store Sum
```

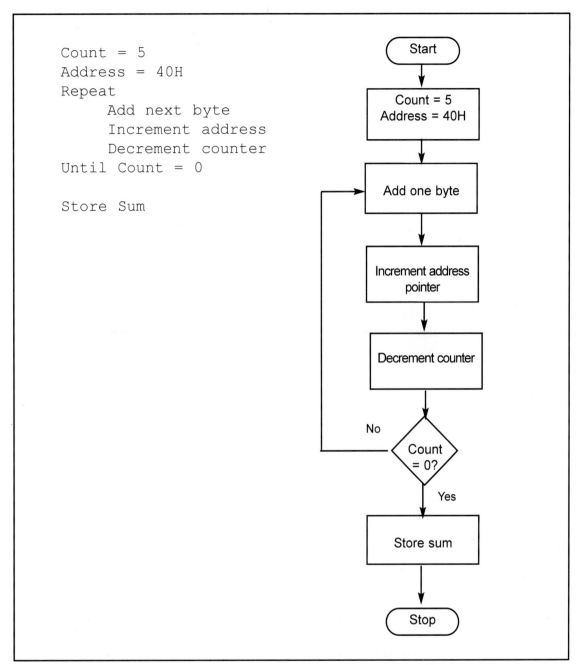

Figure D-7. Pseudocode versus Flowchart for Program D-1

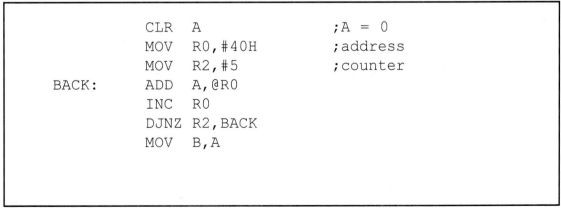

```
            CLR  A              ;A = 0
            MOV  R0,#40H        ;address
            MOV  R2,#5          ;counter
BACK:       ADD  A,@R0
            INC  R0
            DJNZ R2,BACK
            MOV  B,A
```

Program D-1

APPENDIX E

8051 PRIMER FOR X86 PROGRAMMERS

	X86	**8051**
8-bit registers:	AL, AH, BL, BH, CL , CH, DL, DH	A, B, R0, R1, R2, R3, R4, R5, R6, R7
16-bit (data pointer):	BX,SI,DI	DPTR
Program Counter:	IP (16-bit)	PC (16-bit)
Input:	MOV DX,port addr IN AL,DX	MOV A,Pn ;(n = 0 - 3)
Output:	MOV DX,port addr OUT DX,AL	MOV Pn,A ;(n = 0 - 3)
Loop:	DEC CL JNZ TARGET	DJNZ R3,TARGET (Using R0-R7)
Stack pointer:	SP (16-bit)	SP (8-bit)
	As we PUSH data onto the stack, it decrements the SP.	As we PUSH data onto the stack, it increments the SP.
	As we POP data from stack it increments the SP.	As we POP data from stack it decrements the SP.

Data movement:

From the code segment:

    `MOV AL,CS:[SI]`   `MOVC A,@A+PC`

From the data segment:

    `MOV AL,[SI]`   `MOVX A,@DPTR`

From RAM:

    `MOV AL,[SI]`   `MOV A,@R0`
    (use SI, DI or BX only)   (Use R0 or R1 only)

To RAM:

    `MOV [SI],AL`   `MOV @R0,A`

APPENDIX F

ASCII CODES

Ctrl	Dec	Hex	Ch	Code	Dec	Hex	Ch	Dec	Hex	Ch	Dec	Hex	Ch
^@	0	00		NUL	32	20		64	40	@	96	60	`
^A	1	01	☺	SOH	33	21	!	65	41	A	97	61	a
^B	2	02	●	STX	34	22	"	66	42	B	98	62	b
^C	3	03	♥	ETX	35	23	#	67	43	C	99	63	c
^D	4	04	♦	EOT	36	24	$	68	44	D	100	64	d
^E	5	05	♣	ENQ	37	25	%	69	45	E	101	65	e
^F	6	06	♠	ACK	38	26	&	70	46	F	102	66	f
^G	7	07	●	BEL	39	27	'	71	47	G	103	67	g
^H	8	08	□	BS	40	28	(	72	48	H	104	68	h
^I	9	09	○	HT	41	29	)	73	49	I	105	69	i
^J	10	0A	◉	LF	42	2A	*	74	4A	J	106	6A	j
^K	11	0B	♂	VT	43	2B	+	75	4B	K	107	6B	k
^L	12	0C	♀	FF	44	2C	,	76	4C	L	108	6C	l
^M	13	0D	♪	CR	45	2D	-	77	4D	M	109	6D	m
^N	14	0E	♫	SO	46	2E	.	78	4E	N	110	6E	n
^O	15	0F	☼	SI	47	2F	/	79	4F	O	111	6F	o
^P	16	10	►	DLE	48	30	0	80	50	P	112	70	p
^Q	17	11	◄	DC1	49	31	1	81	51	Q	113	71	q
^R	18	12	↕	DC2	50	32	2	82	52	R	114	72	r
^S	19	13	‼	DC3	51	33	3	83	53	S	115	73	s
^T	20	14	¶	DC4	52	34	4	84	54	T	116	74	t
^U	21	15	§	NAK	53	35	5	85	55	U	117	75	u
^V	22	16	▬	SYN	54	36	6	86	56	V	118	76	v
^W	23	17	↨	ETB	55	37	7	87	57	W	119	77	w
^X	24	18	↑	CAN	56	38	8	88	58	X	120	78	x
^Y	25	19	↓	EM	57	39	9	89	59	Y	121	79	y
^Z	26	1A	→	SUB	58	3A	:	90	5A	Z	122	7A	z
^[	27	1B	←	ESC	59	3B	;	91	5B	[	123	7B	{
^\	28	1C	⌐	FS	60	3C	<	92	5C	\	124	7C	|
^]	29	1D	↔	GS	61	3D	=	93	5D	]	125	7D	}
^^	30	1E	▲	RS	62	3E	>	94	5E	^	126	7E	~
^_	31	1F	▼	US	63	3F	?	95	5F	_	127	7F	⌂

APPENDIX G

ASSEMBLERS, DEVELOPMENT RESOURCES, AND SUPPLIERS

This appendix provides various sources for 8051 assemblers and trainers. In addition, it lists some suppliers for chips and other hardware needs. While these are all established products from well-known companies, neither the authors nor the publisher assumes responsibility for any problem that may arise with any of them. You are neither encouraged nor discouraged from purchasing any of the products mentioned; you must make your own judgment in evaluating the products. This list is simply provided as a service to the reader. It also must be noted that the list of products is by no means complete or exhaustive. To suggest other products to be included in future editions of this book, please send your company's name, product name and description, and internet address to the Authors' email listed in the introduction.

8051 assemblers

The 8051 assembler is provided by many companies. Some of them provide a shareware version of their products on the Web. You can download them from their Web Sites. However, the size of code for these shareware versions is limited to 1K (or 2K). Figure G-1 lists some of them.

8051 trainers

There are many companies that produce and market 8051 trainers. Figure G-2 provides a list of some of them.

Franklin Software
www.fsinc.com

Keil
www.keil.com

Dunfield Development Systems
www.dunfield.com/download.htm

Product Language Corp.
www.plcorp.com/51.htm

Figure G-1. Assembler Suppliers

Axiom Manufacturing
717 Lingco Dr. Ste. # 209
Richardson, TX 75081
(972) 994-9676 Fax: (972) 994-9170
www.axman.com

Rigel Corp.
P. O. BoxX 90040
Gainsville, FL 32607
(352)373-4629
http://rigelcorp.com

Urda Inc.
1811 Jancey Street Ste. #200
Pittsburgh, PA 15206
1-800-338-0517 (412) 363-0990
www.urda.com

RSR Electronics
www.elexp.com

Figure G-2. Trainer Suppliers

The following is a Web site for FAQ (frequently asked questions) about the 8051:
www.ece.orst.edu/~pricec/8051/faq

Parts Suppliers

Figure G-3 provides a list of suppliers for many electronics parts.

RSR Electronics
Electronix Express
365 Blair Road
Avenel NJ 07001
Fax: (732) 381-1572
Mail Order: 1-800-972-2225
In New Jersey: (732) 381-8020
www.elexp.com

Altex Electronics
11342 IH-35 North
San Antonio TX 78233
Fax: (210) 637-3264
Mail Order: 1-800-531-5369
www.altex.com

Digi-Key
1-800-344-4539 (1-800-DIGI-KEY)
FAX: (218) 681-3380
www.digikey.com

Radio Shack Mail order: 1-800-THE-SHACK

JDR Microdevices
1850 South 10th St.
San Jose CA 95112-4108
Sales 1-800-538-5000
(408) 494-1400
Fax: 1-800-538-5005
Fax: (408) 494-1420
www.jdr.com

Mouser Electronics
958 N. Main St.
Mansfield TX 76063
1-800-346-6873
www.mouser.com

Jameco Electronic
1355 Shoreway Road
Belmont CA 94002-4100
1-800-831-4242
(415) 592-8097
Fax: 1-800-237-6948
Fax: (415) 592-2503
www.jameco.com

B. G. Micro
P. O. Box 280298
Dallas Tx 75228
1-800-276-2206 (Orders Only)
(972) 271-5546
Fax: (972) 271-2462
This is an excellent source of LCDs, ICs,
keypads, etc.
www.bgmicro.com

Tanner Electronics
1301 West Beltline Rd, Suite 119
Carrollton TX 75006
(972) 242-8702

Figure G-3. Electronics Suppliers

APPENDIX H

DATA SHEETS

The information presented in this chapter is collected from the MCS®-51 Architectural Overview and the Hardware Description of the 8051, 8052 and 80C51 chapters of this book. The material has been selected and rearranged to form a quick and convenient reference for the programmers of the MCS-51. This guide pertains specifically to the 8051, 8052 and 80C51.

MEMORY ORGANIZATION

PROGRAM MEMORY

The 8051 has separate address spaces for Program Memory and Data Memory. The Program Memory can be up to 64K bytes long. The lower 4K (8K for the 8052) may reside on-chip.

Figure 1 shows a map of the 8051 program memory, and Figure 2 shows a map of the 8052 program memory.

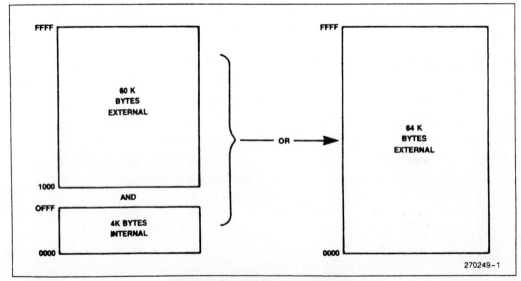

Figure 1. The 8051 Program Memory

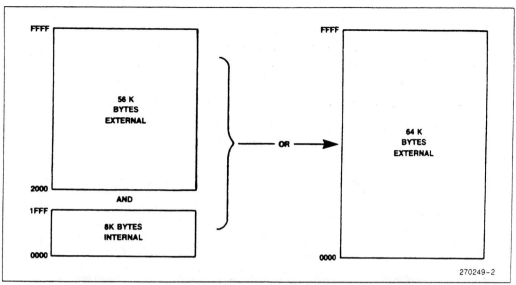

Figure 2. The 8052 Program Memory

Data Memory:

The 8051 can address up to 64K bytes of **Data Memory external** to the chip. The "MOVX" instruction is used to access the external data memory. (Refer to **the MCS-51 Instruction** Set, in this chapter, for detailed description of instructions).

The 8051 has 128 bytes of on-chip RAM **(256 bytes in the 8052) plus** a number of Special Function Registers (SFRs). The lower 128 bytes of RAM can be accessed **either by direct addressing** (MOV data addr) or by indirect addressing (MOV @Ri). Figure 3 shows the 8051 and the 8052 **Data Memory** organization.

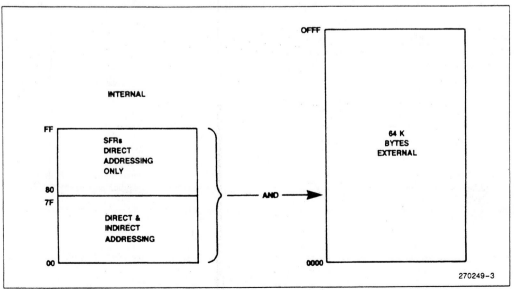

Figure 3a. The 8051 Data Memory

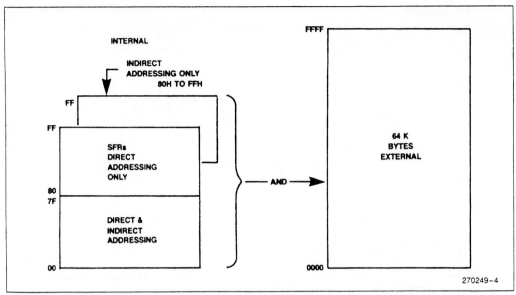

Figure 3b. The 8052 Data Memory

INDIRECT ADDRESS AREA:

Note that in Figure 3b the SFRs and the indirect address RAM have the same addresses (80H–0FFH). Nevertheless, they are two separate areas and are accessed in two different ways.

For example the instruction

        MOV   80H, #0AAH

writes 0AAH to Port 0 which is one of the SFRs and the instruction

        MOV   R0, #80H

        MOV   @R0, #0BBH

writes 0BBH in location 80H of the data RAM. Thus, after execution of both of the above instructions Port 0 will contain 0AAH and location 80 of the RAM will contain 0BBH.

Note that the stack operations are examples of indirect addressing, so the upper 128 bytes of data RAM are available as stack space in those devices which implement 256 bytes of internal RAM.

DIRECT AND INDIRECT ADDRESS AREA:

The 128 bytes of RAM which can be accessed by both direct and indirect addressing can be divided into 3 segments as listed below and shown in Figure 4.

1. Register Banks 0-3: Locations 0 through 1FH (32 bytes). ASM-51 and the device after reset default to register bank 0. To use the other register banks the user must select them in the software (refer to the MCS-51 Micro Assembler User's Guide). Each register bank contains 8 one-byte registers, 0 through 7.

Reset initializes the Stack Pointer to location 07H and it is incremented once to start from location 08H which is the first register (R0) of the second register bank. Thus, in order to use more than one register bank, the SP should be intialized to a different location of the RAM where it is not used for data storage (ie, higher part of the RAM).

2. Bit Addressable Area: 16 bytes have been assigned for this segment, 20H-2FH. Each one of the 128 bits of this segment can be directly addressed (0-7FH).

The bits can be referred to in two ways both of which are acceptable by the ASM-51. One way is to refer to their addresses, ie. 0 to 7FH. The other way is with reference to bytes 20H to 2FH. Thus, bits 0–7 can also be referred to as bits 20.0–20.7, and bits 8-FH are the same as 21.0–21.7 and so on.

Each of the 16 bytes in this segment can also be addressed as a byte.

3. Scratch Pad Area: Bytes 30H through 7FH are available to the user as data RAM. However, if the stack pointer has been initialized to this area, enough number of bytes should be left aside to prevent SP data destruction.

406

Figure 4 shows the different segments of the on-chip RAM.

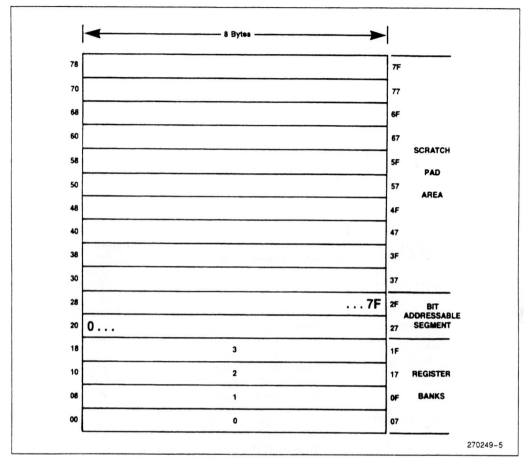

Figure 4. 128 Bytes of RAM Direct and Indirect Addressable

2-7

MCS®-51 PROGRAMMER'S GUIDE AND INSTRUCTION SET

SPECIAL FUNCTION REGISTERS:

Table 1 contains a list of all the SFRs and their addresses.

Comparing Table 1 and Figure 5 shows that all of the SFRs that are byte and bit addressable are located on the first column of the diagram in Figure 5.

Table 1

Symbol	Name	Address
*ACC	Accumulator	0E0H
*B	B Register	0F0H
*PSW	Program Status Word	0D0H
SP	Stack Pointer	81H
DPTR	Data Pointer 2 Bytes	
DPL	Low Byte	82H
DPH	High Byte	83H
*P0	Port 0	80H
*P1	Port 1	90H
*P2	Port 2	0A0H
*P3	Port 3	0B0H
*IP	Interrupt Priority Control	0B8H
*IE	Interrupt Enable Control	0A8H
TMOD	Timer/Counter Mode Control	89H
*TCON	Timer/Counter Control	88H
*+T2CON	Timer/Counter 2 Control	0C8H
TH0	Timer/Counter 0 High Byte	8CH
TL0	Timer/Counter 0 Low Byte	8AH
TH1	Timer/Counter 1 High Byte	8DH
TL1	Timer/Counter 1 Low Byte	8BH
+TH2	Timer/Counter 2 High Byte	0CDH
+TL2	Timer/Counter 2 Low Byte	0CCH
+RCAP2H	T/C 2 Capture Reg. High Byte	0CBH
+RCAP2L	T/C 2 Capture Reg. Low Byte	0CAH
*SCON	Serial Control	98H
SBUF	Serial Data Buffer	99H
PCON	Power Control	87H

* = Bit addressable
\+ = 8052 only

2-8

408

WHAT DO THE SFRs CONTAIN JUST AFTER POWER-ON OR A RESET?

Table 2 lists the contents of each SFR after power-on or a hardware reset.

Table 2. Contents of the SFRs after reset

Register	Value in Binary
*ACC	00000000
*B	00000000
*PSW	00000000
SP	00000111
DPTR	
DPH	00000000
DPL	00000000
*P0	11111111
*P1	11111111
*P2	11111111
*P3	11111111
*IP	8051 XXX00000,
	8052 XX000000
*IE	8051 0XX00000,
	8052 0XX000000
TMOD	00000000
*TCON	00000000
*+T2CON	00000000
TH0	00000000
TL0	00000000
TH1	00000000
TL1	00000000
+TH2	00000000
+TL2	00000000
+RCAP2H	00000000
+RCAP2L	00000000
*SCON	00000000
SBUF	Indeterminate
PCON	HMOS 0XXXXXXX
	CHMOS 0XXX0000

X = Undefined
• = Bit Addressable
+ = 8052 only

SFR MEMORY MAP

8 Bytes

F8									FF
F0	B								F7
E8									EF
E0	ACC								E7
D8									DF
D0	PSW								D7
C8	T2CON		RCAP2L	RCAP2H	TL2	TH2			CF
C0									C7
B8	IP								BF
B0	P3								B7
A8	IE								AF
A0	P2								A7
98	SCON	SBUF							9F
90	P1								97
88	TCON	TMOD	TL0	TL1	TH0	TH1			8F
80	P0	SP	DPL	DPH				PCON	87

↑
Bit
Addressable

Figure 5

Those SFRs that have their bits assigned for various functions are listed in this section. A brief description of each bit is provided for quick reference. For more detailed information refer to the Architecture Chapter of this book.

PSW: PROGRAM STATUS WORD. BIT ADDRESSABLE.

CY	AC	F0	RS1	RS0	OV	—	P

CY	PSW.7	Carry Flag.	
AC	PSW.6	Auxiliary Carry Flag.	
F0	PSW.5	Flag 0 available to the user for general purpose.	
RS1	PSW.4	Register Bank selector bit 1 (SEE NOTE 1).	
RS0	PSW.3	Register Bank selector bit 0 (SEE NOTE 1).	
OV	PSW.2	Overflow Flag.	
—	PSW.1	User definable flag.	
P	PSW.0	Parity flag. Set/cleared by hardware each instruction cycle to indicate an odd/even number of '1' bits in the accumulator.	

NOTE:
1. The value presented by RS0 and RS1 selects the corresponding register bank.

RS1	RS0	Register Bank	Address
0	0	0	00H-07H
0	1	1	08H-0FH
1	0	2	10H-17H
1	1	3	18H-1FH

PCON: POWER CONTROL REGISTER. NOT BIT ADDRESSABLE.

SMOD	—	—	—	GF1	GF0	PD	IDL

SMOD	Double baud rate bit. If Timer 1 is used to generate baud rate and SMOD = 1, the baud rate is doubled when the Serial Port is used in modes 1, 2, or 3.
—	Not implemented, reserved for future use.*
—	Not implemented, reserved for future use.*
—	Not implemented, reserved for future use.*
GF1	General purpose flag bit.
GF0	General purpose flag bit.
PD	Power Down bit. Setting this bit activates Power Down operation in the 80C51BH. (Available only in CHMOS).
IDL	Idle Mode bit. Setting this bit activates Idle Mode operation in the 80C51BH. (Available only in CHMOS).

If 1s are written to PD and IDL at the same time, PD takes precedence.

*User software should not write 1s to reserved bits. These bits may be used in future MCS-51 products to invoke new features. In that case, the reset or inactive value of the new bit will be 0, and its active value will be 1.

INTERRUPTS:

In order to use any of the interrupts in the MCS-51, the following three steps must be taken.

1. Set the EA (enable all) bit in the IE register to 1.
2. Set the corresponding individual interrupt enable bit in the IE register to 1.
3. Begin the interrupt service routine at the corresponding Vector Address of that interrupt. See Table below.

Interrupt Source	Vector Address
IE0	0003H
TF0	000BH
IE1	0013H
TF1	001BH
RI & TI	0023H
TF2 & EXF2	002BH

In addition, for external interrupts, pins $\overline{INT0}$ and $\overline{INT1}$ (P3.2 and P3.3) must be set to 1, and depending on whether the interrupt is to be level or transition activated, bits IT0 or IT1 in the TCON register may need to be set to 1.

ITx = 0 level activated

ITx = 1 transition activated

IE: INTERRUPT ENABLE REGISTER. BIT ADDRESSABLE.

If the bit is 0, the corresponding interrupt is disabled. If the bit is 1, the corresponding interrupt is enabled.

EA	—	ET2	ES	ET1	EX1	ET0	EX0

EA   IE.7   Disables all interrupts. If EA = 0, no interrupt will be acknowledged. If EA = 1, each interrupt source is individually enabled or disabled by setting or clearing its enable bit.

—   IE.6   Not implemented, reserved for future use.*

ET2   IE.5   Enable or disable the Timer 2 overflow or capture interrupt (8052 only).

ES   IE.4   Enable or disable the serial port interrupt.

ET1   IE.3   Enable or disable the Timer 1 overflow interrupt.

EX1   IE.2   Enable or disable External Interrupt 1.

ET0   IE.1   Enable or disable the Timer 0 overflow interrupt.

EX0   IE.0   Enable or disable External Interrupt 0.

*User software should not write 1s to reserved bits. These bits may be used in future MCS-51 products to invoke new features. In that case, the reset or inactive value of the new bit will be 0, and its active value will be 1.

APPENDIX H: DATA SHEETS   **411**

MCS®-51 PROGRAMMER'S GUIDE AND INSTRUCTION SET

ASSIGNING HIGHER PRIORITY TO ONE OR MORE INTERRUPTS:

In order to assign higher priority to an interrupt the corresponding bit in the IP register must be set to 1.

Remember that while an interrupt service is in progress, it cannot be interrupted by a lower or same level interrupt.

PRIORITY WITHIN LEVEL:

Priority within level is only to resolve simultaneous requests of the same priority level.

From high to low, interrupt sources are listed below:

IE0
TF0
IE1
TF1
RI or TI
TF2 or EXF2

IP: INTERRUPT PRIORITY REGISTER. BIT ADDRESSABLE.

If the bit is 0, the corresponding interrupt has a lower priority and if the bit is 1 the corresponding interrupt has a higher priority.

—	—	PT2	PS	PT1	PX1	PT0	PX0

—	IP. 7	Not implemented, reserved for future use.*
—	IP. 6	Not implemented, reserved for future use.*
PT2	IP. 5	Defines the Timer 2 interrupt priority level (8052 only).
PS	IP. 4	Defines the Serial Port interrupt priority level.
PT1	IP. 3	Defines the Timer 1 interrupt priority level.
PX1	IP. 2	Defines External Interrupt 1 priority level.
PT0	IP. 1	Defines the Timer 0 interrupt priority level.
PX0	IP. 0	Defines the External Interrupt 0 priority level.

*User software should not write 1s to reserved bits. These bits may be used in future MCS-51 products to invoke new features. In that case, the reset or inactive value of the new bit will be 0, and its active value will be 1.

2-13

412

TCON: TIMER/COUNTER CONTROL REGISTER. BIT ADDRESSABLE.

TF1	TR1	TF0	TR0	IE1	IT1	IE0	IT0

TF1   TCON. 7   Timer 1 overflow flag. Set by hardware when the Timer/Counter 1 overflows. Cleared by hardware as processor vectors to the interrupt service routine.

TR1   TCON. 6   Timer 1 run control bit. Set/cleared by software to turn Timer/Counter 1 ON/OFF.

TF0   TCON. 5   Timer 0 overflow flag. Set by hardware when the Timer/Counter 0 overflows. Cleared by hardware as processor vectors to the service routine.

TR0   TCON. 4   Timer 0 run control bit. Set/cleared by software to turn Timer/Counter 0 ON/OFF.

IE1   TCON. 3   External Interrupt 1 edge flag. Set by hardware when External Interrupt edge is detected. Cleared by hardware when interrupt is processed.

IT1   TCON. 2   Interrupt 1 type control bit. Set/cleared by software to specify falling edge/low level triggered External Interrupt.

IE0   TCON. 1   External Interrupt 0 edge flag. Set by hardware when External Interrupt edge detected. Cleared by hardware when interrupt is processed.

IT0   TCON. 0   Interrupt 0 type control bit. Set/cleared by software to specify falling edge/low level triggered External Interrupt.

TMOD: TIMER/COUNTER MODE CONTROL REGISTER. NOT BIT ADDRESSABLE.

GATE	C/$\overline{T}$	M1	M0	GATE	C/$\overline{T}$	M1	M0

                 TIMER 1   TIMER 0

GATE   When TRx (in TCON) is set and GATE = 1, TIMER/COUNTERx will run only while INTx pin is high (hardware control). When GATE = 0, TIMER/COUNTERx will run only while TRx = 1 (software control).

C/$\overline{T}$ Timer or Counter selector. Cleared for Timer operation (input from internal system clock). Set for Counter operation (input from Tx input pin).

M1   Mode selector bit. (NOTE 1)

M0   Mode selector bit. (NOTE 1)

NOTE 1:

M1	M0	Operating Mode	
0	0	0	13-bit Timer (MCS-48 compatible)
0	1	1	16-bit Timer/Counter
1	0	2	8-bit Auto-Reload Timer/Counter
1	1	3	(Timer 0) TL0 is an 8-bit Timer/Counter controlled by the standard Timer 0 control bits, TH0 is an 8-bit Timer and is controlled by Timer 1 control bits.
1	1	3	(Timer 1) Timer/Counter 1 stopped.

TIMER SET-UP

Tables 3 through 6 give some values for TMOD which can be used to set up Timer 0 in different modes.

It is assumed that only one timer is being used at a time. If it is desired to run Timers 0 and 1 simultaneously, in any mode, the value in TMOD for Timer 0 must be ORed with the value shown for Timer 1 (Tables 5 and 6).

For example, if it is desired to run Timer 0 in mode 1 GATE (external control), and Timer 1 in mode 2 COUNTER, then the value that must be loaded into TMOD is 69H (09H from Table 3 ORed with 60H from Table 6).

Moreover, it is assumed that the user, at this point, is not ready to turn the timers on and will do that at a different point in the program by setting bit TRx (in TCON) to 1.

TIMER/COUNTER 0

As a Timer:

Table 3

| MODE | TIMER 0 FUNCTION | TMOD | |
		INTERNAL CONTROL (NOTE 1)	EXTERNAL CONTROL (NOTE 2)
0	13-bit Timer	00H	08H
1	16-bit Timer	01H	09H
2	8-bit Auto-Reload	02H	0AH
3	two 8-bit Timers	03H	0BH

As a Counter:

Table 4

| MODE | COUNTER 0 FUNCTION | TMOD | |
		INTERNAL CONTROL (NOTE 1)	EXTERNAL CONTROL (NOTE 2)
0	13-bit Timer	04H	0CH
1	16-bit Timer	05H	0DH
2	8-bit Auto-Reload	06H	0EH
3	one 8-bit Counter	07H	0FH

NOTES:
1. The Timer is turned ON/OFF by setting/clearing bit TR0 in the software.
2. The Timer is turned ON/OFF by the 1 to 0 transition on INT0 (P3.2) when TR0 = 1 (hardware control).

TIMER/COUNTER 1

As a Timer:

Table 5

| MODE | TIMER 1 FUNCTION | TMOD | |
		INTERNAL CONTROL (NOTE 1)	EXTERNAL CONTROL (NOTE 2)
0	13-bit Timer	00H	80H
1	16-bit Timer	10H	90H
2	8-bit Auto-Reload	20H	A0H
3	does not run	30H	B0H

As a Counter:

Table 6

| MODE | COUNTER 1 FUNCTION | TMOD | |
		INTERNAL CONTROL (NOTE 1)	EXTERNAL CONTROL (NOTE 2)
0	13-bit Timer	40H	C0H
1	16-bit Timer	50H	D0H
2	8-bit Auto-Reload	60H	E0H
3	not available	—	—

NOTES:
1. The Timer is turned ON/OFF by setting/clearing bit TR1 in the software.
2. The Timer is turned ON/OFF by the 1 to 0 transition on INT1 (P3.3) when TR1 = 1 (hardware control).

T2CON: TIMER/COUNTER 2 CONTROL REGISTER. BIT ADDRESSABLE

8052 Only

TF2	EXF2	RCLK	TCLK	EXEN2	TR2	C/$\overline{T2}$	CP/$\overline{RL2}$

TF2   T2CON. 7   Timer 2 overflow flag set by hardware and cleared by software. TF2 cannot be set when either RCLK = 1 or CLK = 1

EXF2   T2CON. 6   Timer 2 external flag set when either a capture or reload is caused by a negative transition on T2EX, and EXEN2 = 1. When Timer 2 interrupt is enabled, EXF2 = 1 will cause the CPU to vector to the Timer 2 interrupt routine. EXF2 must be cleared by software.

RCLK   T2CON. 5   Receive clock flag. When set, causes the Serial Port to use Timer 2 overflow pulses for its receive clock in modes 1 & 3. RCLK = 0 causes Timer 1 overflow to be used for the receive clock.

TLCK   T2CON. 4   Transmit clock flag. When set, causes the Serial Port to use Timer 2 overflow pulses for its transmit clock in modes 1 & 3. TCLK = 0 causes Timer 1 overflows to be used for the transmit clock.

EXEN2   T2CON. 3   Timer 2 external enable flag. When set, allows a capture or reload to occur as a result of negative transition on T2EX if Timer 2 is not being used to clock the Serial Port. EXEN2 = 0 causes Timer 2 to ignore events at T2EX.

TR2   T2CON. 2   Software START/STOP control for Timer 2. A logic 1 starts the Timer.

C/$\overline{T2}$ T2CON. 1   Timer or Counter select.

               0 = Internal Timer. 1 = External Event Counter (falling edge triggered).

CP/$\overline{RL2}$ T2CON. 0   Capture/Reload flag. When set, captures will occur on negative transitions at T2EX if EXEN2 = 1. When cleared, Auto-Reloads will occur either with Timer 2 overflows or negative transitions at T2EX when EXEN2 = 1. When either RCLK = 1 or TCLK = 1, this bit is ignored and the Timer is forced to Auto-Reload on Timer 2 overflow.

TIMER/COUNTER 2 SET-UP

Except for the baud rate generator mode, the values given for T2CON do not include the setting of the TR2 bit. Therefore, bit TR2 must be set, separately, to turn the Timer on.

As a Timer:

Table 7

MODE	T2CON	
	INTERNAL CONTROL (NOTE 1)	EXTERNAL CONTROL (NOTE 2)
16-bit Auto-Reload	00H	08H
16-bit Capture	01H	09H
BAUD rate generator receive & transmit same baud rate	34H	36H
receive only	24H	26H
transmit only	14H	16H

As a Counter:

Table 8

MODE	TMOD	
	INTERNAL CONTROL (NOTE 1)	EXTERNAL CONTROL (NOTE 2)
16-bit Auto-Reload	02H	0AH
16-bit Capture	03H	0BH

NOTES:
1. Capture/Reload occurs only on Timer/Counter overflow.
2. Capture/Reload occurs on Timer/Counter overflow and a 1 to 0 transition on T2EX (P1.1) pin except when Timer 2 is used in the baud rate generating mode.

SCON: SERIAL PORT CONTROL REGISTER. BIT ADDRESSABLE.

SM0	SM1	SM2	REN	TB8	RB8	TI	RI

SM0   SCON. 7   Serial Port mode specifier. (NOTE 1).

SM1   SCON. 6   Serial Port mode specifier. (NOTE 1).

SM2   SCON. 5   Enables the multiprocessor communication feature in modes 2 & 3. In mode 2 or 3, if SM2 is set to 1 then RI will not be activated if the received 9th data bit (RB8) is 0. In mode 1, if SM2 = 1 then RI will not be activated if a valid stop bit was not received. In mode 0, SM2 should be 0. (See Table 9).

REN   SCON. 4   Set/Cleared by software to Enable/Disable reception.

TB8   SCON. 3   The 9th bit that will be transmitted in modes 2 & 3. Set/Cleared by software.

RB8   SCON. 2   In modes 2 & 3, is the 9th data bit that was received. In mode 1, if SM2 = 0, RB8 is the stop bit that was received. In mode 0, RB8 is not used.

TI   SCON. 1   Transmit interrupt flag. Set by hardware at the end of the 8th bit time in mode 0, or at the beginning of the stop bit in the other modes. Must be cleared by software.

RI   SCON. 0   Receive interrupt flag. Set by hardware at the end of the 8th bit time in mode 0, or halfway through the stop bit time in the other modes (except see SM2). Must be cleared by software.

NOTE 1:

SM0	SM1	Mode	Description	Baud Rate
0	0	0	SHIFT REGISTER	Fosc./12
0	1	1	8-Bit UART	Variable
1	0	2	9-Bit UART	Fosc./64 OR Fosc./32
1	1	3	9-Bit UART	Variable

SERIAL PORT SET-UP:

Table 9

MODE	SCON	SM2 VARIATION
0	10H	Single Processor Environment (SM2 = 0)
1	50H	
2	90H	
3	D0H	
0	NA	Multiprocessor Environment (SM2 = 1)
1	70H	
2	B0H	
3	F0H	

GENERATING BAUD RATES

Serial Port in Mode 0:

Mode 0 has a fixed baud rate which is 1/12 of the oscillator frequency. To run the serial port in this mode none of the Timer/Counters need to be set up. Only the SCON register needs to be defined.

$$\text{Baud Rate} = \frac{\text{Osc Freq}}{12}$$

Serial Port in Mode 1:

Mode 1 has a variable baud rate. The baud rate can be generated by either Timer 1 or Timer 2 (8052 only).

2-19

USING TIMER/COUNTER 1 TO GENERATE BAUD RATES:

For this purpose, Timer 1 is used in mode 2 (Auto-Reload). Refer to Timer Setup section of this chapter.

$$\text{Baud Rate} = \frac{K \times \text{Oscillator Freq.}}{32 \times 12 \times [256 - (\text{TH1})]}$$

If SMOD = 0, then K = 1.
If SMOD = 1, then K = 2. (SMOD is the PCON register).

Most of the time the user knows the baud rate and needs to know the reload value for TH1.
Therefore, the equation to calculate TH1 can be written as:

$$\text{TH1} = 256 - \frac{K \times \text{Osc Freq.}}{384 \times \text{baud rate}}$$

TH1 must be an integer value. Rounding off TH1 to the nearest integer may not produce the desired baud rate. In this case, the user may have to choose another crystal frequency.

Since the PCON register is not bit addressable, one way to set the bit is logical ORing the PCON register. (ie, ORL PCON, #80H). The address of PCON is 87H.

USING TIMER/COUNTER 2 TO GENERATE BAUD RATES:

For this purpose, Timer 2 must be used in the baud rate generating mode. Refer to Timer 2 Setup Table in this chapter. If Timer 2 is being clocked through pin T2 (P1.0) the baud rate is:

$$\text{Baud Rate} = \frac{\text{Timer 2 Overflow Rate}}{16}$$

And if it is being clocked internally the baud rate is:

$$\text{Baud Rate} = \frac{\text{Osc Freq}}{32 \times [65536 - (\text{RCAP2H, RCAP2L})]}$$

To obtain the reload value for RCAP2H and RCAP2L the above equation can be rewritten as:

$$\text{RCAP2H, RCAP2L} = 65536 - \frac{\text{Osc Freq}}{32 \times \text{Baud Rate}}$$

SERIAL PORT IN MODE 2:

The baud rate is fixed in this mode and is $\frac{1}{32}$ or $\frac{1}{64}$ of the oscillator frequency depending on the value of the SMOD bit in the PCON register.

In this mode none of the Timers are used and the clock comes from the internal phase 2 clock.

SMOD = 1, Baud Rate = $\frac{1}{32}$ Osc Freq.

SMOD = 0, Baud Rate = $\frac{1}{64}$ Osc Freq.

To set the SMOD bit: ORL   PCON, #80H. The address of PCON is 87H.

SERIAL PORT IN MODE 3:

The baud rate in mode 3 is variable and sets up exactly the same as in mode 1.

MCS®-51 INSTRUCTION SET

Table 10. 8051 Instruction Set Summary

Interrupt Response Time: Refer to Hardware Description Chapter.

Instructions that Affect Flag Settings[1]

Instruction	Flag			Instruction	Flag		
	C	OV	AC		C	OV	AC
ADD	X	X	X	CLR C	O		
ADDC	X	X	X	CPL C	X		
SUBB	X	X	X	ANL C,bit	X		
MUL	O	X		ANL C,/bit	X		
DIV	O	X		ORL C,bit	X		
DA	X			ORL C,bit	X		
RRC	X			MOV C,bit	X		
RLC	X			CJNE	X		
SETB C	1						

[1]Note that operations on SFR byte address 208 or bit addresses 209-215 (i.e., the PSW or bits in the PSW) will also affect flag settings.

Note on instruction set and addressing modes:

Rn — Register R7–R0 of the currently selected Register Bank.

direct — 8-bit internal data location's address. This could be an Internal Data RAM location (0–127) or a SFR [i.e., I/O port, control register, status register, etc. (128–255)].

@Ri — 8-bit internal data RAM location (0–255) addressed indirectly through register R1 or R0.

#data — 8-bit constant included in instruction.

#data 16 — 16-bit constant included in instruction.

addr 16 — 16-bit destination address. Used by LCALL & LJMP. A branch can be anywhere within the 64K-byte Program Memory address space.

addr 11 — 11-bit destination address. Used by ACALL & AJMP. The branch will be within the same 2K-byte page of program memory as the first byte of the following instruction.

rel — Signed (two's complement) 8-bit offset byte. Used by SJMP and all conditional jumps. Range is −128 to +127 bytes relative to first byte of the following instruction.

bit — Direct Addressed bit in Internal Data RAM or Special Function Register.

Mnemonic		Description	Byte	Oscillator Period
ARITHMETIC OPERATIONS				
ADD	A,Rn	Add register to Accumulator	1	12
ADD	A,direct	Add direct byte to Accumulator	2	12
ADD	A,@Ri	Add indirect RAM to Accumulator	1	12
ADD	A,#data	Add immediate data to Accumulator	2	12
ADDC	A,Rn	Add register to Accumulator with Carry	1	12
ADDC	A,direct	Add direct byte to Accumulator with Carry	2	12
ADDC	A,@Ri	Add indirect RAM to Accumulator with Carry	1	12
ADDC	A,#data	Add immediate data to Acc with Carry	2	12
SUBB	A,Rn	Subtract Register from Acc with borrow	1	12
SUBB	A,direct	Subtract direct byte from Acc with borrow	2	12
SUBB	A,@Ri	Subtract indirect RAM from ACC with borrow	1	12
SUBB	A,#data	Subtract immediate data from Acc with borrow	2	12
INC	A	Increment Accumulator	1	12
INC	Rn	Increment register	1	12
INC	direct	Increment direct byte	2	12
INC	@Ri	Increment direct RAM	1	12
DEC	A	Decrement Accumulator	1	12
DEC	Rn	Decrement Register	1	12
DEC	direct	Decrement direct byte	2	12
DEC	@Ri	Decrement indirect RAM	1	12

All mnemonics copyrighted © Intel Corporation 1980

2-21

418

Table 10. 8051 Instruction Set Summary (Continued)

Mnemonic		Description	Byte	Oscillator Period
ARITHMETIC OPERATIONS (Continued)				
INC	DPTR	Increment Data Pointer	1	24
MUL	AB	Multiply A & B	1	48
DIV	AB	Divide A by B	1	48
DA	A	Decimal Adjust Accumulator	1	12
LOGICAL OPERATIONS				
ANL	A,Rn	AND Register to Accumulator	1	12
ANL	A,direct	AND direct byte to Accumulator	2	12
ANL	A,@Ri	AND indirect RAM to Accumulator	1	12
ANL	A,#data	AND immediate data to Accumulator	2	12
ANL	direct,A	AND Accumulator to direct byte	2	12
ANL	direct,#data	AND immediate data to direct byte	3	24
ORL	A,Rn	OR register to Accumulator	1	12
ORL	A,direct	OR direct byte to Accumulator	2	12
ORL	A,@Ri	OR indirect RAM to Accumulator	1	12
ORL	A,#data	OR immediate data to Accumulator	2	12
ORL	direct,A	OR Accumulator to direct byte	2	12
ORL	direct,#data	OR immediate data to direct byte	3	24
XRL	A,Rn	Exclusive-OR register to Accumulator	1	12
XRL	A,direct	Exclusive-OR direct byte to Accumulator	2	12
XRL	A,@Ri	Exclusive-OR indirect RAM to Accumulator	1	12
XRL	A,#data	Exclusive-OR immediate data to Accumulator	2	12
XRL	direct,A	Exclusive-OR Accumulator to direct byte	2	12
XRL	direct,#data	Exclusive-OR immediate data to direct byte	3	24
CLR	A	Clear Accumulator	1	12
CPL	A	Complement Accumulator	1	12

Mnemonic		Description	Byte	Oscillator Period
LOGICAL OPERATIONS (Continued)				
RL	A	Rotate Accumulator Left	1	12
RLC	A	Rotate Accumulator Left through the Carry	1	12
RR	A	Rotate Accumulator Right	1	12
RRC	A	Rotate Accumulator Right through the Carry	1	12
SWAP	A	Swap nibbles within the Accumulator	1	12
DATA TRANSFER				
MOV	A,Rn	Move register to Accumulator	1	12
MOV	A,direct	Move direct byte to Accumulator	2	12
MOV	A,@Ri	Move indirect RAM to Accumulator	1	12
MOV	A,#data	Move immediate data to Accumulator	2	12
MOV	Rn,A	Move Accumulator to register	1	12
MOV	Rn,direct	Move direct byte to register	2	24
MOV	Rn,#data	Move immediate data to register	2	12
MOV	direct,A	Move Accumulator to direct byte	2	12
MOV	direct,Rn	Move register to direct byte	2	24
MOV	direct,direct	Move direct byte to direct	3	24
MOV	direct,@Ri	Move indirect RAM to direct byte	2	24
MOV	direct,#data	Move immediate data to direct byte	3	24
MOV	@Ri,A	Move Accumulator to indirect RAM	1	12

All mnemonics copyrighted © Intel Corporation 1980

APPENDIX H: DATA SHEETS   **419**

Table 10. 8051 Instruction Set Summary (Continued)

Mnemonic		Description	Byte	Oscillator Period
DATA TRANSFER (Continued)				
MOV	@Ri,direct	Move direct byte to indirect RAM	2	24
MOV	@Ri, # data	Move immediate data to indirect RAM	2	12
MOV	DPTR, # data16	Load Data Pointer with a 16-bit constant	3	24
MOVC	A,@A + DPTR	Move Code byte relative to DPTR to Acc	1	24
MOVC	A,@A + PC	Move Code byte relative to PC to Acc	1	24
MOVX	A,@Ri	Move External RAM (8-bit addr) to Acc	1	24
MOVX	A,@DPTR	Move External RAM (16-bit addr) to Acc	1	24
MOVX	@Ri,A	Move Acc to External RAM (8-bit addr)	1	24
MOVX	@DPTR,A	Move Acc to External RAM (16-bit addr)	1	24
PUSH	direct	Push direct byte onto stack	2	24
POP	direct	Pop direct byte from stack	2	24
XCH	A,Rn	Exchange register with Accumulator	1	12
XCH	A,direct	Exchange direct byte with Accumulator	2	12
XCH	A,@Ri	Exchange indirect RAM with Accumulator	1	12
XCHD	A,@Ri	Exchange low-order Digit indirect RAM with Acc	1	12

Mnemonic		Description	Byte	Oscillator Period
BOOLEAN VARIABLE MANIPULATION				
CLR	C	Clear Carry	1	12
CLR	bit	Clear direct bit	2	12
SETB	C	Set Carry	1	12
SETB	bit	Set direct bit	2	12
CPL	C	Complement Carry	1	12
CPL	bit	Complement direct bit	2	12
ANL	C,bit	AND direct bit to CARRY	2	24
ANL	C,/bit	AND complement of direct bit to Carry	2	24
ORL	C,bit	OR direct bit to Carry	2	24
ORL	C,/bit	OR complement of direct bit to Carry	2	24
MOV	C,bit	Move direct bit to Carry	2	12
MOV	bit,C	Move Carry to direct bit	2	24
JC	rel	Jump if Carry is set	2	24
JNC	rel	Jump if Carry not set	2	24
JB	bit,rel	Jump if direct Bit is set	3	24
JNB	bit,rel	Jump if direct Bit is Not set	3	24
JBC	bit,rel	Jump if direct Bit is set & clear bit	3	24
PROGRAM BRANCHING				
ACALL	addr11	Absolute Subroutine Call	2	24
LCALL	addr16	Long Subroutine Call	3	24
RET		Return from Subroutine	1	24
RETI		Return from interrupt	1	24
AJMP	addr11	Absolute Jump	2	24
LJMP	addr16	Long Jump	3	24
SJMP	rel	Short Jump (relative addr)	2	24

All mnemonics copyrighted © Intel Corporation 1980

Table 10. 8051 Instruction Set Summary (Continued)

Mnemonic		Description	Byte	Oscillator Period
PROGRAM BRANCHING (Continued)				
JMP	@A + DPTR	Jump indirect relative to the DPTR	1	24
JZ	rel	Jump if Accumulator is Zero	2	24
JNZ	rel	Jump if Accumulator is Not Zero	2	24
CJNE	A,direct,rel	Compare direct byte to Acc and Jump if Not Equal	3	24
CJNE	A, # data,rel	Compare immediate to Acc and Jump if Not Equal	3	24

Mnemonic		Description	Byte	Oscillator Period
PROGRAM BRANCHING (Continued)				
CJNE	Rn, # data,rel	Compare immediate to register and Jump if Not Equal	3	24
CJNE	@Ri, # data,rel	Compare immediate to indirect and Jump if Not Equal	3	24
DJNZ	Rn,rel	Decrement register and Jump if Not Zero	2	24
DJNZ	direct,rel	Decrement direct byte and Jump if Not Zero	3	24
NOP		No Operation	1	12

All mnemonics copyrighted © Intel Corporation 1980

APPENDIX H: DATA SHEETS   **421**

Table 11. Instruction Opcodes in Hexadecimal Order

Hex Code	Number of Bytes	Mnemonic	Operands	Hex Code	Number of Bytes	Mnemonic	Operands
00	1	NOP		33	1	RLC	A
01	2	AJMP	code addr	34	2	ADDC	A, # data
02	3	LJMP	code addr	35	2	ADDC	A,data addr
03	1	RR	A	36	1	ADDC	A,@R0
04	1	INC	A	37	1	ADDC	A,@R1
05	2	INC	data addr	38	1	ADDC	A,R0
06	1	INC	@R0	39	1	ADDC	A,R1
07	1	INC	@R1	3A	1	ADDC	A,R2
08	1	INC	R0	3B	1	ADDC	A,R3
09	1	INC	R1	3C	1	ADDC	A,R4
0A	1	INC	R2	3D	1	ADDC	A,R5
0B	1	INC	R3	3E	1	ADDC	A,R6
0C	1	INC	R4	3F	1	ADDC	A,R7
0D	1	INC	R5	40	2	JC	code addr
0E	1	INC	R6	41	2	AJMP	code addr
0F	1	INC	R7	42	2	ORL	data addr,A
10	3	JBC	bit addr, code addr	43	3	ORL	data addr, # data
11	2	ACALL	code addr	44	2	ORL	A, # data
12	3	LCALL	code addr	45	2	ORL	A,data addr
13	1	RRC	A	46	1	ORL	A,@R0
14	1	DEC	A	47	1	ORL	A,,@R1
15	2	DEC	data addr	48	1	ORL	A,R0
16	1	DEC	@R0	49	1	ORL	A,R1
17	1	DEC	@R1	4A	1	ORL	A,R2
18	1	DEC	R0	4B	1	ORL	A,R3
19	1	DEC	R1	4C	1	ORL	A,R4
1A	1	DEC	R2	4D	1	ORL	A,R5
1B	1	DEC	R3	4E	1	ORL	A,R6
1C	1	DEC	R4	4F	1	ORL	A,R7
1D	1	DEC	R5	50	2	JNC	code addr
1E	1	DEC	R6	51	2	ACALL	code addr
1F	1	DEC	R7	52	2	ANL	data addr,A
20	3	JB	bit addr, code addr	53	3	ANL	data addr, # data
21	2	AJMP	code addr	54	2	ANL	A, # data
22	1	RET		55	2	ANL	A,data addr
23	1	RL	A	56	1	ANL	A,@R0
24	2	ADD	A, # data	57	1	ANL	A,@R1
25	2	ADD	A,data addr	58	1	ANL	A,R0
26	1	ADD	A,@R0	59	1	ANL	A,R1
27	1	ADD	A,@R1	5A	1	ANL	A,R2
28	1	ADD	A,R0	5B	1	ANL	A,R3
29	1	ADD	A,R1	5C	1	ANL	A,R4
2A	1	ADD	A,R2	5D	1	ANL	A,R5
2B	1	ADD	A,R3	5E	1	ANL	A,R6
2C	1	ADD	A,R4	5F	1	ANL	A,R7
2D	1	ADD	A,R5	60	2	JZ	code addr
2E	1	ADD	A,R6	61	2	AJMP	code addr
2F	1	ADD	A,R7	62	2	XRL	data addr,A
30	3	JNB	bit addr, code addr	63	3	XRL	data addr, # data
31	2	ACALL	code addr	64	2	XRL	A, # data
32	1	RETI		65	2	XRL	A,data addr

2-25

422

Table 11. Instruction Opcodes in Hexadecimal Order (Continued)

Hex Code	Number of Bytes	Mnemonic	Operands		Hex Code	Number of Bytes	Mnemonic	Operands
66	1	XRL	A,@R0		99	1	SUBB	A,R1
67	1	XRL	A,@R1		9A	1	SUBB	A,R2
68	1	XRL	A,R0		9B	1	SUBB	A,R3
69	1	XRL	A,R1		9C	1	SUBB	A,R4
6A	1	XRL	A,R2		9D	1	SUBB	A,R5
6B	1	XRL	A,R3		9E	1	SUBB	A,R6
6C	1	XRL	A,R4		9F	1	SUBB	A,R7
6D	1	XRL	A,R5		A0	2	ORL	C,/bit addr
6E	1	XRL	A,R6		A1	2	AJMP	code addr
6F	1	XRL	A,R7		A2	2	MOV	C,bit addr
70	2	JNZ	code addr		A3	1	INC	DPTR
71	2	ACALL	code addr		A4	1	MUL	AB
72	2	ORL	C,bit addr		A5		reserved	
73	1	JMP	@A + DPTR		A6	2	MOV	@R0,data addr
74	2	MOV	A,#data		A7	2	MOV	@R1,data addr
75	3	MOV	data addr,#data		A8	2	MOV	R0,data addr
76	2	MOV	@R0,#data		A9	2	MOV	R1,data addr
77	2	MOV	@R1,#data		AA	2	MOV	R2,data addr
78	2	MOV	R0,#data		AB	2	MOV	R3,data addr
79	2	MOV	R1,#data		AC	2	MOV	R4,data addr
7A	2	MOV	R2,#data		AD	2	MOV	R5,data addr
7B	2	MOV	R3,#data		AE	2	MOV	R6,data addr
7C	2	MOV	R4,#data		AF	2	MOV	R7,data addr
7D	2	MOV	R5,#data		B0	2	ANL	C,/bit addr
7E	2	MOV	R6,#data		B1	2	ACALL	code addr
7F	2	MOV	R7,#data		B2	2	CPL	bit addr
80	2	SJMP	code addr		B3	1	CPL	C
81	2	AJMP	code addr		B4	3	CJNE	A,#data,code addr
82	2	ANL	C,bit addr		B5	3	CJNE	A,data addr,code addr
83	1	MOVC	A,@A + PC		B6	3	CJNE	@R0,#data,code addr
84	1	DIV	AB		B7	3	CJNE	@R1,#data,code addr
85	3	MOV	data addr, data addr		B8	3	CJNE	R0,#data,code addr
86	2	MOV	data addr,@R0		B9	3	CJNE	R1,#data,code addr
87	2	MOV	data addr,@R1		BA	3	CJNE	R2,#data,code addr
88	2	MOV	data addr,R0		BB	3	CJNE	R3,#data,code addr
89	2	MOV	data addr,R1		BC	3	CJNE	R4,#data,code addr
8A	2	MOV	data addr,R2		BD	3	CJNE	R5,#data,code addr
8B	2	MOV	data addr,R3		BE	3	CJNE	R6,#data,code addr
8C	2	MOV	data addr,R4		BF	3	CJNE	R7,#data,code addr
8D	2	MOV	data addr,R5		C0	2	PUSH	data addr
8E	2	MOV	data addr,R6		C1	2	AJMP	code addr
8F	2	MOV	data addr,R7		C2	2	CLR	bit addr
90	3	MOV	DPTR,#data		C3	1	CLR	C
91	2	ACALL	code addr		C4	1	SWAP	A
92	2	MOV	bit addr,C		C5	2	XCH	A,data addr
93	1	MOVC	A,@A + DPTR		C6	1	XCH	A,@R0
94	2	SUBB	A,#data		C7	1	XCH	A,@R1
95	2	SUBB	A,data addr		C8	1	XCH	A,R0
96	1	SUBB	A,@R0		C9	1	XCH	A,R1
97	1	SUBB	A,@R1		CA	1	XCH	A,R2
98	1	SUBB	A,R0		CB	1	XCH	A,R3

APPENDIX H: DATA SHEETS 423

Table 11. Instruction Opcodes in Hexadecimal Order (Continued)

Hex Code	Number of Bytes	Mnemonic	Operands	Hex Code	Number of Bytes	Mnemonic	Operands
CC	1	XCH	A,R4	E6	1	MOV	A,@R0
CD	1	XCH	A,R5	E7	1	MOV	A,@R1
CE	1	XCH	A,R6	E8	1	MOV	A,R0
CF	1	XCH	A,R7	E9	1	MOV	A,R1
D0	2	POP	data addr	EA	1	MOV	A,R2
D1	2	ACALL	code addr	EB	1	MOV	A,R3
D2	2	SETB	bit addr	EC	1	MOV	A,R4
D3	1	SETB	C	ED	1	MOV	A,R5
D4	1	DA	A	EE	1	MOV	A,R6
D5	3	DJNZ	data addr,code addr	EF	1	MOV	A,R7
D6	1	XCHD	A,@R0	F0	1	MOVX	@DPTR,A
D7	1	XCHD	A,@R1	F1	2	ACALL	code addr
D8	2	DJNZ	R0,code addr	F2	1	MOVX	@R0,A
D9	2	DJNZ	R1,code addr	F3	1	MOVX	@R1,A
DA	2	DJNZ	R2,code addr	F4	1	CPL	A
DB	2	DJNZ	R3,code addr	F5	2	MOV	data addr,A
DC	2	DJNZ	R4,code addr	F6	1	MOV	@R0,A
DD	2	DJNZ	R5,code addr	F7	1	MOV	@R1,A
DE	2	DJNZ	R6,code addr	F8	1	MOV	R0,A
DF	2	DJNZ	R7,code addr	F9	1	MOV	R1,A
E0	1	MOVX	A,@DPTR	FA	1	MOV	R2,A
E1	2	AJMP	code addr	FB	1	MOV	R3,A
E2	1	MOVX	A,@R0	FC	1	MOV	R4,A
E3	1	MOVX	A,@R1	FD	1	MOV	R5,A
E4	1	CLR	A	FE	1	MOV	R6,A
E5	2	MOV	A,data addr	FF	1	MOV	R7,A

2-27

Instruction Set Summary

	0	1	2	3	4	5	6	7
0	NOP	JBC bit,rel [3B, 2C]	JB bit, rel [3B, 2C]	JNB bit, rel [3B, 2C]	JC rel [2B, 2C]	JNC rel [2B, 2C]	JZ rel [2B, 2C]	JNZ rel [2B, 2C]
1	AJMP (P0) [2B, 2C]	ACALL (P0) [2B, 2C]	AJMP (P1) [2B, 2C]	ACALL (P1) [2B, 2C]	AJMP (P2) [2B, 2C]	ACALL (P2) [2B, 2C]	AJMP (P3) [2B, 2C]	ACALL (P3) [2B, 2C]
2	LJMP addr16 [3B, 2C]	LCALL addr16 [3B, 2C]	RET [2C]	RETI [2C]	ORL dir, A [2B]	ANL dir, A [2B]	XRL dir, a [2B]	ORL C, bit [2B, 2C]
3	RR A	RRC A	RL A	RLC A	ORL dir, #data [3B, 2C]	ANL dir, #data [3B, 2C]	XRL dir, #data [3B, 2C]	JMP @A + DPTR [2C]
4	INC A	DEC A	ADD A, #data [2B]	ADDC A, #data [2B]	ORL A, #data [2B]	ANL A, #data [2B]	XRL A, #data [2B]	MOV A, #data [2B]
5	INC dir [2B]	DEC dir [2B]	ADD A, dir [2B]	ADDC A, dir [2B]	ORL A, dir [2B]	ANL A, dir [2B]	XRL A, dir [2B]	MOV dir, #data [3B, 2C]
6	INC @R0	DEC @R0	ADD A, @R0	ADDC A, @R0	ORL A, @R0	ANL A, @R0	XRL A, @R0	MOV @R0, @data [2B]
7	INC @R1	DEC @R1	ADD A, @R1	ADDC A, @R1	ORL A, @R1	ANL A, @R1	XRL A, @R1	MOV @R1, #data [2B]
8	INC R0	DEC R0	ADD A, R0	ADDC A, R0	ORL A, R0	ANL A, R0	XRL A, R0	MOV R0, #data [2B]
9	INC R1	DEC R1	ADD A, R1	ADDC A, R1	ORL A, R1	ANL A, R1	XRL A, R1	MOV R1, #data [2B]
A	INC R2	DEC R2	ADD A, R2	ADDC A, R2	ORL A, R2	ANL A, R2	XRL A, R2	MOV R2, #data [2B]
B	INC R3	DEC R3	ADD A, R3	ADDC A, R3	ORL A, R3	ANL A, R3	XRL A, R3	MOV R3, #data [2B]
C	INC R4	DEC R4	ADD A, R4	ADDC A, R4	ORL A, R4	ANL A, R4	XRL A, R4	MOV R4, #data [2B]
D	INC R5	DEC R5	ADD A, R5	ADDC A, R5	ORL A, R5	ANL A, R5	XRL A, R5	MOV R5, #data [2B]
E	INC R6	DEC R6	ADD A, R6	ADDC A, R6	ORL A, R6	ANL A, R6	XRL A, R6	MOV R6, #data [2B]
F	INC R7	DEC R7	ADD A, R7	ADDC A, R7	ORL A, R7	ANL A, R7	XRL A, R7	MOV R7, #data [2B]

Key:

[2B] = 2 Byte, [3B] = 3 Byte, [2C] = 2 Cycle, [4C] = 4 Cycle, Blank = 1 byte/1 cycle

Instruction Set

Instruction Set Summary (Continued)

	8	9	A	B	C	D	E	F
0	SJMP REL [2B, 2C]	MOV DPTR,# data 16 [3B, 2C]	ORL C, /bit [2B, 2C]	ANL C, /bit [2B, 2C]	PUSH dir [2B, 2C]	POP dir [2B, 2C]	MOVX A, @DPTR [2C]	MOVX @DPTR, A [2C]
1	AJMP (P4) [2B, 2C]	ACALL (P4) [2B, 2C]	AJMP (P5) [2B, 2C]	ACALL (P5) [2B, 2C]	AJMP (P6) [2B, 2C]	ACALL (P6) [2B, 2C]	AJMP (P7) [2B, 2C]	ACALL (P7) [2B, 2C]
2	ANL C, bit [2B, 2C]	MOV bit, C [2B, 2C]	MOV C, bit [2B]	CPL bit [2B]	CLR bit [2B]	SETB bit [2B]	MOVX A, @R0 [2C]	MOVX wR0, A [2C]
3	MOVC A, @A + PC [2C]	MOVC A, @A + DPTR [2C]	INC DPTR [2C]	CPL C	CLR C	SETB C	MOVX A, @RI [2C]	MOVX @RI, A [2C]
4	DIV AB [2B, 4C]	SUBB A, #data [2B]	MUL AB [4C]	CJNE A, #data, rel [3B, 2C]	SWAP A	DA A	CLR A	CPL A
5	MOV dir, dir [3B, 2C]	SUBB A, dir [2B]		CJNE A, dir, rel [3B, 2C]	XCH A, dir [2B]	DJNZ dir, rel [3B, 2C]	MOV A, dir [2B]	MOV dir, A [2B]
6	MOV dir, @R0 [2B, 2C]	SUBB A, @R0	MOV @R0, dir [2B, 2C]	CJNE @R0, #data, rel [3B, 2C]	XCH A, @R0	XCHD A, @R0	MOV A, @R0	MOV @R0, A
7	MOV dir, @R1 [2B, 2C]	SUBB A, @R1	MOV @R1, dir [2B, 2C]	CJNE @R1, #data, rel [3B, 2C]	XCH A, @R1	XCHD A, @R1	MOV A, @R1	MOV @R1, A
8	MOV dir, R0 [2B, 2C]	SUBB A, R0	MOV R0, dir [2B, 2C]	CJNE R0, #data, rel [3B, 2C]	XCH A, R0	DJNZ R0, rel [2B, 2C]	MOV A, R0	MOV R0, A
9	MOV dir, R1 [2B, 2C]	SUBB A, R1	MOV R1, dir [2B, 2C]	CJNE R1, #data, rel [3B, 2C]	XCH A, R1	DJNZ R1, rel [2B, 2C]	MOV A, R1	MOV R1, A
A	MOV dir, R2 [2B, 2C]	SUBB A, R2	MOV R2, dir [2B, 2C]	CJNE R2, #data, rel [3B, 2C]	XCH A, R2	DJNZ R2, rel [2B, 2C]	MOV A, R2	MOV R2, A
B	MOV dir, R3 [2B, 2C]	SUBB A, R3	MOV R3, dir [2B, 2C]	CJNE R3, #data, rel [3B, 2C]	XCH A, R3	DJNZ R3, rel [2B, 2C]	MOV A, R3	MOV R3, A
C	MOV dir, R4 [2B, 2C]	SUBB A, R4	MOV R4, dir [2B, 2C]	CJNE R4, #data, rel [3B, 2C]	XCH A, R4	DJNZ R4, rel [2B, 2C]	MOV A, R4	MOV R4, A
D	MOV dir, R5 [2B, 2C]	SUBB A, R5	MOV R5, dir [2B, 2C]	CJNE R5, #data, rel [3B, 2C]	XCH A, R5	DJNZ R5, rel [2B, 2C]	MOV A, R5	MOV R5, A
E	MOV dir, R6 [2B, 2C]	SUBB A, R6	MOV R6, dir [2B, 2C]	CJNE R6, #data, rel [3B, 2C]	XCH A, R6	DJNZ R6, rel [2B, 2C]	MOV A, R6	MOV R6. A
F	MOV dir, R7 [2B, 2C]	SUBB A, R7	MOV R7, dir [2B, 2C]	CJNE R7, #data, rel [3B, 2C]	XCH A, R7	DJNZ R7, rel [2B, 2C]	MOV A, R7	MOV R7, A

Key:

[2B] = 2 Byte, [3B] = 3 Byte, [2C] = 2 Cycle, [4C] = 4 Cycle, Blank = 1 byte/1 cycle

PACKAGE TYPES

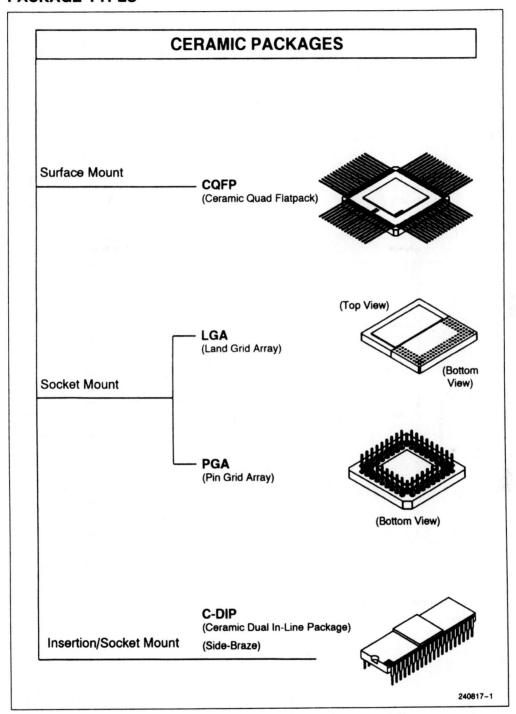

CERAMIC PACKAGES

Surface Mount ──── **CQFP**
(Ceramic Quad Flatpack)

LGA
(Land Grid Array)

(Top View)

(Bottom View)

Socket Mount

PGA
(Pin Grid Array)

(Bottom View)

C-DIP
(Ceramic Dual In-Line Package)
(Side-Braze)

Insertion/Socket Mount

240817-1

1-3

(Reprinted by permission of Intel Corporation, Copyright Intel Corp. 1992)

PACKAGE TYPES (Continued)

PLASTIC PACKAGES
Surface Mount

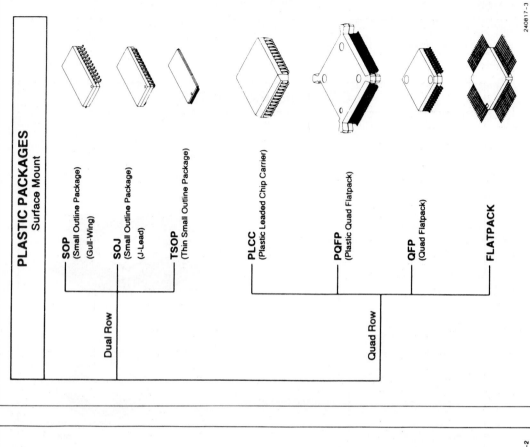

SOP (Small Outline Package) (Gull-Wing)

SOJ (Small Outline Package) (J-Lead)

TSOP (Thin Small Outline Package)

Dual Row

PLCC (Plastic Leaded Chip Carrier)

PQFP (Plastic Quad Flatpack)

QFP (Quad Flatpack)

FLATPACK

Quad Row

240817-3

1-5

PACKAGE TYPES (Continued)

LEADLESS CHIP CARRIER PACKAGES

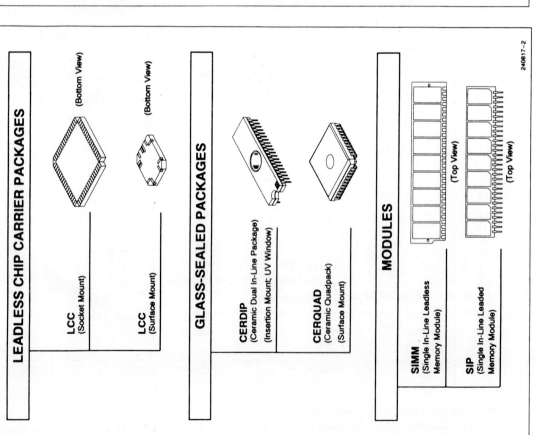

(Bottom View)

LCC (Socket Mount)

(Bottom View)

LCC (Surface Mount)

GLASS-SEALED PACKAGES

CERDIP (Ceramic Dual In-Line Package) (Insertion Mount; UV Window)

CERQUAD (Ceramic Quadpack) (Surface Mount)

MODULES

(Top View)

(Top View)

SIMM (Single In-Line Leadless Memory Module)

SIP (Single In-Line Leaded Memory Module)

240817-2

1-4

428

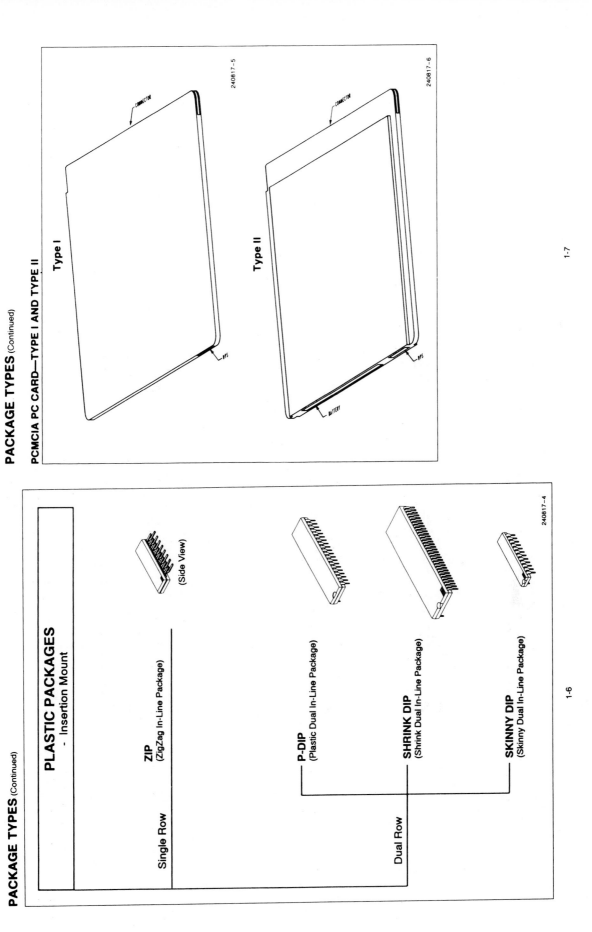

PACKAGE TYPES (Continued)

PCMCIA PC CARD—TYPE I AND TYPE II

Type I

CONNECTOR

PPS

240817-5

Type II

CONNECTOR

PPS

BATTERY

240817-6

1-7

PACKAGE TYPES (Continued)

PLASTIC PACKAGES
- Insertion Mount

Single Row

ZIP
(ZigZag In-Line Package)

(Side View)

Dual Row

P-DIP
(Plastic Dual In-Line Package)

SHRINK DIP
(Shrink Dual In-Line Package)

SKINNY DIP
(Skinny Dual In-Line Package)

240817-4

1-6

INDEX

!

8031 microcontroller   30
    interfacing with ROM   287-291
8051 microcontroller   28
    addressing modes   96, 98
    family members   29, 30
    features of the 8051   28
    history of the 8051   28
    OTP version   32
    pin description   84
    RAM memory map   53
    register banks   53, 54, 55
    register values on RESET   85
    registers   36
    ROM memory map   46, 47
    stack   56
8052 microcontroller   29
8255 chip   304-311
8751 microcontroller   30

A

abs file   42, 43
AC flag   50, 51
ACALL instruction   71, 75, 326
ADC devices   243-251
    connection to 8255   243-245
    testing   246-247
ADC804 chip   243
ADC808 chip   250, 251
ADD instruction   38, 51, 110, 111,
    326, 327, 328
ADDC instruction   112, 114, 329
address bus   14-15
addressing modes   96-98
    direct addressing mode   98, 101
    immediate addressing mode   96
    indexed addressing mode   104, 106
    register addressing mode   97
    register indirect addressing mode
        101, 103
AJMP instruction   329

ALE pin   87
AND gate   9
ANL instruction   128, 329, 330, 331
ASCII   7, 8
    ASCII programming   137
    ASCII table   401
asm file   42, 43
assemblers   40, 402-403
AT89C51 microcontroller   30

B

baud rate   187
BCD number systems   112-113
    BCD addition and correction   114
    BCD applications   137
binary numbers   2
    addition   6
    representation   3
bit   13
buffers   383
bus   14
byte   13

C

CALL instruction   57, 71-75
    See also ACALL, LCALL.
Centronics connector   320-321
checksum byte   281
CJNE instruction   131, 132, 331, 332
CLR instruction   144, 333
comments   41
control bus   14
conversion
    ASCII to packed BCD   138
    binary to decimal   3
    binary to hex   4
    decimal to binary   2
    decimal to hex   4
    hex to binary   4
    hex to decimal   5
    packed BCD to ASCII   138
counters. See timers.

CPL instruction   116, 130, 144, 333
CPU   14-15
crosstalk   394
currents   391
CY flag   50, 51, 111-124, 150, 328

D

DA instruction   113, 138, 334, 335
DAC808 chip 267
DAC devices   266-270
    programmimg   270
daisy chain   367
data bus   14
DB directive   48
DB-25 pins   187, 320
DB-9 signals   187-189
DEC instruction   335
decoders   12
directives   40, 41, 47
DIV instruction   118, 335, 336
DJNZ instruction   66, 336
DRAM   282
DS5000 microcontroller   31
duplex transmission   185

E

EA pin   86
embedded systems   25-26
END directive   49
EPROM   276
EQU directive   48

F

fan-out   382, 384, 390
flag register   50, 110-124
flip-flops   12
flowcharts   396-399

G

gigabyte   13
ground bounce   392

H

half duplex   185
hex file   42, 371-374
hexadecimal numbers   4
    addition, subtraction   7

I

I/O ports   87, 91-93, 385-390
    fan-out   390
    programming   91-93, 306-311
    reading input pin   385
    reading internal latch of output port
        152-154
    reading status of input pin   152-154
    writing to the ports   386
IC technology   376-381
idle mode   391
INC instruction   116, 336, 337
interrupts 208-234
    disabling   210
    edge-triggered interrupts   218-222
    enabling   210
    external hardware interrupts 216-222
    interrupt handler   208
    interrupt vector table   208
    level-triggered interrupt   216-222
    priority   227, 228
    programming   212-215
inverters   10, 377, 378
IP register   228

J

JB instruction   144, 146, 147, 337
JBC instruction   144, 338
JC instruction   67, 338

JMP instruction   338
JNB instruction   144, 146, 147, 338
JNC instruction   68, 338, 339
JNZ instruction   68, 339
jump instructions   66-71
   conditional jumps   67-70, 340-341
   unconditional jumps   69, 341-342
JZ instruction   67, 68, 340

K

keyboards   261-266
   keypress detection   263-266
   programming   265
kilobyte   13

L

labels   41, 49
LCALL instruction   71, 72, 75, 341
LCDs   236-241
   command codes   237
   connection to 8255   312
   data sheet   240
   LCD timing   241
   pin descriptions   236
   programming   238
LEDs   236
linking   43
LJMP instruction   69, 341
LM34/ LM35 sensors   248
logic instructions   128-133
loop   66
   nested loop   67
lst file   42, 43

M

machine cycles   76
machine language   39
MAX232/233 chips   191, 192
MC1408 DAC   267
megabyte   13
memory. See semiconductor memory

microcontroller   24
   choosing a microcontroller   26, 27
microprocessor   24
mnemonic   39
MOV instruction   37, 342, 343
MOVC instruction   344, 345
MOVX instruction   346
MUL instruction   117, 346, 347

N

NAND gate   10
nested loop   78
nibble   13
NOP instruction   347
NOR gate   10
NV-RAM   280

O

obj file   42, 43
one's complement   7
open drain gates   381
open collectors   381
OR gate   9
ORG directive   48
ORL instruction   128, 347, 348, 349
oscillator   85
OV flag   50-51, 122-124, 328
overflow   122-124

P

P flag   50, 51
packed BCD   113
parallel communication   184
PCON register   200
polling   208
POP instruction   56-59, 73, 349
Ports 0 - 3   87-90, 304
power dissipation   391
power down mode   392
printer operation   320

program counter   44, 46, 69-71
PROM   276
PSEN pin   87
pseudocode   396
PSW register   50, 51, 55, 148
PUSH instruction   73, 350

R

RAM   13, 15, 279
   bit-addressable   149
real-time clock   31
registers   36, 358
   bit-addressable registers   147
   SFR registers   99, 100
RESET   44, 85
RET instruction   350
RETI instruction   225, 350
RL instruction   134, 351
RLC instruction   135, 351
ROM   13, 15, 44, 276
   program ROM   44, 45
rotate instructions   134-136
RR instruction   134, 351
RRC instruction   134-135, 351
RS232 standards   187, 188

S

SBUF register   194
SCON register   194, 195
semiconductor memory   274-302
   address decoding   284
   interfacing to 8051   287
   memory space of the 8051   292
   MOVX instruction   293
   organization   274
   speed   275
sensors   243-251
   selection   248
serial communication   184-190, 193-203
   8051 programming   196-203
   asynchronous   185
   data framing   186
   synchronous   185

serial communication port   223-227
SETB instruction   144, 352
SFR registers   99
signal conditioning   249
signed number representation   119
single-bit instructions   144-150
SJMP instruction   69, 352
sleep mode   391
source file   43
SRAM   279
src file   42-43
stack   56-58
   in CALL instruction   73
   popping   57
   pushing   56
stepper motors   256-261
   programming   258
structured programming   397-400
SUB instruction   115
SUBB instruction   115, 116, 352, 353
subroutines   73
   See also CALL instructions.
SWAP instruction   135, 353

T

TCON register   176, 177, 218-222, 227
terabyte   13
thermistor   248
time delays 76, 78, 168
timers   31, 158-173
   clock source   160
   event counter   173-178
   mode 0 programming   169
   mode 1 programming   161-162
   mode 2 programming   169
   registers   158, 159
   interrupt programming   212-215
TMOD register   159, 173
transducers   248
transient current   393
transistors   376, 377
transmission line ringing   394
tri-state buffer   9, 384
TTL technology   379, 380
two's complement 7, 115

U

UART/USART   185
unpacked BCD   112

W

wire-wrapping   366-367
   8051-based system   368-374
word   13

X

XCH instruction   353, 354
XCHD instruction   354
XOR gate   10
XRL instruction   129, 354, 355
XTAL1, XTAL2   85